An Unnatural History of Religions

Scientific Studies of Religion: Inquiry and Explanation

Series editors: Luther H. Martin, Donald Wiebe, William W. McCorkle Jr., D. Jason Slone and Radek Kundt

Scientific Studies of Religion: Inquiry and Explanation publishes cutting-edge research in the new and growing field of scientific studies in religion. Its aim is to publish empirical, experimental, historical, and ethnographic research on religious thought, behaviour, and institutional structures. The series works with a broad notion of scientific that includes innovative work on understanding religion(s), both past and present. With an emphasis on the cognitive science of religion, the series includes complementary approaches to the study of religion, such as psychology and computer modelling of religious data. Titles seek to provide explanatory accounts for the religious behaviors under review, both past and present.

The Attraction of Religion, edited by D. Jason Slone and James A. Van Slyke
The Cognitive Science of Religion, edited by D. Jason Slone and William W. McCorkle Jr.
Contemporary Evolutionary Theories of Culture and the Study of Religion, Radek Kundt
Death Anxiety and Religious Belief, Jonathan Jong and Jamin Halberstadt
The Mind of Mithraists, Luther H. Martin
New Patterns for Comparative Religion, William E. Paden
Philosophical Foundations of the Cognitive Science of Religion, Robert N. McCauley with E. Thomas Lawson
Religion Explained?, edited by Luther H. Martin and Donald Wiebe
Religion in Science Fiction, Steven Hrotic
Religious Evolution and the Axial Age, Stephen K. Sanderson
The Roman Mithras Cult, Olympia Panagiotidou with Roger Beck

An Unnatural History of Religions: Academia, Post-truth and the Quest for Scientific Knowledge

Leonardo Ambasciano

BLOOMSBURY ACADEMIC
LONDON · NEW YORK · OXFORD · NEW DELHI · SYDNEY

BLOOMSBURY ACADEMIC
Bloomsbury Publishing Plc
50 Bedford Square, London, WC1B 3DP, UK
1385 Broadway, New York, NY 10018, USA

BLOOMSBURY, BLOOMSBURY ACADEMIC and the Diana logo are trademarks of
Bloomsbury Publishing Plc

First published in Great Britain 2019

A catalogue record for this book is available from the British Library.

A catalog record for this book is available from the Library of Congress.

Series: Scientific Studies of Religion: Inquiry and Explanation

ISBN: HB: 978-1-3500-6238-2
    ePDF: 978-1-3500-6239-9
    eBook: 978-1-3500-6240-5

Typeset by RefineCatch Limited, Bungay, Suffolk
Printed and bound in Great Britain

To find out more about our authors and books visit www.bloomsbury.com
and sign up for our newsletters.

To all those who tried (in vain) to turn off the lights. Close, but no cigar.

To the builders of the lighthouse, Marco Goso and Amelia Engaddi Cresta, without whom none of this would have been possible.

To my wife Elisabeta and to Philip Davison, Alison Lennox and their team, for showing me that the night is at its darkest just before the brightest dawn, with gratitude.

To all who, against all odds, have succeeded in lighting a candle in the dark.

I owe you. And I love you.

Contents

Figures and Tables

Figures

Tables

Preface:
Ghosts, Post-truth Despair and Brandolini's Law

The sleep of reason produces monsters.

Francisco Goya

This volume is based on a paradox.

How come that, despite centuries of scientific research, the main academic discipline dedicated to the historical study of religion has been – and still is – so blindly devoted to an apologetic study of its research subject?

The paradox lies in the fact that nature, in the guise of deep-historical, impersonal, aimless and meaningless evolutionary dynamics, has provided us, *Homo sapiens*, with a *software* that, for the mere purpose of its *hardware* self-maintenance, survival and replication, is inclined to produce immediate, personal, intentional and purposeful answers which are intuitively at odds with evolutionary and scientific thinking (e.g. Evans 2000; Kelemen 2012; for nonhuman animals and their cognitive abilities, see de Waal 2013 and de Waal 2016).[1] When confronted with science and evolution, just as the famous episode of the TV animated sitcom *Family Guy* where Carl Sagan's *Cosmos* had been overdubbed to be broadcast in the US Bible Belt, believers, religionists and theologians are likely to answer 'God'.[2]

And yet, no goddess, god, spirit, ghost or superhuman being has ever walked among us mere mortals (however, Carl Sagan did). No divine intervention of some sort has ever interrupted our daily existence. No miracle has ever been ascertained beyond reasonable doubt. Apparently, the era of miracles is closed, and rivers of theological ink have been shed to justify this absence. Today, miracles are within the domain of science. The very perception, diffusion and reception of miracles have been demystified by neurosciences and cognitive science. Today, science is so close to magic that the two are almost indistinguishable. Immunotherapy and CRISPR/Cas genome editing techniques promise to offer a solution for devastating conditions, new fossils are continually improving our understanding of the 4.5 billion years of life evolution on this pale blue dot we call home, and Mars Exploration Rovers on a planet 220 million kilometres away from Earth (on average) are gathering and sending us data almost on a daily basis. And yet, what might appear as the unquestionable triumph of rational approaches to understand life, the cosmos and everything in between, crumbles down before 'alternative facts' and fake news. By sheer volume of shared online data, this will be probably remembered as the era of digital manipulation ('have you seen that shadow of an *alien* on the latest NASA pics?'), of imaginary conspiracies ('it's a cover up!'), of extinct animal sightings (wouldn't it be great to observe a *living* thylacine, by the way?), of Moon-landing hoaxes, of chemtrail and anti-vax movements. This is the era in which the blood liquefaction miracle of St Januarius coexists with de-extinction research aimed at bringing back the charismatic Pleistocene and Holocene megafaunae (are we going to see a *living* thylacine, after all?), the era in which millions of smartphones and tablets all over the globe have failed to record a single, incontrovertible instance of a divine or miraculous sign, and yet we are constantly flooded by tearful stories of miracles of every sort on our social media networks (and kittens playing the piano too, of course). This is the era of *post-truth*. Move away, 'Spooky' Mulder. We *all* want to believe.

Spoiler alert: there might be no happy ending to this story. This is rather unsurprising. For most of our history, the vast majority of human beings have been almost naturally insensitive to critical thinking and scientific research, and even more prone to confuse technology with magic, which is just one step away from being miraculous. Once upon a time, writing technology was reputed a sacred device to practise magic, fix the words of the gods, or cast *magical* spells to bind or summon demons. Today, tap on the screen of your smartphone and, *magically*, you can almost telepathically communicate from Reykjavìk to Canberra, scroll the sacred texts of all the major religious traditions past and present, and still bind and summon virtual demons in an online MMORPG. Electric grids, communication satellites, repeaters, digital signals, are all obliterated, nay, happily ignored, and the triumph of applied scientific technology – the result of centuries of interconnected research – turns into a nightmarish maelstrom of *magic 2.0*, where videos and pictures of weeping statues of Madonnas, thirsty simulacra of Ganesha, and Abrahamic angels of all kinds are shared millions of times (cf. Eco 2008: 102–1). Online, traditional, visceral, committed, non-negotiable beliefs translate into strong emotions which, in turn, filter out, or distort, everything that does not conform to someone's assumptions, and the result is that *you see what you want to see*. If critical thinking is applied at all, it is usually hijacked by biases and marred by fallacies. Specialists and scholars are themselves derided as slaves of the system. 'There must be a governmental conspiracy on 9/11, here are the clues!' (9/11 might be easily substituted with the death of Elvis, Paul McCartney's döppelganger, alien visitors from outer space, reptilians, you name it.) Once those committed beliefs set in, and a support network is created, a rational change of heart is difficult, if it is possible at all (Prothero 2013; D'Ancona 2017).

All of this is *post-truth*. Despite the superficial differences, the overarching *trait d'union* is the perception of 'truthiness', that is, the appeal to one's own emotions just because it feels right to believe in that particular something. Therefore, preference is accorded to the likeliness of something held to be 'true' over factual, objective truth. There is a clear echo of the most extreme forms of postmodernism in this definition. However, if post-truth refers to 'circumstances in which objective facts are less influential in shaping public opinion than appeals to emotion and personal beliefs', then this phenomenon is not just something *old*, but also *recurrent* (Oxford Dictionaries 2016a). Already in 1647, Spanish Jesuit Baltasar Gracián y Morales (1601–1658) alerted his readers to

> *take care when gathering information.* We live mainly on information. We see very little for ourselves and live on others' testimony. Hearing is truth's last point, and a

lie's first. Truth is normally seen and rarely heard. It rarely reaches us unadulterated, especially when it comes from far off. It is always tinged with the emotions through which it has passed. Passion tints everything it touches, making it odious or pleasing.

Gracián 2015: 14; original emphasis

Three centuries later, during the apogee of Interwar totalitarianism, George Orwell rightly identified the modern origins of truth manipulation in the combination of emotional attachment and reactionary ideologies (D'Ancona 2017: 4–5). When ideologies, prejudices, and intuitive beliefs take over institutional and collective knowledge, truth itself is always the first victim. But if that is the case, then it is quite easy to recognize that post-truth has *always* gone hand in hand with politics, academic research – *and religion*. The digital era of social media and shared online content is only acting as a real-time quantitative amplifier and accelerator, the likes of which, to be sure, have never been witnessed. It is difficult to stay critically focused when bombarded by more information than one can handle in one's entire lifetime – all of it just one click away. The usual and most comfortable defence against challenging ideas and cognitive overload is to cling onto one's own cherished point of view. Although apparently paradoxical, this is understandable (the devil you know is better than the devil you don't, anyway), and this is why science is so difficult. When push comes to shove, curious and intelligent people find curious and intelligent ways to dodge criticism and justify their theses – or beliefs. Which is a great way to advance knowledge, if you stick to the rules of critical thinking; if not, that is the perfect recipe for a post-truth disaster (Chatfield 2018). The Humanities and the social sciences have been a hotbed for the intelligent defence of belief systems and, among their ranks, historians of religions have been the unquestionable masters of post-truth.

Magic powers, paranormal events, anti-democratic advocacy, theological intrusions, ethnocentric chauvinism, anti-scientific stances, creationist support, all abound in the writings of some of the most renowned historians of religions of the past. Why is that? The usual defence, whether interested or just lazy, is that everyone during their age, whether recent or remote, thought like that. This excuse, however, is based on the 'everyone does it' fallacy (Warburton 2007: 64) and, as such, has no bearing on the point raised above. At most, it can be objected that everyone did or thought so *within* the environment of that discipline, which begs the question of why such an environment had been colonized by such scholars or, alternatively, why this academic space has been capable of exerting such cultural pressures on its denizens. Which are exactly the kinds of question that I try to tackle in this book, and their relevance is self-evident to the archaeology of the post-truth era.

Looking back in time, institutional religion has always been the main actor in the arena where science and pseudoscience have been long since duelling. Since Darwin published his groundbreaking work on the *Origin of Species* in 1859, the evolutionary debunking of the argument from design in a world deprived of any theodicy has risen to symbolize *the* godless, and religiously dangerous, scientific research. Admittedly, various accommodationist stances have tried, and are still trying, to assuage this conflict, albeit to no epistemological avail. Already a century before Darwin, in the continuing and obvious absence of any miracle or superhuman sign, David Hume advocated a *Natural History of Religion* (1757) which, in a sardonically dissimulated style, accounted for the natural causes behind complex sets of beliefs and behaviours (such as cognitive anthropomorphism, terror management and wishful self-preservation), while acknowledging the futility of religion for morality and its penchant for bringing out and justifying the worst violent behaviours (topics also tackled in his *Dialogues Concerning Natural Religion*, 1779; Hume 2008a). But long before that, more than 2,000 years earlier, Greek philosopher Xenophanes led a 'full frontal attack' against traditional theology, by famously stating not just that every human culture has imagined its gods with characteristics which typically followed that of the people in question (thus undermining any universalizing claim), but even that 'if cows and horses or lions had hands, or could draw with their hands and make things as men can, horses would have drawn horse-like gods, cows cow-like gods, and each species would have made the gods' bodies just like their own' (Waterfield 2009: 22, 27; cited fragment DK21B15, KRS 169). And these are just a few notable examples. As anyone sufficiently acquainted with ancient historiography should know, critical thinking, atheism and agnosticism, in many different forms, have always coexisted and wrestled with religious institutions and dogmas (Minois 1998).[3] But if before 1859 tension between science and pseudoscience existed in the form of a zero-sum guerrilla war punctuated by the occasional outburst of religious intolerance, the three-fold combination of the slow scientific accumulation of knowledge after the Scientific Revolution, the European Enlightenment, and Darwin's works, resulted in the veritable outbreak of a world war. One of the main battlefields of such conflict in the academy has been the history of religion(s) (HoR, henceforth).

HoR is a very complicated academic discipline, in that, *mirabile dictu*, it has no established agreed-upon charter. From a historiographical point of view, the discipline shifts back and forth from the universal (i.e. a single, innate *religion*) to the particular (i.e. the many historical *religions*) while trying to understand or justify their relationship on ontological, metaphysical, cultural and historical levels (Smith 2004: 184–6).[4] What can be safely stated is that the HoR is based on comparison writ large. But, as remarked earlier, the very basic cognitive mechanisms and biases that intuitively allow for the comparative endeavour (such as essentialism and teleological thinking, which we will tackle later on) are an obstacle for a correct, historical understanding of the impact of evolutionary processes on culture and religion. As it appears, ordinary human cognitive construals that provide the scaffolding for intuitive explanations find the very logic of evolution counterintuitively baffling (Blancke and de Smedt 2013; Girotto, Pievani and Vallortigara 2014; Coley and Tanner 2015). Therefore, and unsurprisingly, the most appealing and successful study of religion(s) is that which has answered positively to, and piggybacked on, these immediate, agentive, essential and teleological non-evolutionary and virtually anti-scientific tenets. Institutions have exploited this penchant, and most scholars in the field, eager to manipulate socio-cultural evolutionism, have provided supportive ethnocentric, theological and/or racial interpretations to the HoR as a whole. However, there is epistemological justice after all. Daniel Lord Smail has written that 'natural history, obviously, does not require an awareness of history on the part of the subjects' (Smail 2008: 70). Historians of religions may have thought that their ideas about how to conceptualize the HoR were right, but most of the times, as the cases analysed in the book will show, they were not. Sometimes they rejected natural history as meaningful for their studies (and some do so even today). And yet, given that those historians of religions were, until proven otherwise, mammals with an evolved brain and evolved cognitive biases, natural history explains both their ideological inclinations and their disciplinary leanings.

However, readers might opine that, if those intuitive biases are natural, and religion is fundamentally a product of those cognitive biases, religious stances might be *natural, right* and *naturally good*. Unfortunately for those who entertain such beliefs, there is a catch. Religious sets of beliefs and behaviours may be cognitively natural, i.e. the natural products of evolutionary pressures (McCauley 2011), but this does not mean that they are inherently good or reliable. Let us indulge in a slight variation on the Humean *is–ought* problem. Heart attacks and strokes due to smoking, obesity and poor dietary habits are the *natural outcome* of an ultimate evolutionary cause (i.e. addictive behaviour and human propensity to eat high-caloric food, a valuable – if occasional – source of energy in nature) and a proximate historical cause (that is, market economy and psychotropic consumerism), but this does not automatically make heart attacks or strokes *naturally good* (cf. Smail 2008; Diamond 2012; Diamond 2015). An *unnatural history of religions* such as the one conducted within the HoR, for want of a better parallel, would be like having cardiac surgeons, maybe even affected by heart attacks and strokes because of such causes, actively downplaying, minimizing or misinterpreting the impact of the ultimate and proximate causes behind cardiovascular issues, encouraging the usual bad habits which lead to such health problems as though nothing has happened, and judging the very causes of such tragic outcomes as *naturally* non-problematic or unavoidable. This is the core of what I have dubbed herein as the academic history of an intrinsically anti-Humean *unnatural religion* as the one pursued within the HoR.

Let me elaborate on this for a moment. Science and religion are similar and, at the same time, profoundly different. Theological knowledge can be as much elaborated and hard to grasp as cutting-edge science itself. Compare for instance the discovery of a subset of homeobox genes that guide the sequential development of physiological structures, and whose activation and functioning are the same across the animal kingdom, with the Christian Holy Trinity, which fuses three distinct religious agents, or hypostases (the Father, the Son and the Holy Spirit), in a single nature, at once consubstantial, co-eternal and equal. However, while cutting-edge science is based on the grasping of more basic but still counterintuitive processes which are not arrived at via merely intuition but through critical thinking and the accumulation of epistemically warranted knowledge, theology builds on and depends upon immediate, widespread and highly intuitive ideas which may arise independently in each cultural setting. This set of cognitively uniform core beliefs and templates has been called 'popular (or folk) religion', which explain the almost panhuman cultural presence, formation and diffusion of anthropomorphic agents constrained by human-like features such as ghosts, dead ancestors living in the otherworld, talking animals as messengers and more. Taken together, and notwithstanding innumerable projects of accommodationism and harmonization, theology and popular religion remain at odds with the tenets of scientific critical thinking and evolution (Coyne 2015). The immediate appeal of intuitions which feed into religious beliefs can be shrugged off only with constant, onerous, effortful, institutionally supported education, which makes science inherently fragile (cf. Ambasciano 2017a).

There is also a willingness to believe, an emotional attachment, an affective investment to believe in beliefs themselves, an addiction that might feed into our intuitions, in a self-reinforcing loop. As philosopher Stephen Law has recently summarized, our cognitive machinery is vulnerable to successful cyberattacks by beliefs involving counterintuitive, divine hidden agents, special places, charismatic objects and miraculous or paranormal actions – which Law has collectively labelled as 'X-claims' – when subjective experiences and testimony are added to the mix (Law 2018). Religions and pseudoscience literally thrive on X-claims. Just think about this for a second: *our computational brainpower is limited.* We can always be deceived. Cognitive predispositions once useful to strengthen group solidarity and individual reliability can be easily hijacked. Not to mention that 'our computational devices are only just good-enough, and they make us prone to false positives in the detection of meaningful patterns, from anthropomorphism to pareidolia, from distorted national myths to imaginary conspiracies. Our neural networks devoted to gathering and processing data into meaningful patterns (no matter how authentic) are unrelentingly at work' (Ambasciano 2015b: 242–3; cf. Gottschall 2012: 96). Do not get me wrong: emotions and cognition are interconnected, and the myth of a perfectly rational, emotionless Vulcan mind is, indeed, a sci-fi caricature. No emotions, no rationality (Galef 2011; Galef 2016). However, emotions *do* get in the way of effective decision-making if no sufficient rational literacy has been provided. It takes brains, a lot of guts, a bit of luck and decades of education and literacy to counter the effects of our potentially fallacious intuitions. And even this might not be enough. We are social animals, and we depend on others to survive. Academic research would be unthinkable without institutional resources, tools, support and funds. And here is the crucial point: it just takes some very simple tweaks in the institutional hierarchy to control funds and scholarships so as to control subtly the academic study of religion(s). The educational costs of a scientifically disengaged HoR are far too high, for higher education shapes the cultural and mental environment of our fellow citizens of tomorrow, those who will vote and be voted for, those who will propose, implement or support policies that will affect us all. In the light of this, what are the long-term, collective, socio-political damages by governmental legislation aimed at endorsing the teaching of the supernatural, of creationism, of theological dogmas, as positive, institutionally endorsed facts? And just like the aforementioned example, who watches the watchmen when the supervisors themselves are prone to be institutionally controlled, cognitively biased, ideologically compromised, and prejudicially manipulated?

As psychologist James Leuba rightly foresaw more than a century ago, 'the advent of psychological analysis and explanation should bring about a crisis more painful, because more profound, than the one due to the less recent appearance of the comparative history of religions and the literary criticism of sacred writings' (Leuba 1912: ix–x). Indeed, both the evolutionary and cognitive revolutions have brought about a 'crisis more powerful' than anything else in the HoR. The explanatory power born of those revolutions has proved to be immense. For the first time ever, scientific tools and critical thinking can support evidence-based, scientifically sound research to produce solid and epistemically warranted knowledge. However, as we will see, even these tools and frameworks can be easily rejected or manipulated to justify apologetic or accommodationist ends. Indeed, most historians of religions, immunized from external or internal criticism within their safe echo chamber, are successfully resisting – and counterattacking. Science, it should be recalled, can be easily lost. The resistance, the disdain and the contempt against science and evolution are quite diffused within the Humanities, especially since the spread of postmodernism, but within the history of the HoR they reach unimaginable levels. Institutions are not helping. As I am writing these lines, the Republican US administration, led by post-truth champion Donald Trump, has reportedly advised the *Centers of Disease Control and Prevention* (CDC) against the use of a list of words or syntagmas from their budget documents, among which, quite expectedly, there are 'evidence-based' and 'science-based'. There is a war, indeed, and it is raging on (Helmore 2017).

This is not a neutral academic volume. This is a book with an agenda, and a strong one. Apparently paradoxical, the sense will become clear only by reading the volume and following its argumentations until the very last sentence. For those who cannot wait, here is the 'too long; didn't read' version: for the sake of the survival of the social-scientific study of the history of religions, the discipline called HoR must go (cf. Lincoln 2012: 135). To resolve the apparent paradox stated at the very beginning of this introduction, we need to think out of the box, and we need a serious discussion. Something new, something different, in the wake of the evolutionary and 'cognition-based social-scientific model [of] historical research' (L. H. Martin 2014: 273) already prefigured thanks to the work of a small and brave scholarly community of forerunners, should take its place. Unfortunately, the current resurgence of pseudoscience, misinformation, deception and inaccurate analyses has infected the field of the HoR to such a degree that no therapy is available to cure it. Brandolini's Law states that 'the amount of energy needed to refute bullshit is an order of magnitude bigger than to produce it' (Williamson 2016), and the current HoR is replete with *bullshit*, i.e. the disregard for truth and the willingness to engage in fakery and postmodern 'instant revisionism' for the sake of it – and for prestige and fame as well (Latour 2004: 228; see Frankfurt 2005, and Pigliucci and Boudry 2013). When this combines with genuine, but equally fallacious, fideism, the result is explosive (Ambasciano 2015a; Ambasciano 2016a; Ambasciano 2016b).[5]

To borrow a much-heard slogan from the 2008 global financial crisis, the HoR is apparently 'too big to fail'. The sheer magnitude of such disciplinary bullshit shields it from disconfirmation. Deconstructing and dismantling the discipline to rebuild it on firmer grounds, while rescuing everything that could be scientifically updated, might seem the most viable option here, but I am quite sure that such a project will *never* be accepted nor be entirely successful – and not just because of the Herculean penance involved. Because of the very existence of those evolved cognitive biases, obscurantist resistance will *always* win. Cutting-edge, counterintuitive science will always be misunderstood by laypeople when popularized for instant fruition or translated into easy terms (cf. Asprem 2016). Intuitive logical fallacies will *always* trump effortful evidence-based research in the field. Miraculous fake news, quackery, fideistic distortions will *always* claim their share. Para-institutional money from religious organizations will *always* exert some sort of bottleneck selection with regard to the research they want to fund (cf. Martin and Wiebe 2016: 221–30).

However, this does not mean that we cannot build a new edifice. For all my existentialist dread and utter pessimism, hope is unquenchable. Modern democracy might have seemed an archived social experiment under the post-Napoleonic Restoration. The extension of civil rights might have appeared like a mirage under the Confederate States of America. The establishment of universal suffrage before the breakthrough of the Suffragettes might have sounded like a hopeless goal. The victory of the Axis powers might have seemed inevitable before Churchill's speech or Roosevelt's intervention. Reaching the Moon or Mars might have looked like the fancy stuff of Jules Verne's novels. Real-time, worldwide, relatively cheap audio-video communication, MRI scans, genetic screening, vaccines, virtually infinite archives of literary and scientific data, were once the subjects of sci-fi adventures or Borgesian novels. In a virtuous loop, technological breakthroughs allow for previously unthinkable answers, which in turn lead to new questions in academic research – and so on. Thanks to one of such serendipitous interactions between contemporary cognitive sciences, philosophy of mind, evolutionary psychology, neurosciences, Big and Deep historiographical research, and other cross-disciplinary contributions the demystified, and demystifying, study of religions, *all religions*, could at last be resolved within a Humean, natural *and* historical science (cf. Bulbulia and Slingerland 2012). Sometimes, *real* change does *really* happen. Moral progress and the advancement of science go hand in hand, because, in both cases, whenever institutional space and support for epistemically warranted criticism and refutation is provided, experimentation and logical thinking flourish to the benefit of all (bar the bullshitters, of course). Prove my pessimism wrong. If you want something to believe in, please *believe* in scientific *hope* for improvement and learn from past errors, engage the discipline critically, 'bring the sword of criticism to [postmodern] criticism itself' (Latour 2004: 227), and fight back post-truth with the arms of critical thinking. As you will discover by reading this book, you are not alone.

Acknowledgements

In 1974, Italian historian Arnaldo Momigliano (1908–1987) wrote that 'when faced with uncertainty, the serious historian consults his colleagues, above all those who have proven themselves to be notoriously skeptical and relentless. Tell me who your friends are, and I will tell you what kind of a historian you are' (Momigliano 2016: 42).

This book would not have been written without the outstanding support of two 'notoriously skeptical and relentless' gentlemen, Luther H. Martin and Donald Wiebe, who have helped me since the darkest days of my PhD, giving me strength to carry on in spite of all the odds. Their implacable critical thinking and exceptional scientific open-mindedness have set a very high bar for me and, in judging this volume, I expect from them nothing less than their usual painstaking scrupulousness.

The network of brilliant scholars they put me in contact with over the years has been oxygen for my exhausted mind. In a few cases, scholarly companionship and formal acquaintance have slowly morphed into close friendship. Panayotis Pachis, Nickolas Roubekas and Aleš Chalupa are splendid examples of such a transformation.

In a series of fortunate events, Aleš kindly invited me to the Department for the Study of Religions, Masaryk University, Brno (Czech Republic) for an Autumn 2016 visiting lectureship. The present volume expands the classes on 'The Historical Study of Religion in the Twentieth Century' I had the pleasure to teach in a friendly environment during that magnificent and colourful Czech autumn. My heartfelt thanks go to all the colleagues, friends and scholars that I had the opportunity to meet there while teaching: David Zbíral, Dalibor Papoušek, Šárka Vondráčková, Tomáš Glomb, Tomáš Hampejs, Vojtěch Kaše, Jana Valtrová, Milan Fujda, Martin Paleček, František Novotný, Juraj Franek, Nimrod Luz, the interdisciplinary teams at GEHIR and LEVYNA and, last but not least, all my students.

The hard core of the Masaryk University classes, and of the present book as well, had been previously delineated on two occasions: (1) a 2016 article entitled 'Mind the (Unbridgeable) Gaps: A Cautionary Tale about Pseudoscientific Distortions and Scientific Misconceptions in the Study of Religion'; (2) a paper presented during the conference 'Relazioni pericolose. La storia delle religioni italiana e il fascismo' organized, and helmed notwithstanding a spiteful anti-scientific contestation, by Roberto Alciati and Sergio Botta, and held at Sapienza Università di Roma (Italy) on 3–4 December 2015. I would like to thank in particular Sergio, Roberto, Marianna Ferrara, Caterina Moro and Sonia Gentili for their unfailing support during and after that alarming and worrisome episode.

Way back in 2013, a discussion with a brilliant student (Gianluca Chiesa), when I was halfway through my PhD at the Università degli Studi di Torino (Italy), had already convinced me of the necessity of such a book. Over the years, innumerable lively discussions, *cappuccini d'orzo*, and good chats with my dear friends in Turin have contributed to shape and refine many of the ideas I present here: Roberto Alciati, Francesco Cassata, Maria Dell'Isola, Enrico Manera, Roberto Merlo, Andrea Nicolotti and Emiliano Rubens Urciuoli. Looking forward to having more *cappuccini* with you all.

Finally, I would like to gratefully acknowledge, in no particular order, the help of Florinela Cazan, Emina Hadzifejzovic, Giovanni Pasquali, Alessandro Tavecchio, Thomas J. Coleman III, Michele Luzzatto, Mac Linscott Ricketts, Mauro Mandrioli, Marco Ferrari, Giuseppe Liberti, Anders K. Petersen, Peter J. Richerson, Jeppe S. Jensen, Maurilio Orbecchi, Davide Bonadonna and Mike Keesey. Last but not least, I am particularly indebted to Derek Gillard for his masterful reading and useful suggestions, and to my friends Francesco Lodone, Daniele Bellavia, Daniele Navone and Andrea Trento for being there through all the good times and the bad times.

Despite all the amazing help and careful checking, the usual disclaimer applies: *any remaining error in the book is entirely my fault.*

Oxford, 26 February 2018

Note on Text

All the translations in this book are my own, except where otherwise noted.
All the figures and tables are my original works, except the artwork listed below:

Figure 1 The disciplinary landscape: major issues in the history of religions visualized as a geological map © Davide Bonadonna, 2018. Reproduced with permission.

Figure 6 'What Evolution Looks Like' © Mike Keesey, 2018. Silhouettes from *PhyloPic* open database (*Phylopic.org*). CC licence BY 4.0. Reproduced with permission.

Figure 7 Silhouette credits, from left to right: *Proconsul* (by Mateus Zica, modified by T. Michael Keesey); *Dryopithecus, Australopithecus, Homo neanderthalensis* (by T. Michael Keesey; silhouettes from *PhyloPic* open database, *Phylopic.org*; CC BY-SA 3.0); Neolithic farmer adapted from Zdeněk Burian's painting *Člověk v mladší době kanmenné*, printed by Státní pedagogické nakladatelství, Prague, 1951; Robert Liston. Printed silhouette. Credit: Wellcome Collection. CC BY 4.0; warrior from J. Shooter (1854). *The Kafirs of Natal and the Zulu Country*. London: 1857, 201. Not to scale. Please note that the first three taxa were not known in Victorian times, as their presence is reported here for the sake of simplicity.

Figure 13 Silhouette credits, from left to right: businessman with briefcase, own work; Augustus as *augur* with *lituus* from Wikipedia, https://commons.wikimedia.org/wiki/File:Augur,_Nordisk_familjebok.png; Bronze figure of Kashmiri in Meditation by Malvina Hoffman. Credit: Wellcome Collection. CC BY 4.0; shaman from Wikipedia, https://commons.wikimedia.org/wiki/Category:Shamanism_in_Russia#/media/File:Shaman_Buryatia.jpg, public domain; *San Giuseppe da Copertino in estasi* di Felice Boscaratti (*ca.* 1762), chiesa di San Lorenzo, Vicenza (Italy). By Didier Descouens, from Wikipedia. CC BY-SA 4.0, https://commons.wikimedia.org/wiki/File:Chiesa_di_San_Lorenzo_a_Vicenza_-_Interno_-_San_Giuseppe_da_Copertino_in_estasi_di_Felice_Boscaratti.jpg Didier Descouens CC BY-SA 4.0.

Figure 15 Silhouette credit: NASA, public domain, from *Phylopic.org*.

Epigraphs are reprinted by kind permission of the respective copyright holders (with the exception of the epigraphs for Chapters 2 and 5, for which my efforts to contact the publishers remain unanswered), and excerpted from the following sources:

Chapter 1   Martin, L. H. and D. Wiebe (eds) (2016). *Conversations and Controversies in the Scientific Study of Religion: Collaborative and Co-authored Essays by Luther H. Martin and Donald Wiebe*. Leiden and Boston: Brill, 224.

Chapter 2   Popper, K. (1994). *The Myth of the Framework: In Defence of Science and Rationality*. Edited by M. Notturno. Abingdon and New York: Routledge, 95.

Chapter 3   Darwin, C. R. (1838). *Notebook M* [Metaphysics on morals and speculations on expression]. CUL-DAR125. Transcribed by K. Rookmaaker, edited by P. Barrett, corrected by J. van Wyhe. RN3. *Darwin Online*, http://darwin-online.org.uk/.

Chapter 4   Gould, S. J. (1986). 'Evolution and the Triumph of Homology, or Why History Matters'. *American Scientist* 74(1): 63.

Chapter 5   E. Ionesco to T. Vianu, 20 February 1943, translated from Vianu, M. A. and V. Alexandrescu (eds) (1994). *Scrisori către Tudor Vianu II (1936–1949)*. Bucharest: Minerva, 233.

Chapter 6   Dennett, D. C. (2013). *Intuition Pumps and Other Tools for Thinking*. London: Penguin, 45.

Chapter 7   Darwin, C. R. (1871). *The Descent of Man, and Selection in Relation to Sex*. London: John Murray. Volume 2. 1st edition, 385. Scanned by J. van Wyhe 1.2006; transcribed (double key) by AEL Data 9. 2006, corrections by van Wyhe. RN3. *Darwin Online*, http://darwin-online.org.uk/.

Epilogue   Momigliano, A. (1979). 'A Hundred Years after Ranke'. In *Primo contributo alla storia degli studi classici e del mondo antico*, 373. Rome: Edizioni di Storia e Letteratura.

Finally, I would like to thank Brill for their permission to reprint, update and revise a significant part of the following works for this book:

'Mind the (Unbridgeable) Gaps: A Cautionary Tale about Pseudoscientific Distortions and Scientific Misconceptions in the Study of Religion'. *Method & Theory in the Study of Religion* 28(2): 141–225.

'Politics of Nostalgia, Logical Fallacies, and Cognitive Biases: the Importance of Epistemology in the Age of Cognitive Historiography.' in A. K. Petersen, I. S. Gilhus, L. H. Martin, J. S. Jensen and J. Sørensen (eds), *Evolution, Cognition, and the History of Religion: A New Synthesis. Festschrift in Honour of Armin W. Geertz*, 280–96. Leiden and Boston: Brill.

1

An Incoherent Contradiction

A history of the development of Religious Studies as a scientific enterprise in the modern university is an incoherent contradiction that reveals tensions between putative claims to academic status and the actual reality of continuing infiltrations of extrascientific agendas into the field.

Luther H. Martin and Donald Wiebe

A geological divide

'History of religions' is an academic label as ambiguous as any. The preposition 'of', as Bruce Lincoln once remarked, 'is not a neutral filler. Rather, it announces a proprietary claim and a relation of encompassment' (Lincoln 1996: 225). However, the relational nature of this non-impartial filler is not entirely clear. What does '*of* religions' really mean? Is history thought of as the main driver of religious phenomena (*objective genitive*)? Are 'religions' supposed to provide the framework, the tools, the sense and the meaning of historical research (*subjective genitive*)? As anyone might guess, this is no linguistic trifle: the two meanings underlie two radically different approaches (see Table 1). Then again, a curious interlocutor – let us say, a prospective student – might feel prompted to ask, does the objective understanding of this not-so neutral label really differ from the study of religions as it is commonly practised in historiographical studies? Is the meaning implied by the subjective genitive all that different from theology? What's the point of having such a *doppelgänger* of other disciplines? What are, then, if any, the main features of this weird academic discipline?

For all her doubts, our curious interlocutor might think that, after all, if such a discipline is part and parcel of the contemporary academic panorama, then it should also possess a precise disciplinary charter – which might look something like this. History of religions (HoR, henceforth) has been and still is usually conceived of as the study of religion (singular) and/or religions (plural) via comparative methods aimed at recovering and enhancing similarities (and, more rarely, differences) among religious beliefs and practices from the ancient past to the present day. The discipline aims at gaining an insightful and precise classification of religious phenomena which, in turn, is supposed to enlighten the religious contents of human cultures from the first written documents to the newest religious movements, from the ethnographic accounts of bold nineteenth-century explorers to the religious vagaries of the Internet and the

Table 1 History of what, exactly?

	Stress on	Method and theory	Adopted by	Theology
Objective genitive	history	historical	Historicism	✗
Subjective genitive	religion(s)	religious	Phenomenology	✓

spiritual, digital-age miscellanea available online. The available academic handbooks and the most important works from the past usually enlist a series of features deemed:

1. to strictly delimit and define the independence of the discipline from other apparently similar fields;
2. to point out the existence of a specific *modus operandi*;
3. to single out the major accomplishments and benefits deriving from the adoption of such m.o.

And yet, all of this is merely a façade. For all the relief and satisfaction that our curious interlocutor might experience, we should admit, for the sake of professional ethics, that the HoR has instead a rather convoluted history, and an even more complicated academic ID. When we look at the history of the discipline, we can notice that, quite astonishingly, a widespread, shared consensus among the researchers about the nature and scope of the HoR is difficult, if not impossible, to recognize. Indeed, the history of the discipline reveals that the grammatical divide between the objective and the subjective genitives runs like a geological fault across the field, with many shifts and cracks that stretch along the surface. Therefore, in order to navigate this rugged and unstable landscape without getting lost, we could definitely use a map.

Mapping the problematique

As Daniel L. Smail has aptly remarked, 'metaphors do much of our thinking for us. Evoking whole fields of thought, they communicate complex ideas and images with extraordinary efficiency' (Smail 2008: 78). In our case, a metaphorical map can be exploited as a cognitive device by which a handy shortcut to highlight the main features that lie unseen is readily provided. The geological divide that characterizes metaphorically the disciplinary landscape is the result of historiographical tension and friction between the two forms of genitives implied in the label 'history of religions'. As we will see in more detail shortly, during the twentieth century this tension climaxed in the objective genitive being adopted by historicism, while the subjective genitive had been mainly embraced by phenomenology (see Table 1). The various unsuccessful attempts to reconcile these two trends might be imagined as cracks that characterize the fault surface of our metaphorical map. Each one of those labyrinthine cracks implies a specific subset of points of view, ideas, issues and – most of all – problems.

By the juxtaposition of metaphor and reality, when we look at the fault zone depicted on our imaginary geological map, we can identify three main layers or discontinuities

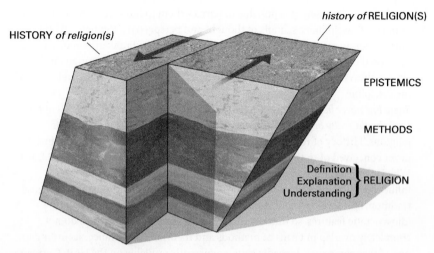

Figure 1 The disciplinary landscape: major issues in the history of religions visualized as a geological map

which represent what can be probably considered as the main issues of HoR *qua* discipline (see Figure 1):

1. As hinted at earlier, there *has never been a disciplinary consensus on what the discipline should exactly be.* In a recent volume, Swiss historian of religions Philippe Borgeaud has underscored that 'history of religions is a branch of knowledge that has its own specificity, but it is a branch that presupposes a trans-disciplinary curiosity. In fact, history of religions is dependent (among others) on history [...], philology [...], sociology, [...] ethnology and anthropology, [...], psychology' (Borgeaud 2013: 33–5). In one of the most remarkable accounts of the discipline from the first half of the 1960s, Mircea Eliade, possibly the most influential historian of religions of the past century, wrote that the 'mission' of the HoR is indeed 'to integrate the results of ethnology, psychology and sociology. Yet, in doing so, it will not renounce its own method of investigation or the viewpoint that specifically defines it' (Eliade 1964: xiii). To be fair, Eliade's statement may sound a bit confusing since it leaves our curious interlocutor wondering about what this specific 'viewpoint' should be when so many different disciplines are involved. It is not just our interlocutor who happens to be confused, for the historians of religions themselves have had their fair share of hassles and headaches to figure out what this discipline should be about. In a remarkably comprehensive panorama of the whole history of the discipline published in 2010, Italian historian of religions Natale Spineto frankly admitted that HoR has an 'uncertain epistemic foundation' and that, 'indeed, the boundaries of the discipline are not very clear: depending on the disciplinary trend, it is possible to expand them to include substantially every field dedicated to the study

of religion. Conversely, it is possible to narrow them to indicate a very restricted number of disciplinary schools of thought and perspectives' (Spineto 2010: 1256). This 'epistemic patchwork' definition of HoR as a sort of Frankenstein discipline composed of many heterogeneous pieces, so to speak, is quite common and it is almost always highlighted as something necessary and useful for the wellbeing of the discipline.

2. *There has never been a shared disciplinary consensus on what the methods of the HoR should really be.* Basically, there are as many methods as historians of religions. 'History of religions', according to historian Ioan P. Culianu, 'is almost never conceived of in the same way by two different scholars' (Culianu 1978: 18). Why is that? Basically, every major national school of HoR, more or less influenced by the presence of each pre-existing, locally predominant theological tradition, has developed independently a set of specific peculiarities and idiosyncratic features, so much so that the most important branches do not completely overlap in terms of methods and theories. For instance, according to one of the major encyclopaedic sources currently available on the HoR, German *Religionswissenschaft* is not quite the same as French *histoire des religions*, and the Italian *storia delle religioni* does not coincide exactly with the English *history of religions* (Casadio 2005a; cf. Wheeler-Barclay 2010: 247). Furthermore, there seems to be a lack of interest in resolving such issues. For instance, here is what Christian K. Wedemeyer wrote back in 2010: 'For some, the study of religion today is precisely attractive because of its intellectual indeterminacy or, to give it a more respectable moniker, its "interdisciplinarity"' (Wedemeyer 2010: xxv). So, the fact that the historical study of religion is an autonomous branch of academia should be accepted at face value while abiding by the 'intellectual indeterminacy' of its methods – which is something weird at best, reprehensible at worst, considering that such disciplinary identity is not the result of any epistemological reflection about cross-disciplinary integration (e.g. O'Rourke, Crowley and Gonnerman 2016). The result is something that, from an institutional point of view, might appear to fly in the face of any disciplinary acceptability.

3. *Finally, there has never been a disciplinary consensus on what religion is, must be, or should be.* This final point might even be considered the root of all disciplinary evil, as it were, for it is the source of all confusion and problems. Most importantly, this is the reason why the HoR has served so many non-neutral, heterogeneous, and ideological purposes and aims, while losing epistemic momentum in endless discussions concerning definition, explanation and understanding (Penner and Yonan 1972). In most of the major works of the past century, it is possible to find appreciative accounts of theological ideas; also, it is not uncommon to see the discipline hailed as the indispensable intellectual device by which the spiritual awakening of a new era could finally begin (von Stuckrad 2014: 113–77). In 1969, for instance, Eliade wrote that 'by means of a competent hermeneutics [that is, the symbolic interpretation of religious contents from a religious perspective], history of religions ceases to be a museum of fossils, ruins, and obsolete *mirabilia* and becomes what it should have been from the beginning for any investigator: a series of "messages" waiting to be deciphered and understood' (Eliade 1984: 2). In this

sense, comparison is used to postulate a universal belief in god(s) as the distinctive trait of humanity. Eliade also openly acknowledged HoR as a 'saving discipline' with a remarkably religious tone (Eliade 1989a: 296, 2 March 1967). The method by which this salvation could be achieved is called 'creative hermeneutics', triumphantly hailed as the 'royal road of the history of religions' capable of leading its practitioners to the spiritual meaning behind and beyond historical events and, ultimately, to the renewal of human creativity (Eliade 1971: x). It goes without saying that those 'messages' are supposed to help in unveiling a meta-theological, fideistic worldview (Juschka 2008), which inevitably blurs the boundary between academic, social, and scientific analysis and active advocacy for certain practices and beliefs. Indeed, a sympathetic perspective on such a controversial topic is not uncommon in the fields that study religion(s) (Stausberg 2014).

As we will see in the next chapter, Eliade's idea highlights the kind of confusion between objective description and subjective redescription that is so typical of classical HoR. The *consensus gentium*, here intended as the agreement of peoples with regard to the ontological existence of a supernatural dimension, is exploited as the ultimate justification for the existence of the discipline. Which, as our curious and now a bit exhausted interlocutor might remark, is a bit like saying that, because people believed in sirens, there must be sirens somewhere and, therefore, something like a sirenological discipline – before having ascertained beyond any reasonable doubt the presence of sirens in the real world. *There are sirens*, indeed: the *Sirenia*, a group of marine mammals that includes manatees and dugongs, but I guess that this is not the mythological kind of sirens you might be thinking of. Moreover, as we will see in detail later on, this idea is fatally affected by the logical fallacy of the bandwagon effect (also known as *argumentum ad populum*), that is, 'an argument that derives its force from the popularity of beliefs on which it is based' (Fellmeth and Horwitz 2009: 39; cf. Ambasciano 2015a). No matter how sincerely you might hope, believing in sirens, and sharing this belief with your family, friends, and colleagues, won't make them any more real than unicorns. If you replace sirens with things like 'levitating shamans', there you have it: hermeneutical HoR in a nutshell (I have not made this up: there is indeed such a belief in the HoR; see Chapter 5).

Maps, compasses and bricks

All things considered, the preliminary survey of our imaginary geological map of the HoR reveals a fault zone at risk of earthquakes. The continuous friction between the rigid rocks which form the two sides of the fault has been building up a lot of stress within the very structure of the fault, which might be unexpectedly released as seismic activity. Under the umbrella term of HoR, 'history' and 'religion(s)' are in constant tension, with 'religion(s)' exerting the most powerful and disruptive force (Lincoln 1996).

Now, if you are really living near an active fault, learning a bit of geological and scientific information will do you no harm and will surely empower your decisions, provided that the information you have is trustworthy. Here is what you should do.

First of all, a map without a compass is useless: learn to orientate yourself, sort out your sources and gather as much data as possible. Find out which are the most reliable and recent. Do your homework. It may turn out that, even though the fault is active, you can cope with it up to a certain extent. Depending on the seriousness of the situation, you might decide to build a better, earthquake-resistant, state-of-the-art designed house, or structurally reinforce the one in which you already live. You can opt for moving and relocating elsewhere, where institutional intervention has already invested in earthquake-resistant buildings or look for places where there is a lower seismic risk. Alternatively, should you be a stubborn denier of the import of such scientific research, you might also decide to ignore the Cassandras who cry wolf and keep on living where you live – at your own risk. As despicable as the latter choice might be, you might also feel drawn to active propaganda to deny that there is a serious geological risk, and that geologists are biased because of some inscrutable hidden agenda.

It is not difficult to see what these metaphors really imply: if HoR is the active fault, then knowledge comes from the history of the discipline (as well as from adjacent disciplines), the compass is the rational process of scientific enquiry, and the Cassandras are the updates from contemporary social and scientific research. Disciplinary denialism, finally, might be understood as the consequences of when 'a well-entrenched belief system comes in conflict with scientific or historic reality, and the believers in this system decide to ignore or attack the facts that they do not want to accept' (Prothero 2013: 4). The problem is that HoR and the other disciplines that study religion(s) offer an overload of information usually characterized by a significant amount of noise. This complicated situation – and the possible way out – has been known to scholars since the very beginning of the discipline.

In 1912, psychologist James Leuba (1868–1946) concluded his volume entitled *A Psychological Study of Religion, Its Origin, Function, and Future* with an appendix that listed 48 interdisciplinary definitions of religions (actually, more than 50 because some entries included more than one definition) (Leuba 1912: 339–61). Leuba defined the rationale for his appendix as follows: 'I trust that the perusal of these forty-eight definitions will not bewilder the reader, but that he will see in them a splendid illustration both of the versatility and the one-sidedness of the human mind in the description of a very complex yet unitary manifestation of life' (Leuba 1912: 339). More specifically, Leuba's intent was to offer what today we would call 'supplementary data', i.e. an additional corpus of data that served as the raw material upon which Leuba systematized four definitional super-categories. As reported in the second chapter of *A Psychological Study of Religion*, Leuba's arrangement of this array of disciplinary definitions consists of the following categories:

1. *intellectual*, a group encompassing definitions focused on metaphysics and natural theology;
2. *emotional*, including the 'feeling of value' of the sacred;
3. *relationary*, with regard to 'the consciousness of our practical relation to an invisible spiritual order';
4. *biological*, i.e. the 'evolutionary and dynamic conception of mental life as opposed to the pre-Darwinian static conception' (resp., Leuba 1912: 29, 45, 38, 42).

Leuba decried the lack of definitional cogency and heuristic force of most definitions included in the first three categories (that is, the lack of convincing classification and explanatory advantage): '[a]lmost all of the definitions that have been reviewed attempt to say what religion is. According to them, it may be almost anything one pleases: a belief, a feeling, an idea, an attitude, a relation, even a faculty' (Leuba 1912: 42). The first three categories, Leuba continued, were 'completely out of harmony' with the fourth class, the newest one, thought to provide the 'most consequential change of point of view in contemporary psychology' (Leuba 1912: 42). Prefiguring Jeppe S. Jensen's interaction between *e*-religion (*external*, i.e. social or individual behaviours) and *i*-religion (*internal*, i.e. mental mechanisms) (Jensen 2014: 1, 42), Leuba wrote that 'in its objective aspect, *active religion* consists, then, of attitudes, practices, rites, ceremonies, institutions; in its *subjective* aspect, it consists of desires, emotions, and ideas, instigating and accompanying these objective manifestations' (Leuba 1912: 53; my emphasis). Finally, fully embracing a Darwinian perspective, Leuba concluded that 'the reason for the existence of religion is not the objective truth of its conceptions, but its biological value. This value is to be estimated by its success in procuring not only the results expected by the worshipper, but also others, some of which are of great significance' – such as the pro-social and morally binding role of such beliefs as high gods (resp., Leuba 1912: 53 and 14; on Leuba's rationalist attitude, see Sharpe 1986: 104). As we will see in detail in the following chapters, Leuba's lucid recommendation was to remain a dead letter for decades.

Commenting on this list almost 80 years later, Jonathan Z. Smith – whose critical works had a groundbreaking impact on contemporary HoR and Religious Studies – has famously noted that religion is a

> term created by scholars for their intellectual purposes and therefore it is theirs to define. It is a second-order, generic concept that plays the same role in establishing a disciplinary horizon that a concept such as 'language' plays in linguistics and 'culture' plays in anthropology. There can be no disciplined study of religion without such a horizon.
>
> Smith 2004: 193–4

Smith did not mean that every definitional effort is equal to another one; like Leuba did before him, Smith considered epistemological evaluation and biology as paramount (Smith 2004: 19–25, 193). Unfortunately, Smith's comment has been used, and abused, to foster a social constructionism deprived of any link to human biology and cognition. Moreover, the lucid awareness or 'self-consciousness of the researcher' hailed by Smith (Jensen 1993: 120) has been rarely cited as a universal glue for the discipline and, as a matter of fact, many of those interested in a self-conscious and epistemologically critical approach have migrated, as it were, to other, relatively new fields, i.e. Religious Studies (RS from now on)[1] and Cognitive Science of Religion (CSR henceforth).

Indeed, critical proposals coming from a small minority of historians of religions, concerning a close re-examination of the most basic assumptions of the discipline from an epistemological perspective (Penner and Yonan 1972), paved the way for the very foundation of both RS and CSR. These two disciplines led the long-awaited conceptual deconstruction or reverse-engineering of the very tenets of classical HoR.

Notwithstanding their quite controversial relationship (see Arnal and McCutcheon 2013: 91–101), these disciplines at their core are much more similar than usually thought (cf. L. H. Martin 2014).[2] They both have contributed to radically re-conceptualize and reorganize the analytical toolbox in the wake of the recognition of the disunity of the concept of 'religion', which, as some CSR scholars claim, is composed of many building blocks, like a box full of Lego bricks, with each brick as a detachable unit having a different evolutionary provenance, function and structure (see Martin 2015: 127 for an 'instructionless' Lego set; cf. Taves 2009 and Barrett 2017). RS, in turn, has focused on the socio-political agendas which the very concepts of religion, as historically stratified and coercive tools to exert hegemonic control do convey and reinforce (e.g. Arnal and McCutcheon 2013: 27–30).

An updated current definition of what the concept of 'religion' is or, perhaps better, refers to, is provided by cognitive anthropologist Harvey Whitehouse:

> arguably religion is not a single coherent entity but only a loose assemblage of patterns of thinking and behaviour that has been conceptualized very differently over time and across different language groups and cultural traditions. Recent research in the cognitive sciences suggests that many of the features commonly associated with the 'religion' label in Western scholarship and popular discourse are not the outcome of a single coherent cluster of causes but, rather, are the by-products of disparate psychological systems that evolved to address very different kinds of problems.
>
> Whitehouse 2013a: 36[3]

Given their non-exclusive cognitive nature, each one of the components of the religious toolkit may be exploited for other non-religious purposes (e.g. 'secular rituals' such as national anthems, sports fanaticism, etc.), and actively co-opted as part of what Italian scholar Furio Jesi (1941–1980) called the *mythological machine*, i.e. the dynamic use of mythographical discourses to reproduce power structures and engender an institutional system of authority within an 'imagined community' (see Manera 2012; cf. Lincoln 2000, and Anderson 1991). What is panhuman is the 'stickiness' and reoccurrences of such mythological and societal structures. We will further explore these disciplines in the following chapters.

Sui generis crisis

The difference between the concept of religion as an instructionless, building-block, modular infrastructure and the tenets of classical HoR highlighted in the opening paragraphs could not be greater. Classical HoR was built around the concept that religion is a 'thing', that religion has a specific identity because it is an autonomous subject. Consequently, religion was thought to have an independent existence as a subject of enquiry. Therefore, religion could be compared across space and time and can be classified into smaller units. This is the so-called *sui generis* idea of 'religion', meaning that religion is conceptualized as belonging to a *specific* genus which, by virtue of its essential features, is ontologically not reducible to any other cultural, social, political or scientific taxonomic rank. This idea was not neutral or impartial; it had complex extra-epistemic reasons that, among other things, were the results of ideological and political strategies (Jensen and Geertz 1991). This is how RS scholar Russell T. McCutcheon has summarized the whole concept in a recent interview:

> *sui generis* religion – i.e. unique, uncaused by something else, irreducible, one of a kind, pre-political, pre-social, the presumption that our object of study is a pristine experience that cannot quite be put into words; here 'words' stands in for some kind of a contingent, historical, public thing that is somehow not fully capable of expressing the prior, interior, pre-political, unmediated experience.
>
> McCutcheon 2014

Whenever this baffling circularity is openly endorsed as the main analytical tool for HoR, a constitutionally religious human being, or *homo religiosus* according to a fancy Latin label, provides nothing more and nothing less than a 'model for a new natural theology', thus exiting a proper, non-confessional, empirical, epistemologically grounded research (Penner and Yonan 1972: 132). As some scholars have argued, this whole *sui generis* concept had been quite successful in supporting the modern establishment of HoR as an autonomous academic discipline in the United States and in Western academia during the Cold War, as the academically legitimized *longa manus* of anti-communist and pro-religious politics (e.g. McCutcheon 1997; Doležalová, Martin and Papoušek 2001; Arnal and McCutcheon 2013: 72–90).

As a result of this prevalent confessional aspect, it is unsurprising to note, as some historians of religions have highlighted themselves, that contemporary HoR is facing a profound epistemological and methodological crisis of identity (e.g. di Nola 1977a: 291–300; Sharpe 1988: 256; Lawson and McCauley 1993 in McCauley 2017: 53–73; Lincoln 2012; Martin and Wiebe 2016: 221–30). The way by which historians of religions arrive at their conclusions and the ideas that support those methods are increasingly under attack. While some historians of religions are really trying to update scientifically their own field, others are strenuously resisting any effort to update and improve their approaches.[4] Resistance is strong, and some national schools are standing up against change more stubbornly than others, even when their pivotal tenets and central ideas seem to lack any sufficient support.

The arms race of natural theology

Probably, the most disturbing aspect to note at this point is that the future problems of the discipline were already present right from the start (cf. Wheeler-Barclay 2010). As the example of Leuba has shown, at the same time many exit strategies to escape from these issues have been devised over the course of the past centuries. This puzzle is *the* constant of the entire history of the history of religions and, in order to understand this situation, we have to contextualize the cognitive foundations of human reasoning within the scaffolding provided by the social history of science.

If 'natural history is the careful observation of nature' which provided 'the foundation for the discovery of evolution' (Travis 2009: 754), then the persistently confessional aspect of the *sui generis* approach of the HoR might be considered as something akin to natural theology, i.e. the identification of a creator, designer and (mostly) benevolent god thanks to intuitive reasoning and the observation of the living world (De Cruz and De Smedt 2015). In a sense, *homo religiosus* is but a folk-social and folk-psychological subset of a natural theology directed at proving the existence of such a god. Ironically, natural theology, in its efforts to accumulate a consistent encyclopaedia of natural knowledge, had been instrumental in the development of evolutionary theory. Even though the fine anatomical organization of such organs as the eye was incessantly hailed as proof of a creation by a caring, careful, designer god, the more natural theology strove to collect precise facts and elaborate complex classifications and explanations of the natural world, the wider the available knowledge that runs foul of any reductionist theological explanation became. Innumerable variations of optic organs apt to detect light and movement, from the more rudimentary to the most advanced, were being added as palaeontological research progressed. And, most of all, theodicy remained theologically challenging, to say the least (Ruse 2009; cf. Dawkins 1986).

Naturalist extraordinaire Charles R. Darwin (1809–1882), who in his youth had been enticed by the explanatory and confirmatory power of natural theology, grew sceptical until he renounced both natural religion and scriptural evidence in favour of an agnostic (and sometimes atheistic) scientific epistemology (Moore 1989; Pievani 2013a). In 1860, Darwin wrote explicitly in a letter to American botanist Asa Gray (1810–1888) that there was no rational way to cope with the problem of theodicy from within the perspective of natural theology:

> but I own that I cannot see, as plainly as others do, & as I shd wish to do, evidence of design & beneficence on all sides of us. There seems to me too much misery in the world. I cannot persuade myself that a beneficent & omnipotent God would have designedly created the *Ichneumonidæ* [a family of parasitoid wasps] with the express intention of their feeding within the living bodies of caterpillars, or that a cat should play with mice. Not believing this, I see no necessity in the belief that the eye was expressly designed.
>
> Letter of Darwin to Asa Gray, 22 May 1860,
> *Darwin Correspondence Project*, DCP-LETT-2814

Likewise, as we shall explore in the third chapter, the Victorian intellectual arms race to explain the most astonishing details of the natural world as the results of a divine project, supported by what was considered as scriptural evidence, eventually led to the birth of evolutionary biology. The same process, characterized by the accumulation of evidence for justifying the existence of *homo religiosus*, led inevitably to a non-theological approach in the historical study of religion(s).

The analogy between natural theology and the history of the HoR ends here: while natural theology has been falsified as an untruthful epistemological enquiry and relegated to pseudoscientific neo-creationism (*contra* Roberts 2009), HoR is atypical in that, contrary to its predecessor (the Victorian science of religion which we will tackle in the next chapter), it has been triumphantly accepted at the High Table of academia (Ginzburg 2010). In the process, HoR has also become a safe harbour for those natural theologians who work in the Humanities (Ambasciano 2014; Ambasciano 2015a; Ambasciano 2016a). This is, in a nutshell, the 'incoherent contradiction' of the HoR (Martin and Wiebe 2016: 224). We will delve deeper into these topics in the following pages, and in the final chapter as well. Meanwhile, let me conclude by underscoring that a history of the HoR might be preliminarily framed as a mix of cyclical advance and regression in which counterintuitive, but trustworthy, scientific research vies for survival against cognitively intuitive, but fallacious, confessional approaches that are responsible for the 'continuing infiltrations of extrascientific agendas into the field' (Martin and Wiebe 2016: 224). If we would want to simplify further this historiographical process, at the risk of losing the fine-grained texture of microscopic precision, we could say that every time a progressive, scientific movement concerned with natural approaches (i.e. scientific, psychologically and/or biologically based) gains momentum, a regressive reaction stops it, and so on, *ad libitum*, each time with unnatural responses (i.e. fideistic, theologically or spiritually based) replacing and neutralizing any previous scientific attempt. This apparently causes similar approaches and perspectives to recur repeatedly and recurrently. The HoR, if we want to return one last time to our metaphor, is a geological hotspot, an active fault zone where pro-science activists and anti-science denialists have fought almost incessantly about the nature, tempo and mode of geological risks themselves. And so far, thanks to the intuitive appeal of folk religion (McCauley 2011), anti-science denialists have had the upper hand, successfully repelling any pitch invasion from pro-science supporters. It is time for us to leave both our imaginary geological maps and our worn-out interlocutor and delve deeper into the unnatural history of HoR so that we can fully understand the legacy of the discipline.

2

The Deep History of Comparison

Science, we may tentatively say, begins with theories, with prejudices, superstitions, and myths. Or rather, it begins when a myth is challenged and breaks down – that is, when some of our expectations are disappointed.

Karl Popper

A rusty toolbox

The comparative study of religion typical of the HoR, whatever its features may be, assumes that there must be some religious entities in time and space that are somehow comparable. And if there are things that can be compared, it means that those things do really share similarities according to which they can be classified. Therefore, *classification* is the first and most important tool of the comparative study of religion. The implicit risk is that religion might encompass at once the agent by which classification is allowed and the items classified. As J. Z. Smith has written, 'religions are not only the objects of classification, they are themselves powerful engines for the production and maintenance of classificatory systems' (Smith 2000: 38). Religious classification, as well as any major attempts at classifying human cultures, was (and still is) mainly driven by *essentialism*, that is, quoting from an article by Benson Saler, the 'idea that an object is what it is because it has certain unchanging and necessary properties or qualities. [...] Essences [...] are not always accessible to perception, but they are held to be determinative both of genuine identity and of true-to-nature behaviour' (Saler 2008: 379). A consequence of this heuristic method of epistemic knowledge (in this case, the intuitive adoption of immediate and unmediated ways to discern similarities in the social and natural world) is that accidents, variants and contingent factors are deemed superfluous. In Greek Platonism, for instance, defects were ascribed to the fact that human beings' ability to perceive the nature of things was limited or, alternatively, imperfections were attributed to the inevitable corruption of the planet Earth against the backdrop of the incorruptible perfection of the universe (Leeds 1988: 467).

According to essentialism, a necessary set of features beyond the appearance of things is reputed to define and delimit the clustering of classes. The underlying epistemology of essentialism, as noted by Anthony Leeds, rests on an *a priori* intuition plus an active filtering: '[o]ne can arrive at a knowledge of some postulated Thing, untroubled by variants and variations, not by induction concerning the variation, but by contemplatively disregarding the variation in order to find the essence' (Leeds 1988: 468). Invariably, essentialism posits that such features are constant, notwithstanding the possibility of immanent, potential developments (e.g. an acorn has the developmental potential to become an oak). This idea provides the cornerstone of many classifications of classical HoR. As we shall see in the next paragraph, this is exactly the opposite of what a scientific, Darwinian epistemology should strive for (i.e. looking for variants and imperfections as the hallmark of history). In any case, essentialism provides an intuitive interpretive pattern for human reasoning and categorizing in general, and not just for philosophical disquisitions. In 1966, Paul McCartney had an accident while riding a moped, which resulted in a cut on the upper lip promptly disguised under a moustache. Because of this change, fans believed that the real Paul was dead and that he was being replaced by an actor. A mythology of sort grew along this premise, with fans working restlessly to uncover the conspiracy while decrypting imaginary clues in both the Beatles' lyrics and cover artwork (MacDonald 2008: 311–12). A sudden change in the appearance is enough to trigger an essentialist suspicion, and today the Internet is rife with conspiracy theories about doppelgängers or clones impersonating supposedly long-dead celebrities (Cresci 2017).

Another critical feature is the *emic/etic dilemma*. 'Emic' is an anthropological term that refers to the semantic level and which defines the insiders' description and/or interpretation of their own worldviews. 'Etic', instead, denotes the systemic dimension and refers to the specialists' description and/or interpretation of the insiders' own description/interpretation, basically positing a meta-interpretation of an interpretation (Jensen 1993: 124–5; Purzycki and McNamara 2016: 155–6). So, in our case, things might seem quite obvious: emic refers to the believers and practitioners themselves plus their own religious systems as defined by themselves, while etic refers to the scholars who are studying, describing and interpreting the descriptions provided by the believers. Yet, HoR does not seem to have a consistent record of getting this qualitative difference right. Before and during the twentieth century, the scholars who studied the classifications provided by their subjects of study were biased believers themselves, and in doing their research they often blurred the distinction between participation and observation. To put it simply, and with the benefit of hindsight, the result was, from a scientific and academic perspective, quite confusing.

When looked at from a big-historical perspective, essentialism and emic/etic perspectives have been two fundamental tools behind the construction and the study of religious classifications (Figure 2). Each past and present human culture has modulated the use of these tools according to its own sensibilities (unless a hypothetical and complete geopolitical isolation prevented such culture from doing so). For instance, during the Roman period, integrative comparison was a culturally codified way to include systematically foreign gods and deities by *conjectural assimilation*, meaning that a local god was assimilated to an apparently similar Roman deity on the basis of some common features. What we know today about ancient Celtic and German gods is mostly due to Roman writers who assimilated those goddesses and gods to similar Roman deities (i.e. respectively, Caesar in his *De Bello Gallico* and Tacitus in his *De*

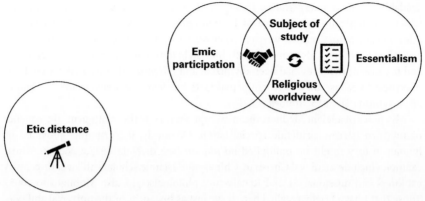

Figure 2 The study of religious worldviews between essentialism and the etic/emic dilemma

origine, situ, moribus ac populis Germanorum, more commonly known as *Germania*). This mechanism implied more than continuity between different beliefs and different gods; it implied an underlying translatability, a common identity.[1] Once this common ground had been established, different comparative enterprises could have been performed. Early Christian writers, for instance, favoured dissimilarities over similarities and distinguished between 'true' religion and 'false' religion – the 'true' religion being theirs, of course.[2] From Late Antiquity onwards, different apologetic and theological answers were advanced to explain away both the presence and the shocking similarities of pagan religions with Christianity (di Nola 1977a: 269; Sharpe 1986: 7–13; Borgeaud 2013: 37–65; Bettini 2014):

1. demonic origins;
2. plagiarism by ancient philosophers or demons;
3. divine condescendence towards some pre-Christian tendencies;
4. natural revelation and/or subsequent degeneration.

Yet, notwithstanding their preference for discontinuity and their unwillingness to engage in a fully detached etic analysis (their interpretation being overshadowed by their own beliefs), early Christian writers remained firmly within the boundaries of a comparative endeavour: even to refute someone else's beliefs, the knowledge of those beliefs is paramount.

Long before that, Greek thinkers and philosophers questioned the validity of myths and set out to describe the rational and comparative interpretations of religions. Cultural and political relations between the Greek world and its neighbours (e.g. ancient Egypt, Persia, India, Mesopotamia) sparked a huge ethnographic, geographic and comparative industry, today mostly available as fragmentary descriptions, keenly dedicated to religious customs and beliefs (for an overview see di Nola 1977a: 263–8; Sharpe 1986: 3–7). The same interpretative account described above was already available as *Greek interpretation* (*interpretatio Graeca*), championed by Herodotus, which was basically the ancestor of Roman conjectural assimilation. Other equally relevant explanatory models were *euhemerism*, which took its name from Euhemerus, according to whom gods were originally human beings that did great things in the past and because of that they had been divinized; and *allegories*, that is, myths, were to be interpreted because they were not actual facts but fantastical adventures (cf. Hawes 2014; Whitmarsh 2015; Roubekas 2017).[3]

My choice of ancient Mediterranean examples is merely the contingent consequence of my own narrow academic specialization. Obviously, the examples provided by human history might be multiplied *ad libitum* (see di Nola 1977a: 268–86, whose examples include medieval Christian, Chinese and Islamic scholars, thirteenth-century explorers and missionaries, and Renaissance philologists; cf. also Momigliano 2005). Those that I have briefly recalled here serve just as instances of the universal ability to trace common patterns between religious matters in order to compare them – or to disassemble them via a critical approach (e.g. L. H. Martin 2014: 66–79; see also Minois 1998; Geertz and Markússon 2010; Jensen 2014: 21). Indeed, the human capacity for social and cultural categorization, exemplified by the existence of classificatory systems about religion(s), might be considered as old as *Homo sapiens* itself, that is, at least 300,000 years, if not older (cf. Geertz 2013). Then again, the mental toolbox for categorizing and classifying is old and rusty, as it has been in use since before the evolution of our primate lineage. Nonhuman animals need to intuitively classify the items in their environments to map their surroundings, in order to act in one way or another, to discriminate between in-group members and outsiders, etc. (Shettleworth 2010: 167–209). Therefore, classification and categorization can be evolutionarily considered as good-enough, imperfect, cognitively intuitive features of animal cognition that can be co-opted for in-group social policy and which are nevertheless prone to maladaptive misfiring (e.g. Ambasciano 2016a: 188–9).

As recently summarized in a reworked excerpt from neuroendocrinologist and primatologist Robert Sapolsky's latest book,

> humans universally make Us/Them dichotomies along lines of race, ethnicity, gender, language group, religion, age, socioeconomic status, and so on. And it's not a pretty picture. We do so with remarkable speed and neurobiological efficiency; have complex taxonomies and classifications of ways in which we denigrate Thems; do so with a versatility that ranges from the minutest of microaggression to bloodbaths of savagery; and regularly decide what is inferior about Them based on pure emotion, followed by primitive rationalizations that we mistake for rationality.
> Sapolsky 2017a; see Sapolsky 2017b: 387–424[4]

Donald Wiebe has recently pointed out that, in the deep history of *H. sapiens*, such ontological categorizations might have been evolutionarily co-opted to constrain extended cooperation and implement a hazard-precaution system, as suggested by the potential feedback loop between the following features (see Martin and Wiebe 2016: 62–89):

1. the role played by evolved, panhuman cognitive mechanisms on the transmission and learning of shared beliefs, resulting in the fallacious, folk-biological and ontological categorization of different human communities as if they were almost different species (Gil-White 2001; see Schaller and Murray 2008, and Schaller and Murray 2010 for a correlation between genotypes and cultural openness);
2. the existence of universal epistemological fallacies tied to in-group self-deception to boost self-confidence (Munz 1985; Trivers 2011);
3. the possible benefits of post-hoc religious justifications for limiting out-group solidarity, in order to strengthen an intuitive disease-avoidance strategy (Villareal 2008; Fincher and Thornhill 2008; Fincher and Thornhill 2012).

However, what had probably been effective as a 'behavioural immune system' in the deep past can be a dangerous and maladaptive hindrance in modern times (Martin and Wiebe 2016: 61). With regard to religion itself, Wiebe asserts that the continued existence of religiously justified violence, ethnocentrism and xenophobia in the globalized network of intercontinental cooperation might be understood as the development of an auto-immune disease with a negative cost/benefit ratio for both in-groups and out-groups (Martin and Wiebe 2016: 62–3). However interesting and thought-provoking such hypotheses might be, since the focus of the volume is just the modern discipline of HoR, in what follows I will humbly limit my exploration to the modern and historical study of religion.

Birth certificate(s): Ultimate origins

Although any identification of a precise historical beginning invariably creates problems of inclusion and exclusion, especially those concerning cultural watersheds (Smail 2014), contemporary disciplinary categorization in academia calls nonetheless for a date of birth. There are two main perspectives on the origins of the HoR. According to some scholars, the modern and comparative study of religions began officially after the Darwinian revolution, in the late nineteenth century, when the organization of modern universities, i.e. national and public institutions dedicated to the accumulation and advancement of shared empirical knowledge, was systematically fine-tuned and the discipline began to be accepted as an autonomous field. Alternatively, it is held that the roots of the discipline lay down into the age of Discovery prior to the Enlightenment (roughly between the late fifteenth and early eighteenth centuries), when the encounter with unexpected human populations scattered worldwide and the exploration of new continents forced European explorers to reconsider the received wisdom on everything they knew and cherished – most of all religion.

However, this is a false dichotomy, since one point of view does not disprove the other. In particular, it is undeniable that during the age of Discovery, amidst an endless catalogue of moral virtues, religion became one of the quintessential traits of human self-depictions: 'as early as the mid-seventeenth century, religion had been identified as a, if not the, uniquely human trait' (Day 2008: 49–50). But it is also true that it was the Darwinian revolution that sparked the academic interest in a truly global history of religions. And yet, as we have seen in the previous paragraph, instances of comparative endeavours between (and within) ancient cultures are also well attested. A way out of this conundrum entails a two-fold recognition. We could say that the *ultimate origins* of the discipline are to be found in the various encounters and intellectual confrontations that took place between different cultures since the dawn of human civilization. As a consequence of this big-historical provincialization, the age of the exploration of the Americas and Australasia becomes a subset of a much larger process of categorization. The *proximate origins* of the discipline, instead, are identifiable in the post-Enlightenment era, with the co-occurrence of three additional factors (Figure 3):

1. the Scientific Revolution and the diffusion of rational, empirical methods and theories to investigate both human beings and the natural world (Ferrone 2015);
2. the increase in ethnographic accounts of human populations that had never been in contact with Western cultures before;
3. the establishment of modern academic institutions – and corresponding growth of international scholarly networks – by the emerging European and (later) American nation-states.

Let us now delve deeper into this reorganization. It is commonly held that the cultural shock waves produced during the age of Discovery put in motion an unstoppable domino effect. Historian of religions Guy Stroumsa has recently claimed that the origins of the HoR are to be traced in the centuries between 1600 and 1800, which saw the 'implosion' of the previously agreed-upon concept of 'religion'. As he writes,

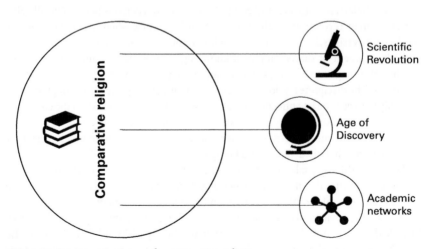

Figure 3 Proximate origins of comparative religion

the intellectual and religious shock caused by the observation of formerly all-but-unknown religious rituals and beliefs in the Americas and Asia provided the trigger without which the new discipline could not have been born [...] The newly discovered continents and cultures were slowly becoming part of the 'cultural landscape' [...] of European intellectuals. Books were now printed in classical or 'exotic' languages, such as Hebrew or Arabic, and the world of European Christendom had been torn asunder. This new knowledge of the diverse religions practiced around the world entailed the urgent need to redefine religion as a universal phenomenon.

Stroumsa 2010: 2–3

It was, in Stroumsa's words, 'a paradigmatic change' (Stroumsa 2010: 2–3). The birth of comparative religion, according to Stroumsa, is something strictly related to the same cultural environment that gave rise to Newtonian physics: religion was 'discovered', in the sense that the confrontation with previously unknown religions and cultures from Asia and the New Worlds prompted, if not a critical reconsideration, at least a thoughtful reflection on the basic common themes behind these 'new' religions (I have put 'new' in quotation marks because they were not new at all, it was just that they were unknown to Europeans). And again, as happened once in Graeco-Roman antiquity, the recognition that each and every society had a more or less institutionalized, or at least an organized, form of religion (in other terms, a *mythological machine*) was something destined to have a major impact on comparative religion. 'Beyond the multiple forms of religion, including the most barbarian forms of idolatry, such as the human sacrifices practiced by some American peoples', continues Stroumsa, 'all religions reflected the unity of humankind. As the Dominican friar Bartolomé de las Casas [...] would say, *Idolas colere humanum est* ("idol worship is human")' (Stroumsa 2010: 7). Even though the interpretive filter provided by the scholars' own religious beliefs was to distort any unbiased understanding (cf. Nongbri 2013), classification, comparison, essentialism and translatability were once again in the spotlight.

This 'discovery' (in quotation marks for the same reason explained above) had a two-fold and quite paradoxical effect: it prompted intellectuals to question the presumption of a unique ethical, theological and cosmological system of religious values, and ultimately led to the acknowledgement of religion as 'a single concept', as 'part of humankind's collective identity' (Stroumsa 2010: 7). This process of slow rethinking was facilitated by the fact that the Western ethical, theological and cosmological system behind religious values had been already (and inescapably) eroded by the European wars of religion between the sixteenth and seventeenth centuries, fought both physically and intellectually between Catholics and Protestants, and within different strains of Catholic and Protestant faith – not to mention the everlasting theological and political confrontation between the three so-called Abrahamic religions, that is, Judaism, Christianity and Islam, whose ultimate 'truths' mutually disconfirmed the others', each one paradoxically reclaiming unique and divine revelation or interpretation (Preus 1987: 205; on the problematic concept of Abrahamic religion, see Hughes 2012).

As a result of such confrontation between old and new problems, the modern idea of *natural religion* was born, an emic concept already in vogue during antique and medieval times to define the fact that religion appeared to be a *sine qua non* feature, an indispensable trait of the human being *qua* human being (Stroumsa 2010: 11), defined by a specific 'basic and minimal list of truths' (Preus 1987: 205–6). In the words of Ivan Strenski, 'natural religion embodies the belief that religion is an innate, built-in "common" feature of being human. It is therefore "natural", because it is a "normal" part of who we are. Natural religion reflects the belief that all people are born with a capacity or talent for being attracted to the ultimate reality' (Strenski 2015: 11).

Most importantly, natural religion was supported and reinforced by *sacred history*, i.e. the view that 'historians writing in the Judeo-Christian tradition' held as true and that 'located the origins of man in the Garden of Eden', that is, somewhere in the Fertile Crescent and sometime around mere thousands of years BCE (Smail 2008: 12). Since the earliest philosophical, cultural and religious contacts, the Near and the Far East have always profoundly influenced and ignited the Mediterranean and European imagination. In ancient Greece, some philosophical currents whose basic tenets were destined to recur later, held that immemorial and primordial sacred wisdom was located in the exotic ancient Egypt or in the far East – dating equally far back in time (cf. Momigliano 1993: 146–7). The equation is simple: the more distant, the better; the more ancient, the more prestigious. And prestige was readily translated into social and cultural terms of originality and purity, and thus embedded in political discourses (Jensen 2014: 101). Modern natural religion simply took this process one step further, globalizing the search for the most ancient traces of revealed sacred wisdom.

Homo religiosus, i.e. 'religious man', can be considered as the final elaboration and ultimate product of both natural religion and sacred history in that it indicates that the human being is religious deep down inside, even though s/he might be atheist or agnostic.[5] More precisely, *homo religiosus* expresses the concept that the human being is literally imbued with a divine transcendence not accessible via ordinary analytical tools and which should be understood and studied tautologically, that is, according to the very religious inner sense that is postulated. In other words, this idea was the precipitate and the apogee of emic reflections on religion (see Figure 4). As we will see

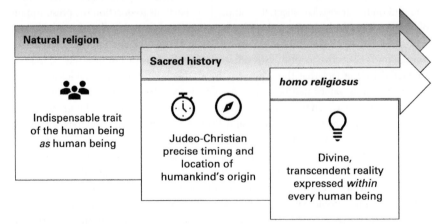

Figure 4 Diachronic and cumulative development of the following concepts: natural religion; sacred history; *homo religiosus*

in the next chapter, *homo religiosus* proved to be a most powerful idea, destined to have an unprecedented impact on the HoR as an academic discipline.

A preliminary note on imperialism, postmodernism and science

Notwithstanding a fideistic attitude, the recognition of such a common ground among human cultures also paved the way for a profound philosophical reassessment of religious beliefs, leaning towards a more rational study of religion. Comparison entails the recognition that any 'absolute truth' of a specific tradition might be reconsidered alongside other absolute truths, and in the long run this reconsideration might emerge into a more or less radical criticism of those very absolute truths as not so absolute as those belonging to that particular tradition might claim them to be. As a result, gradually, naturalistic accounts of religion, that is, accounts that tried to eschew and avoid emic explanation in favour of natural and social (i.e. etic) explanations, were developed. The modern European process of 'explaining religion' in rational terms, thus gradually renouncing the emic claims of absolute truth and rejecting or reworking the various explanations provided by believers, was by no means exclusive to modernity since it had previous and brilliant precursors in ancient Greece and Rome, as we have already and briefly seen. As a matter of fact, the European Renaissance saw the recovery and the renewed study and circulation of ancient works on the nature of religion, igniting an interest in those ancient philosophical works that framed or criticized different religious beliefs by comparing them. Thus, it can be said that the rescue, restoration and circulation of past Graeco-Roman interpretations of religions, which provided a filter by which difference and interpretations were projected not only far away in space but also far away in time, constitutes an integral part of the modern age of religious discovery and confrontation (cf. Bod 2015).

More specifically, what set apart modern Europe from similar historical parallels was not just the breadth, depth and magnitude of a truly global research, but the intimate connection with nascent scientific, rational, empirical and non-confessional approaches born out of the 'horror and the vast and dramatic consequences' of the religious wars of the sixteenth and seventeenth centuries (Ferrone 2015: 100; cf. Martin and Wiebe 2016: 222–3). As James Turner has remarked, other strictly related circumstances had been equally important, such as the mid-nineteenth-century rise of 'an increasingly liberal Protestant theology' interested in explaining in an emic way similarities between different religions as a means to recover a sort of super- or meta-theology with ethical concerns (Turner 2014: 370). Of course, when seen as a whole, those were far from being purely disinterested processes. As was the case with other imperialistic enterprises and politics of the past, the creation of a comparative *corpus* of knowledge was made possible by another 'horror', i.e. the same societal power structures that were later to exploit this accumulation of knowledge as a tool to implement a mercantilist and capitalistic dominance hierarchy which allowed for colonialism, missionizing and, basically, the strategic control of subordinate peoples (Wheeler-Barclay 2010: 253; Chidester 2014). Such imperial organization focused on the exploitation of manpower and natural resources allowed for the development of the huge networks of intellectual and practical knowledge that led to, and supported, the Scientific Revolution (Bowler and Morus 2005: 215–18). As to the academic study of religion, as we shall see in more detail in Chapters 4 and 5, approximately between the 1890s and the Interwar period, an anti-Enlightenment vendetta against etic and scientific approaches came to dominate the Continental HoR with the (sometimes explicit) two-fold aim of offering intellectual support to European imperialism and colonialism, as well as of legitimizing an aggressive interstate nationalism which underlay an equally violent politics of expansion and domination within the same European confines (Leertouwer 1991, and Gandini 2003: 195–202, respectively on Dutch and Italian HoRs and colonialism; cf. also Junginger 2008 and Segal 2002a for an overview on myth and politics). However, this outcome cannot be equated with most of the *preceding* comparative and scientific efforts to study religion, nor can it be generalized, for even during the heyday of the extreme right-wing politicization of the HoR some intellectuals, albeit admittedly isolated, managed to express their critiques and doubts.

At the same time, it is undeniable that, once properly contextualized, scholars interested by this radical transformation differed wildly in their closeness to political or legislative power, and their influence varied accordingly. Without denying in any way the violent, shameful, and painful legacy of modern European imperialism and its ties to science and technology (e.g. Palladino and Worboys 1993), distinctions should always be provided lest an over-generalizing, postmodern tendency to equate science with oppression phagocytize and neutralize the epistemological, rational advances produced by the Enlightenment and the Scientific Revolution (Ambasciano 2014: 24–7; cf. also Malik 2017 on the decolonization of university curricula; on the distinction between science and technology, see Mesoudi *et al.* 2013). Likewise, it would be extraordinarily naïve – not to say historiographically wrong – to equate *tout court* (non-Western) religions with freedom, for power relations of intra-group, inter-group, and gendered violence or domination are more or less explicit in every past and present human social structure dominated by religion (e.g. Lincoln 2005; Bremmer 2011; Murphy 2011; Jerryson, Juergensmeyer and Kitts 2013; Sela, Shackelford and Liddle 2015). On the other hand, we should equally pay attention to the inter-cultural, emic, confessional encounters that resulted in a two-way reconsideration of 'religion'. Indeed, as was the case with Hellenistic and Roman reinventions and renegotiations allowed by imperial expansion, modern and contemporary intercontinental contacts prompted a mutual, eclectic and sometimes radical, re-elaboration of religion both in the dominant and in the dominated parties – scholars and historians of religions included (Bentlage *et al.* 2016).

Each one of these topics would need an entire library to be fully covered with the depth and attention it deserves – and I cannot pretend to cover all of them in a few lines here. As I refer the interested reader to the bibliographical resources recalled above, I wholeheartedly invite those readers interested in the relation between postmodernism, science and the HoR to bear with me until we reach together Chapter 6. Meanwhile, I must return to the focus of the present chapter. What follows is a run-through of some major thinkers that helped to shape the historical and naturalistic study of religion(s), a historiographical enquiry started by Samuel Preus' (1987) volume *Explaining Religion*, in which criticism and comparison, even when born within the confines of theology, set the stage for further scientific approaches. It is by no means an exhaustive list, as it merely provides a quick-and-dirty chronological and partial summary of the complicated historical process that led to the HoR.

Rationality as cumulative by-product of comparison: From Jean Bodin to Edward Burnett Tylor

To recap, the influence of classical sources and the presence of theological tenets were two of the most important drivers of comparative religion during the early modern period. Although working within a Christian framework, intellectuals like Jean Bodin (1530–1596), who adopted the classical Graeco-Roman dialogical form of exposition between a fictional group of men, dethroned Christianity from its pedestal to compare it with other beliefs, highlighting how contemporary religious rivalries were a product of Adam's original sin. Bodin's *Colloquium heptaplomeres de rerum sublimium arcanis abditis* (1588) offered a 'rational' answer to such a problem (rational, of course, between scare quotes, for it still was a fideistic one): the solution was the recovery of the best religion, i.e. the most ancient and prestigious, the one originally granted by God in the Garden of Eden. In any case, as Luther H. Martin and Donald Wiebe have highlighted, 'by debating the fundamentals of religion', Bodin's imaginary 'disputants br[ought] religion into doubt and suggest[ed] the need for tolerance', and by doing so, they implicitly undertook and endorsed a comparative analysis of religion(s) (Martin and Wiebe 2016: 222). English diplomat Herbert of Cherbury (1583–1648) engaged to carry out a similar comparative endeavour. Herbert extracted the most basic religious features (or, in modern terms, innate ideas) behind those major religions' tenets he was acquainted with, thus paving the way for modern deism by appealing to a sort of innate human predisposition. As summarized by Preus, this set of features entailed 'that there is one God; that ought to be worshipped; that virtue is the principle part of Religion; that we ought to repent of our sins, that there are punishments in the next life' (Preus 1987: 28 n. 14). These are clearly features found in monotheistic Abrahamic religions (quite a limited survey, then), but the most important point of Herbert's reflection was that his set of basic features recalled above was not just to be found in the Sacred Books of those religions, but was to be understood as something deeply rooted in human conscience *tout court*.

Generally speaking, the road towards a scientific and etic enfranchisement from both a confessional normative framework and emic tenets was far from being straight and smooth: both perspectives co-existed in rather unexpected ways – sometimes out of institutional respect or expediency. Frenchman Bernard Fontanelle (1657–1757), for instance, elaborated a universal and naturalistic account of ancient mythologies in which he highlighted that ancient myths were explanations of natural phenomena limited by the knowledge available at that time and which, consequently, included frauds and errors. He also rejected the demonic origins of ancient oracles as they were merely the products of priests who wanted to dupe and control the population. While Fontanelle 'left the intelligent reader to complete his suggestions and extend the critical consequences to every religion' (Minois 2012: 184), he excluded Christianity from any critical observation, opting for an explicit immunization of its own theological tradition (Preus 1987: 53). The same sort of intellectual analysis was pursued also by Italian philosopher Giambattista Vico (1668–1744) who, like Fontanelle, worked from within a strictly theological perspective, as he intended to offer a more 'rational' theodicy, i.e. the theological and ultimate explanation of good and evil, justified by God's will itself. According to Vico, history had a divine sense, a Providence willed by God, which through its laws accounted for the development and succession of historical facts (cf. Leeds 1988: 461). And yet, since the organization of human institutions like religion had a social history and a societal development, they were created by human beings and were not the products of inscrutable superhuman entities. Moreover, even though inscribed into a providential scheme, changes in mentality were ascribed to a feedback loop of contextual changes in society and culture over time (Preus 1987: 81). We are undoubtedly in the same religious and theological horizon of the previous thinkers, a horizon evident in both the natural theology of a human being provided with a 'natural instinct for divinity' and a teleological, divine history willed by a God who knew already the ultimate end of history (Preus 1987: 77). However, the recognition that the development of historical facts itself is a human affair gave a significant boost to study human institutions as something conceived and built by human beings alone. In the way that natural theology led to evolutionary biology, it could be said that rational approaches in the study of religion(s) were the inevitable yet originally unintended by-product, as it were, of the gradual accumulation of comparative enterprises (Wiebe 1999: 3). Perhaps Popper's dictum, that 'science [...] begins with theories, with prejudices, superstitions, and myths', is more valid for the HoR than for any other social science or field in the Humanities (Popper 1994: 95).

The critical point of no return was reached when the psychohistorical accounts of religion became gradually detached from apologetics, that is, the search for theological justifications in the history of religious facts, which characterized Herbert and Vico and prevented Fontanelle from being more open in his critique. This intellectual turn became finally fully fledged with Scottish philosopher David Hume (1711–1776), whose criticism highlighted that previous explanatory attempts were merely justifications to the continued existence of religion itself, in a vicious circle of self-reinforcing belief. On the basis of a conception of human mental functioning as driven by mechanisms that today we would call cognitive (cf. Forrest 2013; Guthrie 2013), Hume advanced two groundbreaking ideas:

1. world religions are too diverse and varied to trace their natural ascendance to a single source identified in a universal, religiously innate, inner sense;
2. religious ideas arise from the fear and anxiety of human beings (especially with regard to the future) as well as from the anthropomorphic intuitive perception of agents in nature (e.g. pareidolia, that is, the intuitive recognition of human-like faces in the moon, on Mars, in the shape of rocks, etc. (Guthrie 1993).

Starting from these natural foundations, and building on the wealth of new ethnographic data, Hume rejected the argument from design and criticized beliefs in miracles as well as the ideas of divine revelation and natural religion, thus challenging Herbert of Cherbury's idea of an inner monotheism (Wheeler-Barclay 2010: 24; Strenski 2015: 23).

The key concept of anthropomorphic personification applied to religion, and arranged into an evolutionary framework, was to be labelled *animism* by Edward Burnett Tylor (1832–1917), the first holder of a chair of Anthropology at Oxford University (1896), where he previously served as a lecturer from 1884 to 1895. Animism – from *anima*, Latin for 'soul' – described the belief in the existence of animated invisible spirits that inhabited the world and populated the religions of the scattered peoples that came in contact with European settlers and explorers during late nineteenth-century colonialism and imperialism. Animism, according to Tylor, was a rational, although ultimately fallacious, product of natural enquiry concerning the properties of living bodies, then extended to explain dreams or hallucinations (Wheeler-Barclay 2010: 92). Tylor's anthropological system, whose most accomplished result was *Primitive Culture* (1871), was supported by a loosely Darwinian evolutionary scheme in which cultural progress was considered as the obviously natural continuation of biological evolution. In such a scheme, ancient beliefs of 'ancestral cultures' were doomed to become 'present-day' superstitions, while the underlying rationale was supplied, at long last, by the recognition of the 'psychic unity of mankind' (Desmond and Moore 2009: 365; Wheeler-Barclay 2010: 84). Almost recalling the Darwinian concept of 'vestiges' as 'rudimentary organs', Tylor specifically defined 'survivals' as 'processes, customs, opinions, and so forth, which have been carried on by force of habit into a new state of society different from that in which they had their original home, and they thus remain as proofs and examples of an older condition of culture out of which a newer has been evolved' (Tylor 1871, 1: 15; e.g. Darwin 1859: 452–6; for other influences, cf. Ratnapalan 2008; Saler 2009: 51–7). Although still biased by ethnocentrism, the continuity between present-day Europeans and other non-Western cultures firmly rejected *degenerationism*, i.e. the scriptural belief according to which 'the human race had degenerated from an original state of moral perfection' (Bowler 1989: 330), while affirming the co-existence of different stages of rationally fallacious religious thought even in contemporary (and thus 'modern' or 'advanced') traditions: 'animism for Tylor is the core of *all* religions' (Guthrie 2000: 106; see Wheeler-Barclay 2010: 92).

On the basis of his social evolutionism, Tylor also devised a componential approach to religion in which each religion could be regarded and analysed as a 'bundle of concepts, myths, and ceremonial activities', each one to be singled out by the researcher (Wheeler-Barclay 2010: 90). According to the same perspective, Tylor detached morality from theological pro-social claims, while positing that science was the ultimate heir of animism, in that only science had been able to clarify beyond any reasonable doubt the original animistic confusion between biology and human cognition (Wheeler-Barclay 2010: 85, 94–5). By the same token, he claimed that anthropology was to direct its efforts to the identification of Victorian superstitions (such as spiritualism and astrology) in order to explain them away and dismantle them (Wheeler-Barclay 2010: 87). With Tylor, we witness a complete reversal of the ideas of natural religion and sacred history: not only culture could have been studied scientifically, but theology was not exempted any more from the application of the same critical tools employed for history or politics, like Fontanelle and Vico before him. With the burden of proof shifted to theology, it was theologians who were now expected to provide a sufficient epistemic warrant for their supernatural claims.

Enter comparative mythology: Friedrich Max Müller

In most contemporary handbooks, Tylor is often grouped together with other forerunners of the so-called 'science(s) of religion', a term that has been sometimes employed to indicate a sort of family resemblance, or general similarities, among a set of early approaches that shared a common subject and a common method, i.e. religion and comparison, however differently understood and implemented (e.g. di Nola 1977a: 300–5; Filoramo e Prandi 1997; Waardenburg 1999; Saler 2000; Segal 2004; Wheeler-Barclay 2010; Sfameni Gasparro 2011). Anyway, even though the so-called 'scientists of religion' such as Émile Durkheim or Sigmund Freud were beyond doubt instrumental in the establishment of some specific branches of anthropology, psychology and sociology dedicated to religion (mostly between the latest decades of the nineteenth century and the first two decades of the twentieth century), two 'founding fathers' were ultimately responsible for the establishment of the comparative mythological approaches that had the most astounding resonance in the soon-to-be-established HoR: Friedrich Max Müller and James G. Frazer.

German philologist, orientalist and translator of ancient Vedic texts, Friedrich Max Müller (1823–1900) taught at Oxford from 1858 onwards. In 1856, Max Müller published *Comparative Mythology*, a landmark essay that marked the first modern scientific attempt at explaining the development of religious contents through time (Max Müller 1909). By following the same branching, divergence and diversification of languages through time, he posited that myths originated from a single mistake, so to say, apparently due to the linguistic and expressive constraints that in ancient times led to the improper characterization of stars and planets as human-like entities described by adjectives. Now, the common ascendance of what will be labelled as Indo-European languages had been already proposed in the late eighteenth century by Sir William Jones (1788). Basically, Max Müller applied Jones' philological analysis to religion. He compared a selection of ancient Eurasian male high gods like Greek *Zeus*, Latin *Jupiter* (whose genitive form is *Iovis*), Vedic *Dyaus Pitar*, Germanic *Tiwaz*, Old Norse *Týr*, etc., and he came to the philological conclusion that they all descended from an original and reconstructed **Dyeus*, which in turn was thought to be originally

> the sun in its various phenomenal modes [...]. In myths [Max] Müller saw not simply the personification of the sun, the dawn, the twilight, and so on, but a metaphysical correspondence that human thought and human language drew between the perception of nature and the analogies that the ancient Indo-Europeans had used when communicating what they perceived. The names that people gave to these phenomena, the *nomina* (sing. *nomen*), were later mistaken

for divine beings, or *numina* (sing. *numen*), and myths began to develop around these names to account for their existence.

Stone 2005: 6234

This is the core of the so-called *nomina–numina* thesis, later critically scrutinized because of Max Müller's reliance on a purely philosophical precedence of myth over ritual (see Wheeler-Barclay 2010: 46–7, 116, 230, 249; on the Müllerian distinction between religion and mythology, cf. Segal 2016).

Max Müller rejected the Romantic pan-Sanskritism previously held by Friedrich Schlegel (1772–1829), who considered Sanskrit as the most ancient language (i.e. the *Ursprache*) from which other Indo-European languages had sprung (Wheeler-Barclay 2010: 39; see Villar 2009: 39). Instead, Max Müller posited a proto-historical cultural stage behind all extinct and extant Indo-European languages (a stage he labelled as 'Aryan', today known as Proto-Indo-European), while firmly refusing a corresponding racial profile: the community was created by the power of language, not blood (Wheeler-Barclay 2010: 52; cf. also Tull 1991 for Max Müller's selective approach towards Indian ancient texts). The diffusion and differentiation of languages through time followed the migration of such ancient peoples to Europe and India (Bowler 1987: 57).[6] Notwithstanding the efforts made by Max Müller to counteract a fashionable and fallacious 'quasi-mystical enthusiasm', the fascination for esoteric Indocentrism and religious pan-Sanskritism would have lingered on in the history of HoR, especially when combined with non-Darwinian progressionism stubbornly focused on the identification of exotic religious substrates and 'living fossils' (Wheeler-Barclay 2010: 52).

Notwithstanding the presence of confessional ideas, such as the belief in natural religion and original revelation, and his participation in the 1893 interfaith event World's Parliament of Religions (see Wiebe 1999: 12, 20; Nicholls 2014), Max Müller's programme for comparative religion was nothing short of a scientific breakthrough:

1. Max Müller rightfully acknowledged impartiality, 'a concern for truth, and a commitment to the rules of critical scholarship' as necessary conditions for an academic, historical, scientific study of religion (Wiebe 1999: 16).
2. He separated philosophical and theological speculation from historical analysis (Max Müller 1893: 146) and, accordingly, one's own beliefs from critical, rational, falsifiable enquiry (Max Müller 1898a: 543).
3. He understood the comparative study of historical religion as part of the natural sciences and, consequently, he realized that, just like natural history, classification of religious data (or 'careful collection of facts') according to each religion's 'origin, growth, decay' was necessary (Max Müller 1898b: 27).
4. He was also aware that, given a sufficient amount of data, classification entailed the 'discovery' of common structures which, in turn, required explanation and 'clarification' (Max Müller 1891: 15 and Max Müller 1878: 169; all citations from Wiebe 1999: 16, 21, 28 nn. 9 and 10).

Interestingly, an interdisciplinary *fil rouge* unites Max Müller's comparative philology and Tylor's groundbreaking anthropology. Referring to Max Müller's work, in 1868

Tylor wrote that 'there is in England at this moment an intellectual interest in religion, a craving for real theological knowledge, such as seldom has been known before' (Tylor 1868, cited in Wheeler-Barclay 2010: 71). Earlier, Tylor had also applauded Max Müller's 'consistent and scientific theory of the development of language from a few simple root-words upwards to the most expressive' (although Tylor discarded Max Müller's fideistic idea of language as the expression of a Platonic and divine 'power inherent in human nature'; see resp., Tylor 1866: 423; Max Müller 1864: 402; cf. Davis and Nicholls 2016: 25–30). Notwithstanding the clear differences concerning the underlying assumptions and starting points of their research (i.e. Tylor's non-confessional positivism vs. Max Müller's religious and teleological attitude), both scholars agreed on the epistemological validity and scientific status of the comparative method. Moreover, Max Müller's intervention was instrumental in bringing Tylor to Oxford (Wheeler-Barclay 2010: 81). Notwithstanding their efforts, scientific and etic comparison was soon to be ditched in favour of fideistic criticisms, and even readapted, or manipulated, to conform to some confessional tenets by other late-Victorian scholars in the field of the science of religion (in particular, Andrew Lang, William Robertson Smith, and Jane Ellen Harrison; see Wheeler-Barclay 2010 and the following paragraphs). Only one scholar stood true to the neo-Tylorian scientific programme behind comparative religion, taking such methodology to its most extreme consequences.

Triumph and dissolution of comparison: James George Frazer

James George Frazer (1854–1941), a Scottish anthropologist and fellow at Trinity College, Cambridge, accomplished a daunting task: he single-handedly wrote a much-celebrated encyclopaedia in twelve volumes nominally focused on a single, quite trivial and frankly odd topic. This central theme was the succession and coming to power of a priest devoted to Latin goddess Diana on Lake Nemi, in Central Italy, where a fugitive slave had to kill his predecessor to become king of the sanctuary, or *rex Nemorensis* (*rex*, i.e. 'king', is here a mere priestly attribute, not a political one: that figure was a *sacerdos*, a priest). Frazer's work, initially published in two volumes and entitled *The Golden Bough* (1890), was gradually expanded to accommodate a mind-blowing array of comparisons that extended from classical Antiquity to coeval ethnographic monographs, from philology to folklore, from the Mediterranean to the New Worlds. Frazer's *magnum opus* reached its twelve-volume systematization in 1915, to which Frazer added a compendium in two volumes (1922) and an *Aftermath* in 1936.

Frazer adopted Tylor's progressive evolution with more emphasis on human basic needs and less stress on mere philosophical speculation, and he identified in the continued existence of Tylorian survivals in each and every society a continuous and persistent threat to scientific advancement – a theme which has been recently re-elaborated from a cognitive perspective (McCauley 2011; see Wheeler-Barclay 2010: 192, 209). In Frazer's own words,

> it is not our business here to consider what bearing the permanent existence of such a solid layer of savagery beneath the surface of society, and unaffected by the

superficial changes of religion and culture, has upon the future of humanity. The dispassionate observer, whose studies have led him to plumb its depths, can hardly regard it otherwise than as a *standing menace to civilisation*. We seem to move on a thin crust which may at any moment be rent by the subterranean forces slumbering below.

Frazer 1890, 1: 236; emphasis added

Interestingly, Frazer's combined lucid pessimism and disillusion with an unflinching belief in science and progress, formulating a quasi-cyclical view in which science has to fight unrelentingly not to be reduced to silence by a mix of ignorance, superstition, irrationality, 'clerical resurgence' and mass manipulation by interested parties (Wheeler-Barclay 2010: 192–3). In a sense, Frazer's worldview, profoundly influenced by the socio-economic and political issues that affected Western Europe between the twilight of the *Belle Époque* and the horrors of World War I, provides the most adequate definition of HoR, intended as an ever-failing discipline whose scientific tenets are always prone to be recurrently dismantled by regressive and confessional interests (cf. Martin and Wiebe 2016). And yet, while implicitly acknowledging the fragility of science, Frazer called to arms 'the battery of the comparative method' to 'breach' the 'venerable walls' of the faithful citadel built on past and present mythologies – a 'thankless task' embraced reluctantly (Frazer 1900, 1: xxvi). Frazer, undaunted, also supplied the most grandiose background for what was to represent the ultimate – and the last – 'monument' of comparative mythology (Wheeler-Barclay 2010: 214).

In the wake of Tylorian progressionism, Frazer managed to demythologize almost every kind of global mythology or folkloric theme from the past to the present known at that time by interpreting its material according to a precise developmental process. This process entailed the following chronological and methodological steps:

1. an understanding of magic as a fallacious yet intuitive system of 'primitive' epistemology (i.e. as a set of cognitive and logical devices by which it was possible to obtain knowledge);
2. a redescription of religion as a more systemic and rational development of magic;
3. the identification of science as the only way to gain fully reliable knowledge on the world and the universe.

According to Frazer, and unfortunately for magicians, magic appears to be based on an erroneous system of assimilation between different ontological domains. Frazer further subdivided magic in *sympathetic magic*, which operates by imitation and similarity (the most classic example: a voodoo doll in the shape of a specific person), and *contagious magic*, by which things once in contact are reputed to remain somehow connected even once separated (e.g. hair of a former lover inserted in a voodoo doll). As weird as it might seem, this is the same socio-cognitive mechanism behind the attribution of special status and priceless value to Hollywood stars' memorabilia, like stage costumes and paraphernalia (e.g. Paden 2016: 202). As Tylor had with animism before him, Frazer acknowledged that modern science was the legitimate successor of magic, in that magic assumed as its main principle the uniformity of laws underlying the functioning of the world, while religion assumed as *a priori* the 'elasticity or variability of nature' subjected to agentive manipulation, thus replicating a cyclical pattern rather than a linear process of development (Frazer 1890, 1: 224). Indeed, the *Golden Bough* reflects on a comparative and historiographical level the lucidly pessimistic Frazerian epistemology (Wheeler-Barclay 2010: 199–201).

The evolutionary pattern adopted by Frazer started from the loose theme of fertility to embrace the development of kingship through the 'manipulation of superstition' (Wheeler-Barclay 2010: 202). In this sense, perhaps the most relevant theoretical contribution of Frazer to the comparative study of religion was his idea of the so-called *dying god*, a model which presupposes the violent death and the rebirth of a god, seemingly inspired in agrarian societies of the ancient Mediterranean by seasonal harvesting and by the cycle of life and death (e.g. Horus, Isis and Osiris in Egypt; Adonis and Dionysus in Greece; Jesus Christ in the Mediterranean region and across the Roman empire, etc.). In Tylorian style, Frazer daringly advanced that the European carnival was nothing other than the survival of the ancient Roman *Saturnalia*, and the ritual appointment of a slave or a criminal as a king to be later executed as a relic of the same theme that, mixed with the magical explanation provided by a convenient scapegoat, also justified the appointment of the *rex Nemorensis*: fertility in the guise of a violent succession by which a younger, untarnished vitality replaced an old, worn-out force. The inclusion of Christianity in such a comparative model, now deprived of any theological immunization, proved to be a disciplinary turning point: the theological uniqueness of Jesus was replaced by an entire series of pre-Christian mythological occurrences from the same cultural and geopolitical macro-area (Wheeler-Barclay 2010: 204–8).

With the exception of his insights on magic, and notwithstanding the wealth of materials he gathered to support his theories, almost nothing of Frazer's work has stood the test of his peers (Beard 1992; for a cognitive reappraisal, see Sørensen 2007). Due to his neglect of socio-political factors, the abstract model of the dying god as a unifying theme has been extensively commented upon, revised and expanded. However, no single consensus has ever been reached on this model as a disciplinary category, mostly because of much interference from confessional, emic perspectives (Smith 1990: 85–115; see also the taxonomy proposed in Bianchi and Vermaseren 1982: 4–6, and the critiques in Lincoln 2015: 13–14). It is undeniable that the *Golden Bough* itself was based on unsupported or questionable links between worldwide bundles of different mythologies (Smith 1993: 208–39). And yet, the many contradictions between and within each edition of his *magnum opus* were preventively acknowledged by Frazer himself as a result of a radically honest empiricist epistemology, according to which theories were 'no more than "light bridges" built to connect isolated islands of facts'. As Frazer himself clearly explained, 'I have changed my views repeatedly, and I am resolved to change them again with every change of the evidence' (Wheeler-Barclay 2010: 196; resp., Frazer 1890, 1: xix; Frazer 1910, 1: xiii). In some cases, Frazer did this after being attacked for his attempt to show the similarities between Christianity and other ancient or non-Western religions, which demonstrates that, as a matter of fact, vehement religious reactions from scholars-believers, such as Andrew Lang, impaired the necessary review and epistemological evaluation (see, e.g. Fraser 2009:

xxv–xxvii about the tormented and gradual exclusion of the section entitled 'The Crucifixion of Christ' over the various editions of the *Golden Bough*).

In any case, the sheer success of *The Golden Bough* was immense at that time, and it has had a powerful impact on literature, poetry and cinema alike (Beard 1992). All those items converged in Francis Ford Coppola's Oscar-winning feature film entitled *Apocalypse Now* (1979), substantially based on Joseph Conrad's *Heart of Darkness* (1899), in turn influenced by Frazer's thought (Hampson 1991). In Coppola's film, Colonel Kurtz, interpreted by Marlon Brando, keeps on his desk a copy of the abridged *Golden Bough*. Indeed, the very core of the film was the weird, hallucinatory succession of a self-proclaimed aged king, once a soldier, violently dethroned by another younger soldier.

This does not detract in the least from the provincializing operation that lay at the heart of Frazer's efforts. Once classical antiquity had been dethroned to make room for a global comparative endeavour – one that did not shy away from revealing that the difference between the religious behaviours of the ancient Romans and those of a Papuan tribe was not in kind but merely in degree (to paraphrase Darwin) – there was no turning back. Indeed, the fame of the *Golden Bough* was basically due to a *cultural shock*, that is, to the fact that it was 'something that had not been done before in English: a treatment from the philosophical, evolutionary point of view, delivered in sonorous and untechnical language, of the beliefs and behavior of the ancient Greeks and Romans as if they were those of "primitives"' (Ackerman 2005a: 3191). Moreover, the comparative multiplication of mythological parallels strikingly similar to the Christ life-history (e.g. Indian Krishna, Roman Mithras, Greek Attis) showed unmistakably the non-Christian Roman influence on Western theology, the general recurrence of certain counterintuitive mythical patterns, and the non-uniqueness of the most central divine claims of Christianity (see Strenski 2015: 71–3). Such were the effects of Darwin's (r)evolution and other kind of social progressionism during, and after, the Victorian era.

Whodunnit?

It goes without saying that this is a woefully incomplete list. I haven't mentioned, for instance, Friedrich Engels' (1820–1895) idea of religion as a major factor in social or class conflict,[7] Max Weber's (1864–1920) concept of religion as the main driver behind certain socio-political formations or economic models, or Bronislaw Malinowski's (1884–1942) interest in the 'ways in which religion functions to inform, model, and/or legitimate concrete patterns of action and organization' (Grottanelli and Lincoln 1998: 315; see also Jensen 2013).

I could also have spent more time on the displacement of theological explanation, which came full circle with Émile Durkheim (1858–1917) and Sigmund Freud (1856–1939), both of whom identified the rationale for the existence of religion in something which believers were not (completely) aware of. As a matter of fact, religion was reduced to sociological and psychological urges. Thus, the sacred became a function of the social group looking for its own identity, according to Durkheim (e.g. Paden 2016), and a misplaced, repressed, and half-concealed array of sexual desires and psychopathologies, according to Freud (e.g. Orbecchi 2015).[8] In a bold checkmating move, the needs of the in-group and the unconscious of the individual were put on the front stage, replacing an abstract and emic idea of religion (Preus 1987: 209).

One could easily add to this list two illustrious developmental and comparative theories of animate and inanimate worship, i.e. Charles des Brosses' (1709–1777) *fetishism*, whose roots dated back to ancient Egypt's zoolatry (Wheeler-Barclay 2010: 25–7), and John Ferguson McLennan's (1827–1881) *totemism*, in which primordial human clans were thought to be related through shared ancestry via a common ancestor (usually an animal) – although a sharp, epistemological distinction between these two concepts and Tylorian's animism is not easy to trace (Sharpe 1986: 75–7). More elaborated and radical theories might include Ludwig Feuerbach's (1804–1872) philosophical attempt to reframe in purely etic terms theology as anthropology plus physiology (Feuerbach 1890); Karl Marx's (1818–1883) concept of 'religiosity [as] a symptom of the proletariat's psychological alienation' that results from specific socio-economic conditions (Slone 2013: 56); August Comte's (1798–1857) 'law of three stages' of development (i.e. theology as fictive explanation, metaphysics as philosophical rationality, and positivism as science. The first theological step was further subdivided into fetishism as superstition, polytheism and monotheism; see Pickering 2009: 3); Henri Hubert's (1872–1927) and Marcel Mauss' (1872–1950) cross-cultural and synchronic theory of sacrifice inspired by Durkheim's works (Podemann Sørensen 2016: 50); Lucien Lévy-Bruhl's (1857–1939) study of primitive, 'prelogical' mentality contrasted with a logical, civilized mentality, which led him to postulate an ontological 'law of participation' in which everything and its opposite can coexist;[9] William James' (1842–1910) psychological and philosophical stress on 'personal religious experience' rooted in the 'mystical states of consciousness' (James 1902: 370);[10] or Friedrich Nietzsche's (1844–1900) ultimate, devastating 'God-Is-Dead' critique of religious truths and authority (cf. Jensen 2014: 20–1).

Such a list would go on almost indefinitely. Every HoR handbook of the recent past has provided readers with an extravaganza of more or less famous names from disparate fields. Sometimes the authors of those lists have included or excluded those scholars who advocated a critical approach to Biblical studies (cf. resp., Kippenberg 2002: 66–8 and Strenski 2015: 19–30; Preus 1987 and Wiebe 1999: 6, 11); sometimes they have highlighted the philological roots (Turner 2014: 293–9, 344–80) or the anthropological influences (Morris 1987: 91–140); sometimes they have even reacted against the progressive scientific lineage of such lists by rejecting 'dogmatically secular, social scientific accounts' (King 2013: 139), or by including rather controversial scholars such as Romantic figures Johann J. Bachofen (1815–1887) and Georg F. Creuzer (1771–1858; cf. Jesi 1973; Jesi 2005), or Victorian polymath Andrew Lang (1844–1212), whose anti-positivism, anti-modernism, degenerationism and ambiguous criticism of comparative religion were to echo for decades in the HoR (Wheeler-Barclay 2010: 104–39; cf. Eliade 1984: 24, 44–5, Stavru 2005, and Sfameni Gasparro 2011: 68, 80–3). Not to mention that some of the scholars recalled above were nonetheless advancing a subtle rejection of positivism and scientific approaches from the inside of the contested discipline (e.g. Lang and William Robertson Smith, whose works we will encounter later). Indeed, there are as many variants as one can possibly imagine.

Whatever the case, the reanalysis of the works of such trail-blazing theorists is a necessary step if we want to understand, and correctly re-evaluate, the disciplinary heritage of the comparative study of religion in general (Xygalatas and McCorkle 2013; Jensen 2014).

However, my point is that, as a matter of fact, not one of the aforementioned scholars can be considered a historian of religions *lato sensu*, nor a historian *stricto sensu* (cf. Eliade 1984: 98). A possible exception would be Jane Ellen Harrison (1850–1928), a Cambridge ancient historian whose atheist spirituality resulted in an original mix of mysticism, feminism and the search for Tylorian origins, all applied to the comparative study of ancient Greek religion. And yet, Harrison ambiguously engaged both the adoption of a Darwinian frame (which we will tackle later on) and an anti-positivistic revolt directed against the very tenets of comparative religion (Wheeler-Barclay 2010: 215–42). Another paradoxical example would be the appointment of Tylor, an anthropologist, as honorary president of the third International Congress of the History of Religions held in Oxford in 1908 (Spineto 2010: 1264). Not to mention that, *when the chips are down*, classical HoR would do away with all those scholars that, in one way or another, threaten the emic, fideistic core of the discipline (in particular, Marx, Engels, Weber, Durkheim, Malinowski; cf. Grottanelli and Lincoln 1998). These paradoxes have been clearly recapped by Tomoko Masuzawa as follows: 'each one of these figures stands as something of a maverick, an intellectual nomad at once inspiring and irritating but never truly constitutive of the discipline' (Masuzawa 2000: 217). So, who are the noble ancestors of the discipline, who can be considered as a founding figure? The paradoxical answer is: each one and no one. The very existence of agreed-upon founding fathers is hotly debated (Gilhus 2014; cf. King 2013). Not even Max Müller enjoys such an unquestionable status (see Wiebe 1999: 27–8 n. 3).

Shape of things to come

The sheer diversity and disciplinary heterogeneity of 'founding figures' recalled above exemplifies three issues that could have put the proverbial nail in the coffin of any project dedicated to the unification of the study of religion(s) under a single banner:

1. The *first* is that, as a subject, religion had already been studied and explored by the incipient academic branches of psychology, psychoanalysis, ethnography, anthropology and history alike. Different disciplines entailed different questions, different problems, different answers, different evaluations, different materials and, quite expectedly, different and heterogeneous results. Therefore, while HoR prides itself on being an autonomous discipline, some of the most important tenets of the past study of religion(s) have been elaborated by scholars and intellectuals trained in many different subjects and with a remarkable, centrifugal interest in seeing religion as a part of greater wholes – the human mind, society, culture or history. While methodological pluralism is vital for any scientific or social interdisciplinary research (cf. Geertz 2014a: 256), in the HoR this had not led to epistemological synthesis or integrative pluralism. On the contrary, it led to a

reactionary, unflinching, dogmatic antireductionism associated with methodological cherry-picking from other disciplines (cf. Idinopulos and Yonan 1994). Moreover, any natural history of the history of religion(s) would be inextricably intertwined with the history of critical thinking and atheism (Minois 1998). And yet, the fact that the HoR successfully managed to silence the unequivocal sisterhood with such a unique field so important for the entire course of modern Western philosophy (Jensen 2014: 18–21; Ruse 2015), should give pause to any celebratory plea.

2. The *second* problem concerns the etic/emic dilemma. For the first time, 'the new scientific ethos [...] made it possible for scholars in the mid- to late-nineteenth century to attempt an emancipation of the study of religion from the religious constraints and to institutionalize a new, non-confessional and scientific approach' (Martin and Wiebe 2016: 223). Unfortunately, the first rationalist enterprises, like those heralded by Max Müller and Frazer, failed because of their incautious, unchecked and biased selection of materials and naïve assumptions (Murray 2010), and *not* because of their epistemic proposals on how to lead a rational research within the confines of a new comparative HoR. Max Müller's own Romantic and theological *a priori* was easily falsifiable with more palaeoanthropological data, and Frazer's monumental fantasy could have immediately vanished against the backdrop of more and better ethnographic data. The falsification of their programmes would have been a trivial exercise in normal science: indeed, the positivistic, rational suggestion behind their huge repositories of data, their developmental schemes or their categorizations might have been updated, corrected and salvaged. However, once the criticism spread, the reaction of many scholars was to refute natural scientific or rational explanations *tout court*. The resulting retreat almost unanimously crossed the boundary between scientific analysis and sympathetic acceptance of religious dogmas or spiritual ideas as true facts (whatever the faith), often trying to endow with academic prestige someone's own religious or spiritual views. This is the primary cause behind the aforementioned disciplinary confusion, and an exemplary case of a baby thrown out with the bathwater.

3. The *third* and final problem lies in the fact that, as a consequence of this rejection, the soon-to-be-established HoR mostly endorsed the vision of religion as a *sui generis*, autonomous entity in order to resolve the first problem (i.e. disciplinary unification). The adherence to scientific methods and theories (such as the ones that, as seen above, were dedicated to frame religion as a part of a bigger picture) seemed to explain away and refute this autonomy, therefore making the very existence of the discipline apparently difficult to support on an institutional level. As an academically autonomous discipline, HoR needed nevertheless to adopt, at least formally, the prestige of a scientific approach in order to be accepted. As a result, a tentative agreement was reached, in that HoR was to adopt superficially certain scientific and natural methods of enquiry, merely to confer a certain degree of academic respectability on what might have been perceived as theology in disguise. Mostly, scientific approaches were despised if not openly and harshly criticized.

We could even say that the foundational tenets of the HoR had been laid logically before Hume and scientifically before Darwin. If, as Popper claimed, science really begins only 'when a myth is challenged and breaks down – that is, when some of our expectations are disappointed' (Popper 1994: 95; cf. Masse *et al.* 2007), then the allegiance of religious scholars to the natural theology of *homo religiosus* and their faith in a paranormal or supernatural worldview explains the pre-Humean charter of the discipline (cf. Ambasciano 2015a). Such fideistic epistemology is evident in the chimerical gallery of 'founding fathers', in the rejection of scientific paradigms, and in the parallel cherry-picking of those tools from scholars and disciplines which conformed to a self-confirming, *a priori* vision. Metaphorically speaking, the geological divide within the discipline had been there since the very beginning, and those who tried to implement an engineering plan of bold renovation to safeguard the buildings on the fault zone were misunderstood, marginalized and silenced.

In 1912, Robert R. Marett (1866–1943), Tylor's successor as reader in Anthropology at the Pitt Rivers Museum (University of Oxford), noted the unique epistemological relation between his own field and the explanatory success of Darwin's evolutionary research programme, and therefore he clearly underscored that 'Anthropology is the child of Darwin. Darwinism makes it possible. Reject the Darwinian point of view, and you must reject anthropology also' (Marett 1914a: 8). His proposal was to '[l]et any and every portion of human history be studied in the light of the whole history of mankind, and against the background of the history of living things in general' (Marett 1914a: 10). Marett also noted that, while 'at first, naturally enough, man did not like' to be aligned with other animals as an unexceptional living being, 'now-a-days, however, we have mostly got over the first shock to our family pride. We are all Darwinians in a passive kind of way. *But we need to darwinize actively*' (Marett 1914a: 9; my emphasis. Cf. Sharpe 1986: 48). Unsurprisingly, and notwithstanding all the previous efforts, a significant number of scholars committed to the study of religion did not want their field to be darwinized at all. In short, the Darwinian (r)evolution was the spark that set ablaze the whole debate about which scientific tools – if any – were to be legitimately used to investigate human culture and religion(s).

3

The Darwinian Road Not Taken

Origin of man now proved. – Metaphysic must flourish. – He who understands baboon would do more towards metaphysics than Locke.

Charles R. Darwin

'The greatest historian of all time'

Comparative religion has a very long and eclectic history, tracing its modern origins back to the European Age of Discovery. Perhaps more interesting, and much less known, is the fact that the deep roots of the whole enterprise lie in the species-specific cognitive capacity for inter- and intra-group comparison and categorization. Quite ironically, this deep history is the consequence of evolutionary processes whose scientific recognition and acceptance were misunderstood, fought or ostracized early on in the mid- to late-nineteenth-century academia. Without repeating what I have already briefly recalled in the *Preface*, this explains succinctly why, notwithstanding the bold, iconoclastic attitude of isolated scholars such as Frazer (Strenski 2015: 66), the future of the historical and comparative study of religion as a whole was to be jeopardized by the intuitive appeal of beliefs supported by the very cognitive deep roots of comparison.

This is all the more striking because, while no *stricto sensu* historian can be considered as *the* founder of comparative religion, Darwin worked as a historian in the truest sense of the term and, as we shall see shortly, he also carefully laid the foundations for the ultimate naturalistic study of religion and society. Biologist and historian of science Frank J. Sulloway aptly noted that, once we consider the descriptive methods, the 'dedication to hypothesis testing', the acute self-awareness of the confirmation bias (that is, the cognitive bias according to which we intuitively tend to look for confirmatory evidence in support of our favoured theory, and automatically discard contrary evidence), and the method and theory to avoid it (a peculiar trait which characterized Darwin's works), 'a good case can be made for considering Charles Darwin the greatest historian of all time' (Sulloway 1998: 366). Palaeontologist and historian of science Stephen J. Gould (1941–2002), in turn, recalled the status of Darwin as a 'historical methodologist' whose scientific *modus operandi*, fully exposed in the *Origin of Species*, was characterized by three major features (resp., Gould 1986: 60, 62–4):

1. 'uniformitarianism in extrapolating [. . .] observed results', that is, the assumptions of constant natural laws and the identification of the 'rate and effect' of the process under investigation;
2. 'inference of history from temporal ordering of coexisting phenomena', i.e. the taxonomical classification of causally tied items;
3. the 'panda principle of imperfection', that is, the historical attention to 'oddities' such as useless traits co-opted for new scopes as a result of historical contingencies which, if unrecognized, may lead to identify false taxonomical homologies.[1]

The attention to neglected details, imperfections and futile traits allowed Darwin to reject both typological essentialism, the ancient argument from design, and the theodicy advocated by natural theology (Desmond and Moore 1991: 479).[2] The cornerstone of his evolutionary theory was at once the opposite of essentialism and the epitome of historical processes that bring about change: each individual is unique in a population with which s/he shares specific traits that can be passed on through the generations (i.e. 'population thinking'; Mayr 1991: 79–80). This historical attention upturned the teleological paradigm which made the scholarly study of nature a static and unchanging field, 'for Darwin's theory is but the historicization of biology' (Martin 2013: 179).

The impact of increasingly successful natural explanations and retrodictions on the birth of comparative religion, plus an equally successful politics of scientific, (apparently) non-confessional institutionalization, cannot be overlooked (Wheeler-Barclay 2010: 3). Well acquainted with Darwin's programme of research, Max Müller used to remark incessantly on the liaison of comparative religion with natural sciences, positing that religion was not 'beyond the reach of scientific treatment, or honest criticism' (Max Müller 1898c: 8). Following a trope in vogue in eighteenth-century natural history (Smail 2008: 47), Max Müller vividly described the aim of comparative religion through the perusal of ancient documents as one akin to the decipherment and understanding of the 'geological annals of the earth' (Max Müller 1898a: vi). He also rejected the speculative 'Hegelian laws of thoughts', as well as the 'Comtean epochs', in favour of something not dissimilar from Darwin's 'panda principle', that is, the historical 'tracing of the origin and first growth of human thought' according to careful reconstruction from incomplete logs of facts (Max Müller 1881a: ix; cf. Wiebe 1999: 28 n. 28). In 1878, prefiguring probably the most central and problematic aspect of the HoR, one that will be scientifically tackled only more than one century later by cognitive sciences and evolutionary psychology, Max Müller asked,

how does [religious belief] arise? What is the historical process which produces the conviction that there is, or that there can be, anything beyond what is manifest to our senses, something invisible, or, as it is soon called, infinite, superhuman, divine? It may, no doubt, be *an entire mistake*, a *mere hallucination* to speak of things invisible, infinite, or divine. But in the case, we want to know all the more, how it is that people, apparently sane on all other points, have from the beginning of the world to the present day, been insane on this one point. We want an answer to this, or we shall surrender religion altogether unfit for scientific treatment.

Max Müller 1878: 169; my emphasis; cf. Wiebe 1999: 9–30

And yet, quite astonishingly, most HoR textbooks, handbooks and encyclopaedic entries stop before delving deeper into the epistemological and methodological relevance of Darwin's works for the natural history of comparative religion, as if evolutionary biology stood beyond the *nec plus ultra* of the historiographical Pillars of Hercules (Smail 2008). Generally speaking, this neglect is part of an idiosyncratic academic compartmentalization and a chronically insufficient scientific literacy in non-scientific disciplines (cf. Snow 1961). Evolution is a very misunderstood topic in the Humanities. Even today, as Daniel L. Smail and Andrew Shryock aptly underscore, a fully humanistic comprehension of the significance of Darwinian evolution is far from being sufficiently achieved (Shryock and Smail 2011a: 12). Not long ago, Ina Wunn remarked that 'in biology, evolution is defined as the adaptive modification of organisms through time by means of natural variability and selection. In the study of religion, however, as well as in other disciplines of the humanities, the term evolution is still understood as a process of progressive development' (Wunn 2003: 391). In the early 1970s, Jonathan Z. Smith already noted the conceptual lag between the theoretical and experimental development of evolutionary biology and the belated, misguided or distorted adoption of cherry-picked biological themes within the HoR (Smith 1993: 244 n. 14, 260 nn. 54, 55). Even worse, the general interest towards pseudoscience, born in the first decades of the twentieth century and flourished after World War II (Thurs and Numbers 2013: 134), was destined to have a major impact on the field. Ultimately, the lesson of 'the greatest historian of all time' was lost.

The threat of a scientific and comparative study of religion

The conceptual watershed erected by the Darwinian methodological proposal could not be accepted light-heartedly, for the tenets of the theory were thought to pose a mortal threat to religion itself. Frazer's work is a case in point. The comparative anthropologist set out to explain the development of religion in a cultural environment where Darwin's success had been already assumed as a *fait accompli* (cf. Ackerman 1990: 33): an 'objective, scientific method' was used 'to hammer the last nail into the post-Darwinian coffin of religion, to show once and for all, by bringing together data on myth, ritual and belief from all over the world and throughout recorded time, that religion was a noble but in the end misguided effort on the part of primitive humanity to understand the nature of reality' (Ackerman 2005b: 72). In other words, Frazer's work represented the dreaded materialization of the anti-religious threat posed by a scientific comparative religion.[3]

As far as believers were concerned, the threat was even greater. Historian of science John Hedley Brooke has listed the many Christian theological elements directly challenged by Darwinian evolutionism:

> the nature of biblical authority, the historicity of the creation narratives, the meaning of Adam's fall from grace and (connected with it) the meaning of Christ's redemptive mission; the nature and scope of God's activity in the world; the

persuasive force of the argument from design; what it meant for humankind to be made in the image of God; and the ultimate grounds of moral values.

Brooke 1991: 281–2; cf. Moore 1981: 218

Mutatis mutandis, a similar list of foundational, non-negotiable dogmas might be extrapolated from any Christian denomination or non-Christian religion.

No matter which religion, the result would be very much the same: an unavoidable conflict, to be resolved by declaring the victory of science over religious dogmas, or dealt with thanks to a theological reconfirmation of the traditional *status quo* (e.g. Numbers 2006; Numbers 2009; Blancke, Hjermitslev and Kjærgaard 2014).[4] For all its mistakes and downsides, comparative religion (that is, the academic precursor of the HoR) was set on a course that was leading to the first case, in that a positivistic approach was supposed to overcome and trump any confessional *a priori* in favour of a methodological approach that either separated agnostically 'precept and practice' (such as was the case with Max Müller; Wiebe 1999: 22) or even openly endorsed atheism (as Frazer's). As a project, it collapsed from the inside because ethnocentric infiltrations, naïve theorizing, and confessional, anti-positivistic reinterpretations to neutralize any threatening evolutionary interference, were unavoidable (Wheeler-Barclay 2010: 10). Moreover, external resistance to the comparative religion project was motivated by the fear that any critical theorizing, 'if taken seriously, would have [had] the effect of destroying the pith of "religion" as we know it' (Masuzawa 2000: 217). Indeed, the very first coherent, long-term academic projects for a fully-fledged HoR such as the Dutch and Austrian schools (whose development will be the subject of the next chapter), implemented an institutional answer in line with the second case, quite explicitly refusing to come to terms with evolution. They were not the first anti-scientific reactions to surface in the field, and surely enough they were not going to be the last. But why was Darwinian evolution discarded insofar as the study of the historical roots, the functions and the development of religious thoughts and behaviours was concerned?

Evolutionism, so it is suggested, was abandoned because of its 'theoretical inadequacies' which made the concept of evolution heuristically useless at best and mistaken at worst (Waller, Edwardsen, and Hewlett 2005: 2917; cf. Filoramo and Prandi 1997: 175). More recently, postmodernism and social constructionism in RS have epistemically delegitimized science and evolution by positing the equal importance of non-scientific discourses (von Stuckrad 2014; cf. Ambasciano 2016b). But is it really so? Were evolutionary attempts at understanding and explaining human culture and religion really flawed and 'inadequate'? Or, rather, does a historiographical misunderstanding erroneously conflate pre- or non-Darwinian evolutionism with biological evolution (Smith 1982: 24)? And, perhaps even more importantly, does every explanation have an equal status regardless of its epistemic warrant? In order to understand and contextualize the full extent of the anti-scientific reaction in the study of religion(s), it is necessary to delve deeper into the epistemology of evolutionary theory and into the history of *human exceptionalism*, a multifaceted concept that, as we will see shortly, can be broken down into four simpler ideas: pithecophobia, teleology, orthogenesis and anthropodenial (Figure 5). In order to understand their development and interaction we need to start from the most basic question: what is it that makes us human?

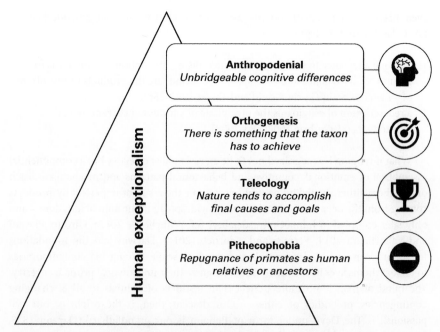

Figure 5 Biases supporting the concept of human exceptionalism

What makes us human

The story of the search for our distinctiveness is as old as human history itself. For a long time before Darwin, naturalists persistently strived to understand, order and classify the natural world (Stott 2012). Around the very end of the eighteenth century, scholars in the natural sciences were trying to make sense of a more or less progressive development in life forms through the deep, geological time of planet Earth (Corsi 2005). The ultimate breakthrough took place in the mid-nineteenth century, when Darwin and co-discoverer of natural selection Alfred Russel Wallace (1823–1913), standing on the shoulders of giants from the past, succeeded in getting the right theoretical toolbox to deconstruct and interpret the relationship between nature, time and change (Mayr 1991). The very concept of 'what makes us human' suddenly became somehow redundant and in need of thoughtful revision. In the words of Matthew Day, '[w]hat, if anything, is a uniquely human trait? [...] If we take the lessons of modern evolutionary theory seriously, it seems as though any claim regarding human uniqueness must either ignore the significance of inter-species continuity or discount the lessons of intra-species variation' (Day 2008: 49–50).

Amid an endless catalogue of moral virtues, religion has always been one of the quintessential traits of human self-depictions: 'as early as the mid-seventeenth century, religion had been identified as a, if not the, uniquely human trait' (Day 2008: 50), but even this revered concept did not manage to save itself from the changing tide. Darwin posited a two-fold strategy:

1. a 'strong allegiance to philosophical materialism – the notion that matter is the ground of all existence and that "spirit" and "mind" are the products or inventions of a material brain' (Luria, Gould and Singer 1985: 585);
2. a redescription of religion as the precipitate of various concurrent mental by-products (cf. Guthrie 2002; Day 2008).

What if religion is recognized not to be an essentialist category but a componential assemblage of emotional, cognitive and behavioural patterns and mechanisms, each one of them either shaped by natural selection or the result of imperfect by-products coming from the deep time of evolution and widespread in the animal kingdom – and only later co-opted in human ultrasociality? (cf. Whitehouse 2013a; Girotto, Pievani and Vallortigara 2014; Ferretti and Adornetti 2014). Darwin laid the foundations for Tylor's componential approach and survival, while forcing his contemporaries to look at themselves in the mirror and recognize their unequivocal primate ancestry: 'the mind of man is no more perfect than instincts of animals to all & changing contingencies, or bodies of either. – Our descent, then, is the origin of our evil passions!! – The Devil under form of Baboon is our grandfather!' (Darwin 1838, *Notebook M*; from Barrett *et al.* 1987: 549–50).

It is quite self-evident to note that, instead, the general assumption that shaped the study of the Humanities, before and during the beginning and establishment of modern academia, and which prevailed until the mid-twentieth century, was one of an undeniable, essential superiority of *H. sapiens*. The original intellectual quest for discovering what makes us human has consequently been led under the aegis of a qualitative difference given for granted on the basis of the essentialist meta-belief known as human exceptionalism (de Waal 2013): we are cognitively bound to intuitively set ourselves apart from other living beings and to cement this divide via our genealogical storytelling (Goldenberg *et al.* 2001; Coley 2007). In a sense, what makes us humans is the neverending effort to define what it is that makes us human.

The *Origin of Species*, 1859: Breaking the chain of being

Human genealogies have been orally or materially translated in various forms, written or visual, since the very inception of our taxon. Kinship is one of the most ancient forms of historical thinking of *H. sapiens*, and innumerable genealogies in the form of trees (frequently focused on male social power and patrilineal descent) crowd the sacred texts of ancient religions and the *codices* of royal dynasties (Shryock, Trautmann and Gamble 2011: 34). Universal histories, religions and mythologies justified kin recruitment in large societies by connecting human beings to mythic ancestors through an imaginary and prestigious common descent (Martin 2000; Christian 2010; L. H. Martin 2014: 80–93, 94–106). The classical idea of the chain of being, which tied every living organism in a fixed formula – with a male individual of *H. sapiens* at the end of the chain – traces its origins in this context of genealogical pedigree (Gould 1989: 28).

Now, Darwin took this intuitively upward progression embedded in kinship thinking and dismantled it, obviously alienating his peers' sympathy. When Darwin's research programme about evolution began to circulate after the publication of the *Origin of Species* (1859), Darwin himself remained for a long time alone in considering that 'higher' or 'lower' were useless heuristic tools for the analysis of the natural world: adaptation is not equivalent to progress (Gould 1977: 34–8). In Darwin's own words, 'natural selection includes no necessary and universal law of advancement or development – it only takes advantage of such variations as arise and are beneficial to each creature under its complex relations of life' (Darwin 1861: 135). To put it differently, evolution has no designed meaning, no ultimate goals and no preordained optimality to achieve insofar as it is simply involved in the tinkering of good-enough solutions (Jacob 1977). Following an argument famously made by Stephen J. Gould, the apparent progressive and upward trend from primeval bacteria to vertebrates is the result of the optic distortion from our present, chauvinistically limited perspective: bacteria outnumber everything else on this planet, and they still reign supreme. What we have is, instead, an increase in variation and complexity through deep time (Gould 1996; cf. Allmon 2002: 49–51, and Sterelny 2007: 90–3).

The firm refutation of ascending progress, teleology, and essentialism was an epistemic consequence of the Darwinian framework. But what did this framework look like? Evolutionary biologist Ernst Mayr (1904–2005) deconstructed and reorganized the bulk of Darwin's evolutionary proposal into five strictly related sub-theories (Mayr 1991: 35–7):

1. *evolution* as the deep-historical, ongoing, contingent, differential development of organisms though time;
2. *common descent* of all living organisms from a common ancestor via a process of continuous branching;
3. *populational speciation*, i.e. the multiplication of species from previous species thanks to a combination of geophysical isolation and presence of individual variations in any given population (i.e. population thinking);
4. *gradualism*, that is, the slow and steady accumulation of diverging anatomical features through time;
5. *natural selection*, i.e. the differential rate of survival of individual organisms due to the relationship between environmental constraints and phenotypic differences (i.e. external and behavioural features).[5] Because of complex factors such as ecological change and genetic recombination occurring in each generation, this relation is ever-changing (resulting in the evolvability of evolution itself).[6]

In 1859, Darwin thought it was inappropriate to reveal the application of such a model to the evolution of humankind, limiting his thoughts to the following laconic, yet groundbreaking, statement: 'in the distant future I see open fields for far more important researches. Psychology will be based on a new foundation, that of the necessary acquirement of each mental power and capacity by gradation. Light will be thrown on the origin of man and his history' (Darwin 1859: 488). And yet, this apparently harmless assertion barely disguised what was almost painfully evident for his audience: the *Origin of Species* was all about humankind, now dethroned and reinserted into the deep-historical fabric of the biological world (Ruse and Richards 2009: xvii). And if that were not enough, the 'distant future' envisaged by Darwin was just around the corner.

The Descent of Man, 1871: Degrees, not kinds

Indeed, a mere 12 years after the publication of the *Origin of Species*, that 'distant future' had finally come. Almost forced by a cultural environment eager to apply evolutionary schemes to human cultures (e.g. Tylor's *Primitive Culture*; see Desmond and Moore 2009), in *The Descent of Man* Darwin set out to show that evolved psychological features critical for the elaboration of religious thought, like agency detection and related emotional responses, are present in different degrees in nonhuman animals. The roots of this interpretation go back to Darwin's famous account of his dog scared by a moving parasol, and the related passage in the *Descent of Man* concluded with the following reflection:

> he [i.e. Darwin's dog] must, I think, have reasoned to himself in a rapid and unconscious manner, that movement without any apparent cause indicated the presence of some strange living agent, and no stranger had a right to be on his territory. The belief in spiritual agencies would easily pass into the belief of one or more gods.
>
> Darwin 1871, 1: 67

In other words, with evolution finally reconnected with anthropological works previously inspired by the *Origin of Species* (such as, once again, Tylor's *Primitive Culture*), Darwin dug deeper into the history of cognition as he highlighted a continuity between the dog's response and the beliefs in animated beings diffused in non-Western societies as identified by Tylor and other anthropologists (Guthrie 2002; Day 2008: 59; Girotto, Pievani and Vallortigara 2014: 391).[7] The analysis of canine psychology was also instrumental in revealing other basic componential blocks for religion, such as the 'deep love of a dog for his master, associated with complete submission, some fear and perhaps other feelings' (Darwin 1871, 1: 68; see Chidester 2009; on emotions, see Darwin 1872a).

The deep-historical continuity revealed by comparative psychology was the direct consequence of Darwin's revolution; any difference between human and nonhuman animals was measurable in 'degrees' (quantity) and not in 'kinds' (quality; Darwin 1871, 1: 105).[8] Darwin also included pro-social morality – i.e. emotions and ideas supporting religious and ideological beliefs such as 'patriotism, fidelity, obedience, courage and sympathy' – as an integral part of natural selection on the basis of their value to promote abnegation and self-sacrifice at the service of the overall fitness of the in-group (Darwin 1871, 1: 166). As David Sloan Wilson has remarked, Darwin envisaged a multilevel process of selection where the diversity of phenotypic differences (in this case, different degrees of moral judgement across and among groups) is the result of the aforementioned materialistic account of mental capacities. According to such perspective, mental traits can be culturally elaborated and passed through generations thanks to their capacity to reinforce social conformity and promote in-group welfare. Then, such traits might be selected via inter-group competition (i.e. warfare; Wilson 2002: 9). As a result, religion is biologically interwoven with culture *lato sensu*: 'the effects of habit, the reasoning powers, instruction, religion, etc.' are able to interact with natural selection and modify the adaptive value of certain behaviours (Darwin 1871, 2: 404; see Pievani and Parravicini 2016 on multilevel selection).

Finally, Darwin's naturalistic account of the continuity between *H. sapiens* and other nonhuman taxa also entailed the deconstruction of previous ideological and socio-political tenets about the uniqueness of *H. sapiens* itself (Day 2008; Desmond and Moore 2009). Ever the conscious and self-aware theorist, Darwin distanced himself from both those commonly, and erroneously, labelled as 'social Darwinists' and progressionist thinkers. Two main features of Darwin's work stand out:

1. Culture might be able to overcome and reverse the blind forces of natural selection thanks to what has been called 'the reversive effect of evolution', according to which the extension of communitarian ethics and sympathy towards disadvantaged individuals (e.g. those affected by ailments or impairments) runs counter to the selective pressure which targets individuals and yet, notwithstanding an immediate loss of fitness in terms of resources, it promotes social cohesion, thus reinforcing in-group solidarity and cooperation which, in a loop and in the long run, promotes in-group fitness (Tort 2008).
2. Competition does not lead unequivocally to better stocks of human groups, nations might lose their advanced status, and war does not improve automatically the fittest individuals (e.g. the youngest, and probably the fittest, individuals are sent to die). Progressive development due to evolutionary forces, in the end, 'is not an iron rule, but a probabilistic one' (Glick 2008: 227). As Darwin himself recalled, 'natural selection acts only in a *tentative manner*' (Darwin 1871, 1: 178; my emphasis).

As far as religion was concerned, the groundbreaking result of Darwin's approach was two-fold: on the one hand, religious intuitions and beliefs were thought to be ultimately (and materially) caused by a mind that had a deep evolutionary history shared with nonhuman animals; on the other hand, religious behaviours finally entered the evolutionary playground as part and parcel of the evolution of sociality and cooperation, making virtually possible the development of any scientific analysis relating to the potentially (non-)adaptive value of religious behaviours and beliefs (for an overview, see Bulbulia *et al.* 2008). It is not difficult to see why Darwin might also be considered the very first modern historian of religions, and it is no wonder that, less than 30 years after the death of Darwin, classicist Jane E. Harrison, while acknowledging forerunners like Hume, attributed the 'creation' of the modern 'scientific study of Religions' to 'Darwinism' itself (Harrison 1909: 494).

The Rubicon: Max Müller versus Darwin

Surprisingly, the first bone of contention between evolutionary biology and comparative religion was centred on an entirely different topic: language. And yet, this choice is not as odd as it might seem at first sight. Language played a considerable part in Darwin's evolutionary account, as it did in Max Müller's own works. As recalled at the beginning of this chapter, Max Müller was an early adopter of Darwin's selective mechanism translated into the domain of language development and differentiation – not just *between* languages, but even *within* each language.[9] Max Müller also conceived myths as pre-rational explanations of natural phenomena (as evidenced by the very names and epithets of the gods he studied), and thus he posited a sort of developmental or evolutionary process of rationality-acquisition by degrees. However, Max Müller was also a staunch supporter of the unbridgeable gap between humankind and animals. As Max Müller wrote, language was

> our Rubicon, and no brute will dare to cross it. This is our matter-of-fact answer to those who speak of development, who think they discover the rudiments at least of all human faculties in apes, and who would fain keep open the possibility that man is only a more favoured beast, the triumphant conqueror in the primeval struggle for life.
>
> Max Müller 1864: 367–8; cited in Richards 1987: 203

Language was inextricably intertwined with conscious thought, and since Max Müller – like many other scholars at that time – considered nonhuman animals as deprived of conscious thought, language and rational thought were pinpointed as a *perfect* and *complex* set of *exclusively* human features, ultimately granted by a divine power. No evolutionary, gradual process was possible. Languages had developed from 'roots' which were tautologically tied to rational thought: 'without roots, no concepts; without concepts, no roots' (Max Müller 1875: 477; from Nicholls 2014: 93). A Cartesian, mechanistic view of animals as automata deprived of conscious thought and language echoed in Max Müller's critique to Darwin's gradual evolution of language. To this background, Max Müller added a Kantian penchant for *a priori* categories, prominent in his *Lectures on Mr. Darwin's Philosophy of Language* (Max Müller 1873; cf. Nicholls 2014; Davis and Nicholls 2016: 90, 93–4).

In the *Descent of Man*, Darwin had previously pointed out that concepts, or a 'long succession of vivid and connected ideas', might be present even in the absence of language (resorting to another example with a sleeping and dreaming dog; Darwin 1871, 1: 58). Like Robert J. Richards recapped, according to Darwin 'complexity [of languages] itself was no sure sign of perfection' (Richards 1987: 204–5). Darwin will respond directly to Max Muller's criticism in the 1874 edition of the *Descent of Man*, positing the co-evolution of mind and thought (Alter 2009). In the wake of the Humean empiricist tradition, and rejecting on the ground of comparative evidence the Kantian approach hailed by Max Müller, Darwin built a compelling evolutionary account in which human proclivities (such as singing) were combined with the sustained use of gradual, onomatopoeic imitation of sounds, both boosted by a cultural environment sensitive to innovations which might have resulted in a deep-historical, loop effect on adaptive fitness (see Darwin 1871, 1: 57–62; Nicholls 2014). When Darwin read the *Lectures*, he duly noted on his copy how aphasia and impairments in the formulation of speech attested to a cogent falsification of Max Müller's hypothesis, declaring 'monstrous sentence – "No thought without words no words without thinking"' (*Darwin Pamphlet Collection*, R240, CUL; from Alter 2009: 46). In the second edition of the *Descent of Man*, Darwin expanded on his negative judgement of Max Müller's *Lectures* (Darwin 1874: 89–90 n. 63). Commenting on Max Müller's ideas according to which 'the use of language implies the power of forming general concepts; and [since] no animals are supposed to possess this power, an impossible barrier is formed between them and man', Darwin pointed out that animals, infants and 'deaf-mutes' are able to 'connect certain sounds' with 'ideas', concluding that it is possible to use sign language to communicate in a perfectly fine way (Darwin 1874: 89).

Notwithstanding the support of linguists and philologists such as Hensleigh Wedgwood, Frederic William Farrar and William Dwight Whitney (Alter 2009; Davis and Nicholls 2016), even Darwin's closest supporters were unwilling to renounce the human Rubicon. Notwithstanding his work on the continuity between human beings and apes (*Man's Place in Nature*, 1863), Darwin's friend and colleague Thomas H. Huxley (1825–1895) remained unconvinced by Darwin's gradualism, advocating instead a saltationist approach (i.e. rapid evolutionary leaps), and thinking that animals were more or less 'mindless automata' (de Waal 2013: 40 on a letter sent by Darwin to Huxley, 27 March 1882, *Darwin Correspondence Project*, DCP-LETT-13744; see also Gould 1991: 129–34; Lyons 1995; Desmond 2009). A third way out of the problematic inclusion of humankind within the fabric of evolution was sought by Wallace, specifically excluding *H. sapiens* from natural selection and inserting it in the grand old scheme of teleological – and supernatural – progressionism (Gould 1980: 47–58; see also Shermer 2002, and Smith and Beccaloni 2008). Wallace, who embraced an 'idiosyncratic theism linked to the spiritualist movement and the growing interest in psychical research' (Wheeler-Barclay 2010: 101), even criticized Tylor's animism for its anti-spiritualistic penchant, claiming that 'it is unsafe to deny facts [elsewhere in his paper thought to be based upon "possible realities"] which have been vouched for by men of reputation after careful enquiry, merely because they are opposed to our prepossessions' (Wallace 1872: 71). For all their differences, Huxley's and Wallace's perspectives were to recur repeatedly in the history of the HoR, where human uniqueness had to be preserved whatever the cost. Primatologist Frans de Waal has recently coined the term *anthropodenial* to define this tendency to highlight some unbridgeable differences between 'us' (*H. sapiens*) and 'them' (nonhuman hominoids; de Waal 2001: 69; cf. Waldau 2013: 155–6), which is the false-negative counterpart of attributing human-like cognitive states to animals, i.e. *anthropomorphism* (Andrews 2011: 473–4). Even in the absence of an openly acknowledged disdain for primate ancestry, the Rubicon was hardly forded at all.

Max Müller tried to combine British empiricism with German philosophical tradition. His rejection of evolution whenever applied to some 'higher' qualities of humankind, on the basis of teleological evolution and Romantic idealism, was nevertheless accompanied by the adoption of evolutionary mechanisms to explain linguistic development and the development of life. As such, his model was doomed to fail. As Angus Nicholls has underscored, 'Müller's explanation of language was ultimately a failure as both Darwinian science and as Kantian *Wissenschaft*' (Nicholls 2014: 94). Max Müller's attempt provides a significant example of the difficult and contrasted relationship between social sciences and natural sciences. And yet it also testifies to the early adoption of scientific methodologies and willingness to engage directly with cutting-edge scientific research. A bona-fide openness to falsification was also remarkable. When Max Müller 'paid a courtesy visit' to Darwin at Down House in 1874, he summarized his views before taking leave, to which Darwin commented by laconically saying '[y]ou are a dangerous man'. Max Müller ventured to reply: 'There can be no danger in our search for truth' (Alter 2009: 48; Max Müller 1899: 203). As Max Müller wrote to Darwin some time later,

> 'more facts & fewer theories' is what we want, at least in the Science of Language, and it is a misfortune if the collectors of facts are discouraged by being told that facts are useless against theories. I have no prejudice whatever against the faculty of language in animals: it would help to solve many difficulties. All I say is, let us wait, let us look for facts, & let us keep *la carrière ouverte*.
>
> Max Müller to Darwin, 13 October 1875; *Darwin Correspondence Project*,
> DCP-LETT-10194

Max Müller was firm in his *a priori* assumptions and religious views, but in the end, loyal to this scientific ethos, he had to acknowledge *bon gré, mal gré* the epistemic shortcomings of his hypothetical and exclusively human 'faculty of faith', he bit the bullet and rejected the idea of a divine revelation and, accordingly, he revised his ideas to embrace both a more empiricist psychology and the 'historical evolution' of societies (Wheeler-Barclay 2010: 56–7; see Max Müller 1878: 30).[10] For all their differences in this regard, the same longing for the adoption of a scientific framework to study religion was shared by Max Müller's Cantabrigian, and die-hard rationalist and atheist colleague, Frazer.[11]

Ladders, progress and pithecophobia

As the confrontation between Darwin and Max Müller reveals, and notwithstanding Darwin's efforts, the distortions that have connected ascending progress (and intuitively positive meaning) to structural complexity have always proved to be much more fascinating, powerful and resilient (Mayr 1991: 35–47). Human exceptionalism has exerted an appealing legitimacy to such anti-Darwinian theories (Shryock and Smail 2011b: 8), to the extent that early Darwinian suggestions and approaches to study religious beliefs and the evolution of religious institutions, in line with the five core sub-theories listed earlier, barely had any echo in the historical and comparative study of religion.

It should not be a surprise that one of the most striking features in past and present widespread ideas of evolution is the *ladder of progress*, which has been visually translated into myriads of forms and styles (Gould 1989; Pietsch 2012) and should not be confused with the Darwinian, quite messy, 'coral of life' and related forms of phylogenetic systematics that help visualize immediately the deep-historical relations between extinct and extant organisms.[12] Unfortunately, each one of those ladders leads the reader on to the slippery slope of an 'iconography of hope' piously devoted to illustrate an appealing, inevitable and linear ascent of *H. sapiens* from the chaotic and primeval slime which dwelt in the oceans some 3.5 billion years ago to the pinnacle of evolution on planet Earth (Pievani 2012). This view crystallized in the *march of progress* which usually portrays a bunch of male hominins walking in a single line from the most primordial to the more advanced – which is, and it goes without saying, *H. sapiens* itself (Gould 1989; Figure 6). Two main fallacious ideas were instrumental in the diffusion of this representation: *orthogenesis*, namely the concept of an inner thrust towards a predetermined evolutionary path (originally reserved for non-adaptive evolution), and *teleology*, that is, the idea that nature tends to accomplish final causes and goals. Both were often infused with meanings about ultimate, transcendent concerns in life. Since then, they have been scientifically disproved, but nonetheless they have remained die-hard ideas in the Humanities (Bowler 1992; Russell 2011). Teleology, in particular, is one of the most important, intuitive pillars that support unnatural religion, so much so that a so-called cognitive 'promiscuity' affects the intuitive application of teleology and design stance to multiple ontological domains (e.g. Kelemen 1999a; Kelemen 1999b; Kelemen and Rosset 2009). As is the case with such immediate and intuitive cognitive mechanisms, a sense of gratifying yet misguided epistemic satisfaction prompts and reinforces their continued use (Buekens 2013). Indeed, Gould outlined that '[t]he familiar iconographies of evolution are all directed – sometimes crudely, sometimes subtly – toward reinforcing a comfortable view of human inevitability and superiority' (Gould 1989: 28). Unsurprisingly, most research in historiography and comparative religion of the past centuries has exploited, and possibly expanded, this commonsensical and fallacious view of life.

Moreover, this goal-oriented view of life successfully promoted a marriage between sacred history and the appealing orthogenesis which inferred evolutionary periods of *progression, regression* (e.g. delinquency and 'racial senility', that is, physical, intellectual, or moral degeneration on both individual and national levels), and *stasis* (i.e. the so-called 'living fossils'), each one coupled with a significant, and preconceived, value judgement (Bowler 1989; Milner 2009: 128–9; see Figure 7). For instance, the newly established twentieth-century academic discipline of historiography was devoted to the faithful application of this framework to the ontogenesis of the national states, providing (directly or indirectly) legitimacy, justification and support in the context of the nation-building process. A more or less invented, prestigious deep past justified the present cultural and political claims, such as missionizing, colonialism and imperialism (Bowler 1989; Smail 2008; Milner 2009: 128–9; Hobsbawm and Ranger 1992). Then, philosophies of history inspired by Hegelian, Spencerian and Bergsonian principles (to recall just a sample from the most influential thinkers of the period) substituted – or accompanied – divine agency in history with the prototype of the Western wealthy male and reinforced the aspirations to a teleological point of view in historiography (Shryock and Smail 2011a: 10).

This is not what evolution looks like.

This is what evolution looks like.

Figure 6 The march of progress (*top*) vs. the bush of evolution (*bottom*)

Although a largely understudied topic, the denial or neglect of the primate ascendance of *H. sapiens* is another important theme that needs to be factored in when evaluating the anti-scientific tendencies of academic historiography. For instance, one of the main pillars of post-Darwinian human exceptionalism in the study of religion was *pithecophobia*, originally diagnosed by zoologist and palaeontologist William King Gregory (1876–1970) as the 'irrational fear of apes and monkeys as potential ancestors [...] brought on by greater knowledge of our own evolution' (Beard 2004: 289–90;

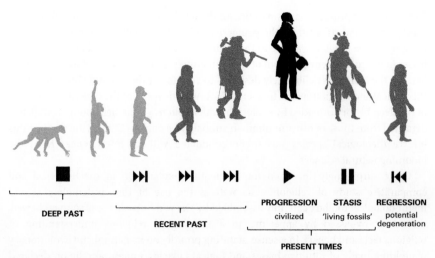

Figure 7 A simplified version of Victorian progressionism

Gregory 1927). In other words, the accumulation of undeniable palaeontological and palaeoanthropological data backfired spectacularly, promoting a directly proportional refusal or denial of their import for human evolution. Max Müller, and even Darwin's closest friends, were at odds with Darwin as far as the gap between human and nonhuman taxa was concerned. Pithecophobia informed a great many works in the Humanities (a topic which definitely needs more dedicated studies) and, as far as the study of religions was concerned, it tinkered with esotericist, orientalist, Indocentric and theological perspectives, and it may still exert an unquestionable charm in some academic quarters. This intellectual malady was also one of the main forces behind the idea that the cradle of humanity, arisen and already formed as Athena from the head of Zeus, lay somewhere in central Asia, a territory which a Romantic vogue endowed with absolute, archaic and mysterious prestige (cf. De Quincey 2013: 72–3). A similar attitude characterized the cultural fascination exerted by the vastly unexplored lands of continental Asia on ancient Greek culture.

As Jonathan Z. Smith has rightly pointed out, a mixture of all these old and new ideas was to characterize precisely the incipient comparative approach to religion, casting a long shadow over the future of the field: 'evolution, as represented by the nineteenth- and early twentieth-century practitioners of anthropology and comparative religions, was an illegitimate combination of the morphological, ahistorical approach to comparison and the new temporal framework of the evolutionists' (Smith 1982: 24).

Original sin

Since the very beginning of modern scientific research, the domains of cognition and human social behaviour were strictly intertwined with religion and evolution. The

Descent of Man proved a baptism of fire for the newborn comparative religion. The test was too hard and stressful for the young discipline. As we will see later on, the failure of comparative religion was due to the intuitive force of religious and anti-positivistic biases in the cultural environment of the Humanities, and Victorian science of religion's heir, i.e. the HoR, was deliberately designed to counteract the scientific invasion of the theological and religious pitch. In a sense, HoR was the result of the conscious rejection of the deep history provided by evolution, with *homo religiosus'* behaviours completely detached from those of other nonhuman animals. In a curious U-turn, the study of this bizarre theological species was to be conducted without interferences from the blooming natural sciences.

Quite surprisingly, the need for a scientific explanation in the historical and comparative study of religion – as well as the use of the epistemological and methodological toolbox carefully assembled by Darwin himself – was soon eschewed, skirted or bypassed by appealing to a tautological religious understanding of religious facts and beliefs. In a sense, standing proudly on its natural and evolutionary scaffolding made of intuitive biases and logical fallacies, unnatural religion declared victory over a humbling but unwelcome natural understanding of religion itself. Consequently, the import of the Darwinian panda principle for comparative religion was eventually forgotten. The panda principle urged researchers to focus on similar environmental forces which might affect the evolution of similar, yet phylogenetically unrelated, traits. Diametrically opposed, the original sin of the whole historiography of comparative religion was the disregard for differential context, which led exactly to the error effortfully avoided by Darwin. One has only to think about the famous comparative work by Jesuit missionary Joseph-François Lafitau (1681–1746), *Moeurs des Sauvages Amériquains, Comparées aux Moeurs des Premiers Temps* (1724), in which Native Americans' customs and religious beliefs were interpreted as homological with (i.e. inherited from) ancient pre-classical (and pseudo-historical) Greek tribes, of which they were thought to be the descendants. As time went by, so Lafitau argued, their primordial and pure religiosity degenerated, resulting in modern-day corrupted customs (Borgeaud 2013: 93–8). Lafitau's homological and fallacious *modus operandi* is exemplary for the field because it was never theoretically amended and falsified, but continuously reiterated, and Lafitau himself has been often included in the innumerable foundational lists of disciplinary forerunners (see di Nola 1977a: 283; Brandewie 1983: 130). Yet, as we have seen, the possession of a comparative method *per se* is a necessary but not sufficient condition: every human being, and nonhuman animal alike, might engage in categorization and systematization. In other words, *epistemology maketh the discipline*, for methodology alone is barely sufficient (cf. Sharpe 1986: 2).

However, as feared by Frazer himself, the fideistic rejection and the accommodationist reaction by the community of colleagues and peers revealed how fragile was the appeal of positivism and how powerful the grip of non-negotiable dogmas. In the long run, Darwin's verdict about the danger of Max Müller's anthropocentric biases justified by religious convictions was to prove correct in the decades to come. As comparative religion 'went on a decline in a post-Frazerian critique of rampant comparative exercises of things stripped of their contexts' (Jensen 1993: 126), its evolutionary methodology was abandoned in favour of functionalism (eagerly adopted by anthropology) and a neo-theological phenomenology (which became the blazon of the new HoR stuck in Lafitau's mistake). In the following chapter, we will tackle the decline of the Victorian science of religion and the inception of the new HoR, with a specific focus on the trail-blazing Dutch, Austrian and Italian disciplinary schools.

4

Goodbye Science

Taxonomy itself guarantees no history.

Stephen J. Gould

The theory of everything

A most famous maxim by Goethe reads '[a] man who has no acquaintance with foreign languages knows nothing of his own' (Goethe 1893: 154, maxim n. 414). Paraphrasing such dictum on language, Max Müller asserted in his first lecture on the 'science of religion' that '[h]e who knows one [religion,] knows none' (Max Müller 1872: 11). Indeed, the Victorian comparative science of religion, the first coherent academic endeavour of this kind, was nothing short of a global, encyclopaedic, trans-disciplinary and cross-cultural comparison among past and present mythologies. The Tylorian search for panhuman trajectories and (pre-)rational explanations embedded in religious thought was the precipitate of the rational and empirical quest for knowledge typical of the Enlightenment. Max Müller's linguistic explanations and Frazer's eclectic systematization epitomized the painstaking categorization of worldwide mythologies according to rational and progressionist patterns of development (Figure 8).

The Victorian science of religion played a major part in the establishment of a truly scientific approach to religion. Hypotheses were advanced and discussed with an unprecedented vigour. The search for the ultimate ancestor of all religious traditions was supported by the psychic unity, and continuity, of mankind: mythologies from Greece, Rome, ancient Egypt, Northern Europe, along with folklore and worldwide ethnographic accounts, were all plumbed to find their greatest common divisor. Pan-Babylonianism and Pan-Egyptianism were also advanced and discussed, and eventually rejected, each one positing an original and pristine culture as *the* unique primordial source for the worldwide diffusion of *all* or *most* religions (in these cases, respectively, Mesopotamian and Egyptian; see Bianchi 1975a: 107–9; Bowler 1987: 57, 217–18; philological Pan-Sanskritism might also be recalled here as a subset of such diffusionist theses). Previous theoretical proposals were keenly examined in a new academic environment, while many seminal theories and concepts related to the ultimate explanation of religion were incessantly advanced. The most famous of such new theoretical counter-proposals was probably *mana*, a label borrowed from Austronesia by Marett. *Mana* was intended as the wedge to dethrone Tylorian animism from the

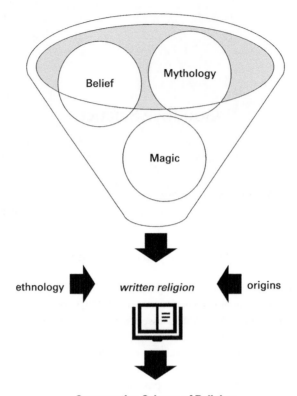

Comparative Science of Religion

Figure 8 Comparative science of religion: original features, influences and main themes

absolute beginning of religion *per se* and reconfigure a primordial phase where 'a sense of the uncanny', an 'impersonal force which [Marett] felt to be present in virtually any unusual object or striking natural phenomenon', was to provide a proto-religious, not-yet or not-strictly animistic feeling or, as the loose translation of *mana* implies, i.e. 'power' (Sharpe 1986: 67).[1]

The main features of all these theoretical paradigms were, as classicist Oswyn Murray listed, 'the beliefs that myth was a secret language disguising a universal earlier stage of humanity, that poetry was the chief literary vehicle for primitive societies, and that ancient art as well as the written record could be used as evidence in the decipherment of this encoded text of a universal lost past' (Murray 2010: 121). Indeed, mythology was on the verge of becoming the humanistic key to solving the theory of everything, from language to social organization, from imagination and folklore to customs and (ir)rationality. And yet, this state-of-the-art approach was to be subverted soon by a rather peculiar Scottish scholar.

A scientific theology? William Robertson Smith's 'dual life'

William Robertson Smith's (1846–1894) multifaceted career, as a Free Church minister, physicist, mathematician, philologist and, last but not least, scholar of comparative religion, is remarkably important in our historiographical account for a number of reasons. *First*, his mix of Higher Criticism (i.e. the philological and critical study of Judeo-Christian sacred texts), Evangelical belief, social evolutionism, social theory and ethnography to create an anthropology of the Middle East dedicated to the recovery of the ancient 'Semitic religious world, close to the vanished world of the Bible' (Strenski 2015: 57, 59), proved to be a groundbreaking proposal. *Second*, he managed to overthrow the previous mythological paradigm by positing that ritual actually came first. *Third*, he was the first successful academic comparativist to advocate an openly accommodationist approach, that is, one dedicated to assuage tension between theological dogmas and science. *Fourth*, notwithstanding his great skills, his approach spectacularly failed to convince his own co-religionists, showing that a compromise was only reachable if epistemology was watered down at the expense of science *tout court* (cf. Sharpe 1986: 79).

Although born in a free and open family environment, Smith[2] was raised within the Free Church, an Evangelical branch that split from the Scottish Church in 1843 to adopt a Calvinist, rigidly traditional and anti-rationalist doctrine based on the human dependence on divine grace and the divine nature of the Bible-as-text (Wheeler-Barclay 2010: 142–3; cf. Black and Chrystal 1912a: 1–31). On a personal level, such a paradoxical mix of a liberal attitude open to scientific, critical discussion and non-negotiable dogmas would constitute the main feature of Smith's work as a scholar. While a believer and a minister of the Free Church since 1870, he adopted German 'believing criticism', in which 'an absolute affirmation of the fact of a supernatural divine revelation' was intertwined with the *non*-divine nature of the texts which recorded that very revelation. One day after being ordained minister, at the age of 24, Smith became professor of Hebrew and Old Testament Exegesis at the Free Church College of Aberdeen (Wheeler-Barclay 2010: 149–50; see also Black and Chrystal 1912a: 114).

Up until 1870, however, Smith had managed to pursue an astonishing and surprising 'dual life' (Anonymous 1894: 557; cf. Wheeler-Barclay 2010: 147). In 1866, Smith was awarded both the prestigious Ferguson scholarship in Mathematics and Classics and the Fullerton and Moir Scholarship. He chose Edinburgh and was admitted to the New College (Beidelman 1974: 4). His examiner, Natural Philosophy professor Peter Guthrie Tait (1831–1901), was so impressed by Smith's scientific acumen that he offered him a position as assistant in his new Physical Laboratory (Anonymous 1894: 557).[3] As recalled by Cargill G. Knott, another assistant of Tait as well as his future biographer, 'when Robertson Smith saw that he could combine the duties of the post with his theological studies at the Free Church College, he accepted Tait's offer; and after training himself in physical manipulation during the summer months of 1868 undertook, the next winter session, the systematic teaching of students in practical physics' (Knott 1911: 71–2).[4] While working at Tait's lab, Smith published on geometry, calculus and electricity (Black and Chrystal 1912b: vi; Livingstone 2014: 46).

An obituary published in *Nature* in 1894, probably written by Tait himself, explicitly lamented that 'unfortunately for Science, and (in too many respects) for himself, his splendid intellectual power was diverted, early in his career, from Physics and Mathematics, in which he had given sure earnest of success' to 'Eastern languages', cultures and – most of all – religions (Anonymous 1894: 557; see Knott 1911: 292). As the obituary recalls, 'what Smith might have done in science' had been already foreshadowed by his early articles (e.g. 'his masterly paper "On the Flow of Electricity in Conducting Surfaces", *Proc. R. S. E.*, 1870') and by his 'splendid service' at the New College (Anonymous 1894: 557).

However, Smith believed in approaching theology as if it was a science, and his accommodationism resulted in the *a priori* logical compatibility of science and religion (cf. Coyne 2015: xii, 97). In a letter dated 29 December 1870, Smith wrote to his friend and mathematician Max Nöther (1844–1921), explaining his views on the matter:

> theology too is a Science & to be successfully handled must be subjected to the methods of all Science. I do not believe in a plurality of Scientific Methods or in any opposition between the mental qualities that fit men for different Scientific pursuits [...]. [T]he ultimate unity of all Science is not a mere distant ideal but a practical fact which every scientific worker will find it profitable to keep constantly before him. And if Science is bound to pay attention to everything real or ideal that forms an actual part of the Universe, there must also be a Science of Religion and especially of Christianity.
>
> SBB Slg Darmstaedter 2d 1870; from Maier 2009: 133

The conclusion of the passage recalled by modern Smith biographer Bernhard Maier exemplifies the perspective of the Scottish scholar: 'and so, my Theological Studies cannot withdraw me from interest in Science in general, but rather stimulate my interest in all those other enquiries in which I recognize an admirable harmony – not with everything that calls itself Theology – but with Theology which is itself scientific' (CUL 7449 A 15; from Maier 2009: 133).

Smith's own 'scientific theology' was the result of a pioneering empirical research in sociology and social evolutionism (see Livingstone 2014: 53 *contra* Kippenberg 2002: 71), intertwined with a firm commitment to Christian dogmas, especially the non-negotiable nature of revelation as real and progressive, and an apparent refusal of merely materialistic and non-teleological biological evolution (Wheeler-Barclay 2010: 147, 169; Livingstone 2014: 51). Accommodationism went hand in hand with 'apologetic purposes': as Smith himself wrote in 1869, 'it is the business of Christianity to conquer the whole universe to itself and not least the universe of thought', so much so that there exists an unresolved tension within the syntagma 'scientific theology' on many respects, not least a dogmatic unwillingness to renounce a historical and teleological *ladder of religious development* oriented towards Christianity (resp., Smith 1912: 135, and Bediako 1997: 310–11). Indeed, when a historical conclusion was to be achieved, such unstable balance was always offset in favour of faith: mere difference in *degrees* between religions became unbridgeable differences in *kind* when Christianity was teleologically concerned (Maier 2009: 269). Even with good intentions, and although he might not have been fully aware of this, Smith's adoption of a comparative approach focused on philology, comparative ethnology and Bible studies belied a critical deconstruction of the comparative science of religion from within, for Christianity retained a privileged place in his social evolutionary analyses, while a stress on the experiential side of religion betrayed his Evangelical roots (Maier 2009: 278; Wheeler-Barclay 2010: 155).

As anticipated above, Smith renounced a promising academic career in physics and mathematics in 1870, when he was appointed to the chair of Hebrew in Aberdeen. And yet, he remained in close contact with Tait and other scientists, corresponding and engaging in a series of quarrels concerning the relationship between science and theology. As a fierce critic of the view that saw science and faith as two opposing forces, in 1874 Smith wrote a newspaper article and harshly criticized both the materialistic, pro-evolutionary approach and the clear-cut separation of science and theology heralded by John Tyndall (1820–1893) in his presidential and militant address to the *British Association for the Advancement of Science* (Maier 2009: 143; Livingstone 2014: 50–1; see Desmond and Moore 1991: 611). One year later, faithful to an accommodationist perspective opposed to any non-teleological biological evolution, Smith helped Tait and fellow physicist Balfour Stewart (1828–1887) write a volume entitled *The Unseen Universe*, in which the two authors set out to show the compatibility between science and religion while asserting that 'recent research in physics rather pointed to the existence of a transcendental universe and the immortality of the soul' (Maier 2009: 143). *The Unseen Universe* was first published anonymously in 1875. One year later, Tait wrote to Smith to ask him whether he would agree to be added as contributor in the forthcoming fourth edition, but Smith refused to comply (Black and Chrystal 1912a: 163–6; Maier 2009: 143). This request coincided with the onset of the most turbulent period of Smith's life.

Smith's heretical accommodationism, 1880s

Smith's editorial participation in the ninth edition of the *Encyclopaedia Britannica* (1875) was to mark the apogee of his efforts in crafting a 'scientific theology'. However, his ideas were to be proven extremely controversial within the Free Church. In 1876, a formal enquiry related to the charge of heresy was started as a direct result of Smith's encyclopaedical entry dedicated to the Bible. In this entry, Smith, notwithstanding his prowess at balancing dogmas and rational academic research, affirmed the non-divine nature of the Bible-as-text and the non-Mosaic nature of the Deuteronomy (Smith 1875). In 1880, after a journey to the Sinai Peninsula and adjacent regions, new entries and articles by Smith were published, and more fuel was added to the theological bonfire (Smith 1880a; Smith 1880b). This time, brilliantly speculative claims which connected totemism with exogamy, polyandry and matrilineal descent, expanded and explained the reconstructed ancient and polytheistic religion of the Biblical peoples (Kippenberg 2002: 69; Maier 2009: 183–4). The Evangelical community deemed such ideas as morally shameful and theologically unacceptable, and 1876 marked the beginning of an excruciating, 5-year-long theological and legal confrontation, culminating in two trials, that ended in 1881 with a compromise. The agreement resulting from the second trial envisaged that, on the one hand, Smith had to be removed from his chair; on the other hand, the Church renounced the official denunciation of his theses (Maier 2009: 150–86, in part. 184; Wheeler-Barclay 2010: 150–5).

The need for moral reassurance during the final year of the trial emerges in a letter sent to Dutch theologian Abraham Kuenen (1828–1891), where Smith, while still avowing his 'personal religious experience on the conception of Christianity as an absolute religion' and rejecting the idea of 'miracle in the medieval sense', notes:

> I am little concerned whether a zoologist can establish an absolute discontinuity between man & the monkey but I have an interest in feeling assured that the human soul & the human race have an absolute vale (greater than the whole Material universe as Jesus teaches us) and are not mere passing phases in the ceaseless fluctuations of the Kosmos.
>
> Maier 2009: 278; BUL BPL 3028

As a consequence of the laborious legal and theological proceedings, a sorrowful Christian existentialism, and a sort of cognitive dissonance between his past experience as a member of the community of the Free Church and his current status as a sort of scientific outsider (see Coyne 2015: 97), were sneaking into Smith's worldview, to which he had to respond by reinforcing his accommodationism: biological evolution was accepted, provided that the teleological Christian project of ultimate salvation was rescued. In any case, the tormented outcome of the trial did not prevent Smith from pursuing successfully his career, as he was able to land a professorship in Arabic at Cambridge a few years later, where he resided from 1883 until his untimely death in 1894, forming a durable friendship with – and mentorship over – Frazer who, in return, dedicated to Smith his *Golden Bough* ('To my friend William Robertson Smith / in gratitude and admiration'; cf. Ackerman 2005b: 20–1, 73; Wheeler-Barclay 2010: 155; Turner 2014: 294).[5]

As illustrated in such pivotal works as *Lectures on the Religion of the Semites* (Smith 1927; delivered in 1887 as Burnett Lectures and originally published 2 years later), Smith's approach was focused on a sociological and evolutionary, albeit ethnocentric and orthogenetic, study of Abrahamic religions, comparing contemporary and nomadic Arabic peoples with ancient, Biblical Hebrews according to the detection of totemic survivals (Wheeler-Barclay 2010: 167). In a passage that closely mimics Spencerian evolutionism, and notwithstanding Smith's own distaste for Spencer (cf. Livingstone 2014: 53), Smith wrote that 'communities of ancient civilisation were formed by the survival of the fittest, and they had all the self-confidence and elasticity that are engendered by success in the struggle for life' (Smith 1927: 260). Earlier, in the same volume, Smith resorted to the old naturalistic trope already exploited by Max Müller, comparing cultural and social evolution to geology:

> the record of the religious thought of mankind, as it is embodied in religious institutions, resembles the geological record of the history of the earth's crust; the new and the old are preserved side by side, or rather layer upon layer. The

classification of ritual formations in their proper sequence is the first step towards their explanation, and that explanation itself must take the form, not of a speculative theory, but of a rational life-history.

Smith 1927: 24; cf. Livingstone 2014: 52–3[6]

Anticipating most sociological explorations of religion, such as Émile Durkheim's (who declared his debt to Smith's work; Maryanski 2014), Smith's evolutionary approach was based on the rejection of the individualistic approach of his colleagues in comparative religion and entailed the following steps (Figure 9):

1. the identification of the basic unit of study in kinship and, therefore, in society;
2. the existence of a primordial totemic stage;
3. the primeval precedence of ritual and practice over mythology;
4. the incorporation of social, lived religion within the confines of economy and politics.

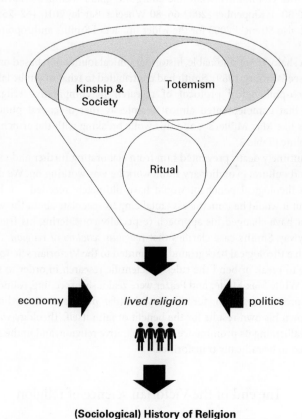

(Sociological) History of Religion

Figure 9 William Robertson Smith's (sociological) history of religion: original features, influences, and main themes

Table 2 Two sides of the same coin: a simplified comparison of Friedrich Max Müller's and William Robertson Smith's method and theory (after Turner 2014: 379–80)

	Friedrich Max Müller	William Robertson Smith
Approach	diachronic	synchronic
Coordinates	temporal	spatial
Philological method	comparative	textual
Focus	mythological	sociological
Aim	uniformity	difference
Framework	genealogical and evolutionary	

This set of interrelated aspects coalesced, methodologically, into the thorough investigation of rituals as a collective manifestation and, epistemologically, into the primacy of sacrifice and ceremonial meals as a socially efficacious way to guarantee the cohesion of the community while including the god/s within the societal network (Sharpe 1986: 80–1; Kippenberg 2002: 66–80; Wheeler-Barclay 2010: 162–75; Livingstone 2014: 53; see also Strenski 2003 for the wider impact of Smith's anthropological theory of sacrifice).[7]

Despite a shaky, if not untenable, historiographical foundation based on ideological prejudices (see Warburg 1989), Smith had contributed to what might be labelled as the first historico-philological protocol of research in comparative religion, with a perspective that complemented almost seamlessly that of coeval philologists and comparatists like Max Müller's – and this while breaking with the armchair tradition of the discipline (Table 2).

Smith's untimely death prevented him from elaborating further and expanding on the social and cultural evolutionary framework he was working on. We do not know whether his theological penchant would have ultimately receded – a bit like Max Müller's – but it would be nonetheless tantalizing to speculate about the way in which Smith might have changed his approach (especially considering his friendship with Frazer). Anyway, Smith's case clarifies that *religious scholars of religion*, that is, those scholars with a theological background committed to the Victorian science of religion, were willing to break or bend the rules of scientific research in order to salvage their own beliefs. While Max Müller and Frazer were zealously exceeding, rounding up their theoretical proposals, Smith, for all his scientific penchant, was ethnocentrically rounding down his own results for the benefit of faith itself. Theology was leaving its mark and reaffirming its prominence on comparative religion, and in the process both were destined to be radically transformed.

The end of the Victorian science of religion

In order to neutralize the threat of scientific reductionism (see also Chapter 3, §*The threat of a scientific and comparative study of religion*), a number of strategies were advanced. For instance, German theologian and church historian Adolf von Harnack

(1851–1930) drily replied to Max Müller's motto about the necessity to know multiple religions by stating that '[w]hoever does not know this religion [i.e. Christianity], knows none, and whoever knows Christianity together with its history, knows all religion' (Harnack 1906: 168). In other words, von Harnack was basically disowning and delegitimizing the very methodological bases of the science of religion, i.e. extra-Christian comparisons (cf. Spineto 2012: 19). It is interesting to note that the slow demise of the Victorian science of religion left a vacant niche for comparative endeavours promptly occupied by eager theologians. German theologian Rudolf Bultmann (1884–1976), for instance, proposed a *demythologization* of the New Testament, which implied an existentialist and symbolic reading of ancient Biblical myths, now conveniently decontextualized from their socio-cultural background to reveal a supposedly universal experience of the world (Segal 2004: 47–51; for the relation between Bultmannian demythologization and future HoR, see also Muthuraj 2007).

As time went by, a more complex dialogue between theology and (an increasingly theology-friendly) HoR began to take place. For instance, American theologian Thomas J. J. Altizer (1927–), well-versed in the phenomenological trends of HoR, resorted to a more radical answer which posited the 'death of God' *tout court*, i.e. the loss of any meaningful link between this world and a transcendent, divine reality. What followed was a frankly paradoxical debate in which HoR became, to a certain extent, more royalist than the king himself, for HoR came to the rescue of positive, orthodox theology, while Altizer's radical theology embraced the comparative and phenomenological eclipse or death of god (see Ricketts 1978; Altizer 2006; cf. also Chapter 5).

These brief historical digressions, just like Smith's trial, reveal how complicated and intertwined the relationship between theology and the comparative study of religion was, and how powerful the theological ballast had always been. While Darwinian evolutionary biology had been able to breach gradually the wall of institutionalization in the United Kingdom thanks to the absence of pre-existing, exclusive disciplinary structures and a lack of reliable scientific paradigms (i.e. natural theology was *the* institutional framework), the comparative science of religion had to fight theological hierarchies and their doctrinal force on their own turf (see Sharpe 1986: 129, and Finkelstein 2000; cf. Ruse 2013). As if that were not enough, the discipline was faltering under the weight of an unchecked and overabundant production of armchair hypotheses and grandiose systematizations. In the absence of an epistemological protocol, and without a consistent peer control, every theory coexisted with its opposite, with scholars unable to falsify each other's claims beyond any reasonable doubt. Any Romantic, (re)invented folkloric poem claimed to have been fortuitously rediscovered was philologically dissected and compared to real historical documents from the past (cf. Kippenberg 2002: 66). Religious traditions were reinvented on the basis of ideological, and political, preconceptions, connecting the dots from scanty archaeological and textual documents to recreate a prestigious past (e.g. Hobsbawm and Ranger 1992; Thiesse 1999; Baár 2010; Merlo 2011).

The Victorian science of religion was to become the victim of its own success. *First*, as shown in the concluding paragraphs of Chapter 2, a disillusion of sorts impinged on the development of such discipline. *Second*, and equally important, the community of peers generally reacted quite harshly to this unprecedented, impudent and unholy hybrid of Humanities, social sciences and natural sciences. Notwithstanding its ultimate adoption for missionizing support and purposes (e.g. Daniélou, Couratin and Kent 1969: 241–70), the comparative project as intended within the Victorian science of religion was equally and strenuously opposed by committed believers, congregations and religious specialists alike.

Eventually, in Britain, the comparative science of religion was displaced by a renewed anthropology less interested in theoretical speculations and more focused on functional aspects and diligent fieldwork. There were still pockets of scholars involved in the historical study of religion(s), of course, but the first British department fully dedicated to the comparative history of religions was inaugurated only in 1967, at Lancaster University, by professor of Religious Studies Ninian Smart (1927–2001) (Turner 2014: 379). It could be said that, at the turn of the century, the concomitant adoption of ritual as the focus of investigation sealed the end of the first wave of scientific approaches to the study of religion (Murray 2010: 126–7). The only way for comparative religion to survive was to renounce science altogether, to leave the social sciences, relocate within the Humanities, and emigrate to Continental Europe (Turner 2014: 378–9; cf. Martin 2016: 34).

Birth certificate(s): HoR's proximate origins

The birth certificate of modern, academic HoR is contested, as we have many beginnings, many false starts, many chairs established across Europe, many different approaches, and way too many scholars to recall here. However, it could be argued that there are at least four main chronological milestones for the discipline:

1. The Higher Education Act of 28 April 1876 (effective by 1877) in the Netherlands, which 'laid the legal and institutional foundations for a new discipline of the science of religion' (Strenski 2015: 80; cf. Platvoet 1998a: 117).
2. The publication of German Catholic priest and anthropologist Wilhelm Schmidt's (1868–1954) *Der Ursprung der Gottesidee* ('The Origin of the Idea of God'), in 12 volumes, published from 1912 to 1955, and preceded by a series of articles (1908–1910) published in French as *L'origine de l'idée de Dieu*.
3. The establishment of the first permanent chair of history of religions in Italy, held by Raffaele Pettazzoni (1893–1959) in Rome since 1923 (preceded by other Italian precursors whose importance for the whole discipline is, with some exceptions, negligible).
4. The foundation of the International Association for the History of Religions (IAHR; initially International Association for the Study of History of Religions) in 1950, with Pettazzoni appointed as second President after the untimely death of Dutch theologian, historian and philosopher of religion Gerardus van der Leeuw.

Many other potential starting points might have been chosen. For instance, a chair dedicated to *Allgemeine Religionsgeschichte* (i.e. 'General History of Religion') was established at the Faculty of Theology, Geneva, Switzerland, in 1874 and dismantled in

1894 because of the dissatisfaction of the faculty, notwithstanding the rather mild pro-theological approach championed there. Another Swiss example worthy of mention here was the series of lectures 'The History of Polytheistic Religions' held by J. G. Müller at the Faculty of Theology at Basel University from 1834 to 1875 (Sharpe 1986: 120). Although overlapping classes were offered in other departments (especially those concerned with Near Eastern history and culture), dedicated courses and chairs soon followed. In the Netherlands, courses about 'Paganism', the 'History of the Idea of God' or about the 'Comparative History of Religions with the exception of Israel's and Christianity' had been taught since 1876 following the institutional elimination of theology from public universities (Spineto 2010: 1258–9). In a few years, the new historico-religious and comparative approach spread all across Western Europe, and other institutions embraced it, although almost always subsumed under various fideistic paradigms: the *Collège de France* and the *Institut Catholique*, both based in Paris (1880), the *Université Libre de Bruxelles* in Belgium (1884), both the Italian universities of Rome and Bologna (respectively, 1886 and 1914), Uppsala University in Sweden (1901), Berlin and Leipzig Universities in Germany (resp., 1910 and 1912), the University of Copenhagen in Denmark (1914), and the University of Oslo in Norway (1915) (Platvoet 1998a: 140 n. 6; Spineto 2010: 1260–3).

In the United States, a lectureship more or less tied to comparative religion *and* theological or ministerial preoccupations began in 1854, at Harvard Divinity School, and then from 1867 to 1904, when the Frothingham Professorship of the History of Religions was finally created. Boston University had a chair dedicated to 'Comparative History of Religion, Comparative Theology, and Philosophy of Religion' established in 1873. Then came the Princeton Theological Seminary (1887), New York University (1887), Cornell University (1891, formally outside of ministerial concerns), the University of Chicago (1892), and the University of Pennsylvania (1894; Turner 2011: 57–9; see Turner 2014: 378–9 for a wider contextualization). More examples might be adduced *ad libitum*, especially if we include Bible Studies, theology, and ancient history, whenever these disciplines engaged in the analysis of religion(s) past and present (Turner 2014). However, a strong case could be made in favour of the pivotal events I have selected above because of the undeniable, long-lasting impact of the Dutch–Austrian–Italian scholarly network and disciplinary reorganization – the effect of which still shapes worldwide, mainstream HoR (Figure 10). As to the usual caveat, I

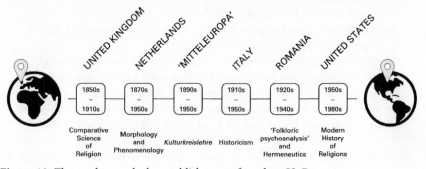

Figure 10 The road towards the establishment of modern HoR

duly remind readers that what follows is a mere geographical selection whose intent is to offer an easy-to-access overview; those interested in an in-depth analysis will find enough material in the in-brackets references.

Netherlands, 1860s: Tiele's (tentative) science of religion

The Dutch context of the late nineteenth century was marked by the establishment of the so-called *duplex ordo* (i.e. the 'two-fold order') via the 1876 law on Higher Education, according to which a formal separation of the academic, theologically free approach to the study of religion(s) and the ecclesiastical, theology-driven pastoral formation, became the law of the Dutch state (Platvoet 1998a: 116; Molendijk 2017: 7). In the long run, this landmark decision, which kickstarted the inception of modern HoR (initially labelled there as 'science of religion'), legitimated an accommodationist approach to re-confessionalize the discipline. This outcome can be considered as the consequence of the academic and cultural environment in which the discipline was set to grow, which remained deeply theological: 'Protestant theologians [were] in a position to make decisions about the shape of university curricula', and they dominated the intellectual and humanistic landscape of that time (Strenski 2015: 79).

However, as the brainchild of Dutch pioneer and Arminian minister Cornelis Petrus Tiele (1830–1902), the Dutch 'science of religion' (*godsdienstwetenschap*) was originally imagined, *first*, as an ally of the natural sciences (Wiebe 1999: 183), and, *second*, as the battering ram with which to break through the citadel of institutional theology (Platvoet 1998a: 116). Like Max Müller, Tiele is often recalled as one of the pioneers, if not *the* founder, of modern HoR (Molendijk 2005: 1–2). Not dissimilar to Max Müller, Tiele too thought of the new discipline as a way to filter spiritual notions and reach a purer and more coherent form of theology and religion (Platvoet 1998a: 118; Molendijk 2004: 323, 347). However, just like Max Müller, Tiele was committed to scientific epistemology, which made him an opponent of the 'obscurantist supranaturalism' of confessional theologies (Tiele 1866a; from Platvoet 1998a: 117; cf. Molendijk 2004: 332) and a staunch advocate of a comparative 'scientific theology' (Tiele 1860: 816). This discipline was based upon the following elements:

1. historical data;
2. the identification of Tylorian survivals;[8]
3. the rejection of the degeneration thesis, i.e. the regressive course of religious development from pristine and divine origins;
4. the abolition of the disconnection between 'natural' (i.e. non-Western and/or non-doctrinal) and 'revealed' (that is, Abrahamic and Eastern) religions;
5. the inclusion of Bible studies as part and parcel of the comparative spectrum;
6. the adoption of general laws describing, and classifying, religious development within a progressive framework (Platvoet 1998a: 118; cf. Molendijk 2004).[9]

This approach was completed by a further distinction between *genealogical* and *morphological* classifications of religion, the former being concerned with historical descent, the latter being dedicated to the identification of developmental stages (Molendijk 2004: 330). From a methodological point of view, Tiele's science of religion was envisioned as the result of the fusion between anthropology and psychology within the natural sciences, with the inclusion of materials coming from aesthetic, historical and systematic theology (Platvoet 1998a: 120).[10]

These Wittgensteinian family resemblances between the late Victorian science of religion and Tiele's own science of religion are not accidental. Direct contacts, acquaintanceship and research invitations strengthened an intellectual partnership started with a letter dated 2 October 1884, in which Tiele was asked by Smith himself to contribute to the ninth edition of the *Encyclopaedia Britannica* (1886),[11] and crowned by the invitation to hold the Gifford Lectures at the University of Edinburgh in 1896 and 1897 (Tiele 1897–1899). Paralleling Smith's working hypothesis, as well as the early Max Müller's vision for the discipline, Tiele's project for a truly scientific theology was nonetheless dominated by a theological *a priori* in the guise of a divine revelation as 'the basic' or 'primeval' force, which instilled a religious sense in human beings 'by nature' (Tiele 1866: 228–30, from Platvoet 1998a: 120), although he recognized that any theorization explicitly based upon the idea of *homo religiosus* as such would be as futile as 'saying that a dog barks because he has the capacity to bark' (Molendijk 2005: 138 n. 50, paraphrasing Tiele 1871: 401). At last, after more than 10 years of relentless attacks against theologians and proposals for institutional reorganization, Tiele's 'strategy of conquest' (Leertouwer 1989) was successful, and in 1878 he became the only professor in the field to be appointed in a faculty of Theology from outside of the Dutch Public Church in a non-ecclesiastical public university (in Leiden; Platvoet 1998a: 117).

Tiele famously stated that the 'science of religion is as distinct from [confessional] theology as astronomy is from astrology, and chemistry is from alchemy' (Tiele 1873: 379; from Platvoet 1998a: 122–3). Indeed, his project, notwithstanding the implicitly unresolved tension between faith and science, could have provided the epistemological basis for a durable scientific agenda (Wiebe 1999: 44, 183). However, just as the coeval Victorian science of religion was doomed, his science of religion proved to be more than permeable to more fideistic infiltrations because of some flaws in the original design itself (but see Wiebe 1999: 39). *First,* Tiele's antireductionistic approach insulated beliefs and matters of faith from further critical scrutiny (Wiebe 1999: 44). *Second,* Tiele tried to dissociate its developmental aspect from Darwinian evolution, insofar as it identified a moral gap between human beings and nonhuman animals (again, a gap embedded in both the inner religiosity of the human being granted by a god, and the direct link between God's love and revelation and human conscience; Platvoet 1998a: 119). *Third,* Tiele showed a certain reluctance, and resistance, to embrace a fully fledged Darwinian approach, for fear of being labelled as reductionist and materialist (see Molendijk 2005: 156, 162–3). *Fourth,* and lastly, the typological development of *religion* through *religions* resulted in Tiele's adoption of the idea of 'mankind as religious by nature', with humankind striving to attain a 'fitter and fuller expression' of a religious devotion seen as 'the eternal working of the divine Spirit' (Tiele 1897–1899, I: 32, 38; for the metaphysical postulates in Tiele's works see Platvoet 1998a: 129). As a result of such premises, the Dutch discipline was bound to adopt a providential way of explaining the course of history, as if a 'divine direction lay behind a movement of history which marched in [...] favour' of whoever was studying it as such (in this case, Dutch theologians; Strenski 2015: 79). This idea suited quite perfectly the belief in teleological directions as described in the previous chapter; naturalistic accounts of religion were abandoned, and evolution *per se* discarded (*contra* Spineto 2010: 1258; cf. Molendijk 2005: 26 on the turn of the century as 'the end of an era').

Netherlands, 1870s: Chantepie's theological reaction

As Donald Wiebe has remarked, Tiele's successors neutralized his attempts, however tentative, to distinguish between theology and science and implement a sort of cultural evolution dealing with religion (Wiebe 1999: 44; Molendijk 2005: 26). His successor at Leiden University, the Norwegian William Brede Kristensen (1867–1953), receded from the scientific project envisaged by Tiele, focusing instead on personal meaning and developing his own anti-Darwinian and anti-evolutionary approach while advocating a sympathetic effort aimed at 'enter[ing] into the subject and mak[ing] the point of view of the people [the historian] is studying his own', so much so that this historian will 'be able to give himself to the other and to unselfishly forget himself; only then is there hope that history will open itself to him and that he will conceive the data in their proper significance' (Kristensen 1954: 73, originally published in 1915; from Plantinga 1989: 176). Thus, so stated Kristensen, historians should strive to become 'Persians in order to understand Persian religion, Babylonians to understand Babylonian religion' – while forgetting that we have access to historical realities only through partial documents which need to be deciphered and interpreted (Kristensen 1954: 77; from Molendijk 2005: 32). Naïvely prone to anachronistic fallacy, Kristensen devised a 'hermeneutics of sympathetic love toward the object of understanding' (Molendijk 2005: 32) which allegedly bypassed the limits of temporal compatibility between different and incongruent historical data (in Kristensen's terms, 'informative comparison'; see Plantinga 1989: 175).

Indeed, after Tiele, this 'stand-in-someone-else's-shoes' attitude would result in two interrelated branches, i.e. phenomenology and morphology. The morphological aspect, recontextualized within an increasingly anti-scientific and theological framework, was to provide a new rationale for the existence of the discipline itself. In the natural sciences, morphology, i.e. the 'science of forms', 'required rigorous, rational sorting of data into classes or kinds, and then further classification of these kinds into more general species and so on until the "tree of life" had fully branched out' (Strenski 2015: 80). However, when used to justify the classification of religious patterns, morphology became a sort of evolution without evolution, so to say, a formally heuristic process deprived of any formal explanatory mechanism – if faith, egotism or ethnocentrism were excluded. This shift revealed the shaky foundations upon which the discipline was built, resulting in an unstable compromise between academic acceptance of the discipline and theological ideas, where the beliefs of the researchers were adopted as the yardstick for judgement and as the implicit criterion by which other religions were to be evaluated.

A more explicit anti-evolutionist stance was advanced by Amsterdam University professor and minister in the Dutch Reformed Church Pierre Daniël Chantepie de la Saussaye (1848–1920). Since the very beginning of his career, Chantepie spearheaded a *theologia religionum* in which all the elements implicitly unresolved in Tiele's thought became endorsed as *the* disciplinary methodology. As summarized by Jan G. Platvoet (1998a), Chantepie's approach was built around three main metaphysical axioms:

1. The relationship between God and men (masculine plural) forms the basis for an 'objective Science of Religion'.
2. The existence of God supports and grants the existence of religion as a subject of study and, consequently, the discipline itself. The acceptance of this 'realm of the unprovable' (Chantepie 1871: 81) entailed that:

   a. God and man are longing for each other;
   b. human origins place humankind above all Creation;
   c. Darwinian evolution is to be firmly rejected (a position based on Max Müller's early writings);[12]
   d. primordial monotheism was the result of God's revelation, the degraded signs of which were still discernible in non-monotheistic religions.

3. Religious history reveals the patterns of human disclosure towards God's revelation as it unfolds in just three ways:

   a. in history itself;
   b. in nature;
   c. in human conscience (Chantepie 1871; from Platvoet 1998a: 123–5; cf. Molendijk 2005: 111–13).

This early manifesto is conceptually enclosed within the boundaries of natural religion and natural theology, with religious beliefs assumed as both method and aim (for a European contextualization of Dutch natural theology, see Mandelbrote 2013: 89–90). With regard to evolution itself, Chantepie held that Darwin's theory was a '"disorderly pile of hypotheses" which even plain common sense must dismiss' (Chantepie 1871: 12–15; from Platvoet 1998a: 124). Humankind, endowed with an inner religious sense, was the pinnacle of Creation, and an unbridgeable chasm separated it from other animals, and this because 'of the very simple rule that a religious being cannot evolve from a non-religious being' (Chantepie 1871: 10–14, 87; from Platvoet 1998a: 124). No wonder that Tiele, despite being somewhat uncomfortable with biological evolution, readily replied to Chantepie's account, defending a scientific approach based on psychology and anthropology, for the defining features of religion, according to Tiele, are to be found in the human mind and not in some supposedly 'objective' external factors like beliefs and rituals, and certainly not in supernatural assumptions (Tiele 1871: 374–80; from Platvoet 1998a: 126; see Molendijk 2005: 112).

Chantepie is officially credited with having implanted into the field the term 'phenomenology' of religion (Chantepie 1887), albeit without providing a clear definition or a specific context and saving any treatment of the concept for a future work which actually was never completed (Molendijk 2005: 28, 30–6, 117–21; cf. Sharpe 1986: 122; Wiebe 1999: 184). The typological classification derived from morphological approaches, lacking a proper epistemic core, was to be more or less closely associated with religiously applied phenomenology, which, as formalized later, became a loose epistemological set of research rules in turn characterized by the following features (cf. Sharpe 1986: 223–4; Penner 1989: 43; Molendijk 2005: 40–7; Strenski 2015: 83; Tuckett 2016a):

1. Religion is something completely different from ordinary reality which deserves a dedicated (i.e. *sui generis* and *sui juris*) approach.
2. Religion cannot be reduced nor scientifically analysed via positivistic methods, lest its core lose its meaning.
3. However, given that the essence of religion *per se* is inaccessible, religious experience is accessible only through its phenomena (i.e. what appears is or *indexes* what lies beyond).
4. In order to gain an eidetic vision on such phenomena (i.e. a factual grasp on the things as they appear), the insider's point of view should be adopted while engaging an empathetic approach. This means that religious phenomena as intended by those who believe in them, whatever they might be, are to be assumed precisely as that, as ontologically believed-in realities. Likewise, the scholar's own religious sensibilities become a heuristic tool.
5. A neutral 'bracketing' or 'suspension' (in Greek, *epochē*) of any critical enquiry on religious beliefs and practices themselves is further practised as to renounce disbelief and value judgement. While apparently this point might seem something 'scientific', it was also an implicit concession to immunize the scholars' own beliefs.[13]
6. In further iterations of historico-religious phenomenology, a subjective, hermeneutic, interpretation was to be assumed as the gold standard of a meta-historical enquiry. 'Understanding' every part of the whole religious system at hand became the primary concern, while interpretation (i.e. the explanation of facts) was mostly discarded (in classical philosophical terms, respectively, *Verstehen* vs. *Erklären*).

In all likelihood, Kristensen was the first scholar to come up with something resembling this set of features, although he failed to promote it as a new method (Molendijk 2005: 33–5). It is certain, instead, that Kristensen's pupil Gerardus van der Leeuw properly developed this approach into a fully independent branch. Anyway, it is important to highlight that such a prototypical scheme, presented here for the mere purpose of simplification, does not do justice to all the various scholars' elaborations (Molendijk 2005: 30–48; Tuckett 2016a). In any case, the fact that many typological and phenomenological projects have been more or less *independently* elaborated by several scholars (see Tuckett 2016a: 77, on Spiegelberg 1982) is the outstanding result of a *parallel* conceptual development favoured by similar cultural pressures and biases within religious environments. And yet, because of the pressure exerted by similar environmental constraints, all those phenomenologies retained a meta-religious Wittgensteinian family resemblance, in which overlapping, recurring and striking features, but no determining genealogical filiation, are discernible (Saler 2000: 158–60; see also Saler 2009: 160–71).[14]

Netherlands, 1930s: van der Leeuw's re-confessionalization

The various strands of typological and hermeneutical phenomenology, when adopted in the HoR, became the intellectual device by which the anti-scientific autonomy of the branch was finally sanctioned – or, in other words, the immunizing strategy by which it became possible to fool the establishment. The adoption of such approaches brought within the new branch a more openly stated critical attitude towards modernity, industrialization, Darwinian and materialistic theories, and the Enlightenment (Leertouwer 1991: 204; Wiebe 1999: 174; Junginger 2008; Tuckett 2016a: 83–4). This reactionary bundle supplied the subtext, and the agenda, of Rudolf Otto's (1869–1937) works, the most important of which was the hugely influential *The Idea of the Holy*, originally published in 1917 (Otto 1923; see Tuckett 2016a: 83–4). In this work, the German Lutheran theologian elaborated on the idea of the autonomy of religion, placing at its core the idea of the perception of the 'numinous', that is, the Latin term which defined the presence of a god in a certain place (*numen*). Religion was not primarily something moral, nor social, and not even rational: religion, according to Otto, was an experience, a particular and very specific experience, something detached from profane normal life, a 'wholly Other' (*das ganz Andere*), a feeling of being in the presence of a '*mysterium tremendum et fascinans*', 'a tremendously powerful yet magnetizing, fascinating mystery' which transcends everything else (Strenski 2015: 86–7). Interestingly, the label 'phenomenology' never occurs in Otto's work, and while many scholars involved in the Dutch phenomenologies of the HoR took great pains to distinguish their own perspectives from Otto's, they nevertheless shared the very same family resemblances (cf. Tuckett 2016a: 84).

Although lack of space prevents me from investigating other equally relevant approaches and phenomenological experimentations (see Molendijk 2005), it is fair to say that the most important project of this kind was that envisioned by Groningen University professor, Egyptologist, theologian, Reformed Dutch Church minister and Minister of Education of the Netherlands for the Labour Party in 1945–1946, Gerardus van der Leeuw (1890–1950). With van der Leeuw's works, the HoR became the modern heir to natural theology via a subversive process of re-confessionalization of the institutional *duplex ordo* (Platvoet 1998a: 131–3; Wiebe 1999: 184). His *Religion in Essence and Manifestation* (1963), originally published in German in 1933, is considered a milestone in the incipient field of the HoR. In it, van der Leeuw states clearly that 'the essence of religion is to be grasped only from above, beginning with God' (van der Leeuw 1963: 679). Characterized by an increasingly obscure methodological jargon and epistemological weakness (Wiebe 1999: 176; Tuckett 2016a; Tuckett 2016b), van der Leeuw's phenomenology of religion was aimed at justifying the role of the typological phenomenology in the HoR as 'theological propaedeutics' (Sharpe 1986: 233;

cf. Platvoet 1998a: 131, 133; see van der Leeuw 1954: 13). In the light of the disciplinary success of van der Leeuw's work, an account of his systematization as summarized by Jonathan Tuckett (2016a; 2016b) might be helpful.

According to van der Leeuw, phenomena are self-evident demonstrations of what appears to be, the structure of which is revealed by a sort of meta-cognitive reflection informed by spiritual apperception. This interpretive mechanism reveals 'structural connections' (van der Leeuw 1963: 673) between similar types of phenomena which are manifested to our phenomenological ability to classify while 'standing aside and understanding what appears into view' (van der Leeuw 1963: 676). In order to accomplish this task, *epochē* must be assisted by the self-aware recognition of the scholar's own inner religiosity. According to van der Leeuw, even though the scholar might be agnostic, there is religion deep down inside every human being (van der Leeuw 1963: 645); consequently, *everything* is religious: 'ultimately, all culture is religious [...] all religion is culture', and 'all significance sooner or later leads to ultimate [i.e. religious] significance' (van der Leeuw 1963: 679, 684; see Cox 2002: 122). The dogmatic acceptance of the existence of God provides the basis for an anti-Darwinian but orthogenetic presence/permanence of the holy (cf. Leertouweer 1991: 200), within which the scholar's own set of religious beliefs retains the peculiar position of *primus inter pares*. The classification of religious typologies via historical examples is brought to the reader via the identification of the underlying models ('prophet', 'priest', 'sacrament', 'saviour', etc., or, in Weberian terms, 'idealtypes'), based on the experiences of the believers themselves but filtered through van der Leeuw's own sensibility (Strenski 2015: 89; Spineto 2010: 1274). Otto's influence is crystal clear in the statement that man (i.e. singular masculine) is drawn to power, and religions are the receptacle of extraordinary potency which cannot be controlled and with which certain objects and human beings are endowed, thus becoming 'sacred' (van der Leeuw 1963: 28). The limiting, *horizontal* power gained from modernity and science can be fallaciously, and religiously, worshipped (Tuckett 2016b), while limiting but *vertical* power from Otto's *ganz Anderes*, the ineffable and numinous 'wholly Other' which cannot be subdued, is properly – and powerfully – religious (van der Leeuw 1963: 681).

Van der Leeuw also formally coined the disciplinary idea of *homo religiosus* to indicate the inner, congenital thrust towards transcendence and religious meaning-making by mastering all aspects of human life, contrasted with *homo negligens*, that is, the man incapable of feeling the awe of the *sensus numinis*, i.e. the feeling of a divine presence (van der Leeuw 1963: 680; cf. Cox 2002: 122). In tautological accordance with his view of religious and theological phagocytosis (Platvoet 1998a: 133), '*homo religiosus* is to be found nowhere else than where "*homo*" himself is found [...] Only he who is not yet human, not yet conscious, he is no "*homo religiosus*"' (van der Leeuw 1937: 160, 165; from Sullivan 1970: 262). Finally, a theologically justified anti-modernism provides a framework for the adoption of Lévy-Bruhl's initial theorization of the *mentalité primitive* according to which, for instance, 'primitive' yet spiritually purer art-like puppet-shows are reputedly superior to 'modern' cinematography (Leertouwer 1991: 204, 210 n. 4; cf. van der Leeuw 1954: 11; see also Westerink 2010 for the role of such theory as an alternative to the evolutionary tenets of comparative religion).

In the decades that followed, the relentless work of some scholars, in particular Fokke Sierskma (1917–1977) and Theo van Baaren (1912–1989), provided an astounding epistemological falsification and methodological deconstruction of van der Leeuw's biased approach, rejecting his *homo religiosus* phenomenology, repudiating the theological subversion of the *duplex ordo*, and adopting a scientific and ethological study of *Homo sapiens* based on comparative psychology which prefigured the most interesting contemporary research branches (see Leertouwer 1991; Platvoet 1998b; cf. Slingerland and Bulbulia 2012). However, the academically isolated role of Sierskma and the quite late change of heart of van Baaren might have contributed to an ongoing neglect; such efforts were in any case to be eventually dwarfed by the intuitive appeal and academically unflinching support enjoyed by van der Leeuw's approach in the Dutch academy (Westerink 2010).

Germany, 1890s: 'Cultural circles', geography and reactionary politics

While the battle for a theology-friendly phenomenology was ravaging the Dutch discipline, a young German seminarist was studying at a recently established Catholic missionary school in the Limburg, the southernmost province of the Netherlands. His name was Wilhelm Schmidt, and the school he attended had been founded in 1875 by German-Dutch Catholic priest Arnold Janssen (1837–1909). The school had been envisaged as the first learning institution of the new religious congregation of the Society of the Divine Word (in Latin, *Societas Verbi Divini*, or SVD, also founded by Janssen in the same year). After the completion of its missionary education, Schmidt was ordained priest in 1892. Three years later, after having studied Semitic languages in Germany, at Berlin University, he was appointed as theology and languages professor at the SVD Saint Gabriel seminary in Mödling, Austria (Henninger and Ciattini 2005). While Schmidt's role as *maître-à-penser* of the new HoR has been sometimes neglected (e.g. Kippenberg 2002; Strenski 2015), his ideas were about to confirm successfully and irremediably the religious-friendly and unscientific course of the nascent discipline.

Expectedly, Schmidt's ideas too were deeply rooted in a peculiar reaction against evolutionary ideas. Curiously, this disciplinary reaction, arisen in Germany in the latest decades of the nineteenth century, was an offshoot of a post-Darwinian interpretation of the development and transmission of cultural ideas, which Schmidt saw as practical and promising enough to tackle the *origin* of religion itself. The movement became widely known as the *Kulturkreislehre*, that is, the theory of the 'cultural circles' intended as geographically enclosed cultural environments marked by a distinctive historical makeup (Andriolo 1979: 133). The concept of *kulturkreis* as 'an area of human civilization within which a uniform material culture can be demonstrated to have existed' (Sharpe 1986: 181) had been elaborated on as a bridge between contemporary ethnology and prehistory mainly by German ethnologists Bernhard Ankermann (1859–1943), Fritz Gräbner (1877–1934) and Leo Frobenius (1873–1938) based on the diffusionist approach to culture advanced by zoologist, ethnologist, journalist, political activist and geographer Friedrich Ratzel (1844–1904).

Influenced by the German reception of Darwinian evolutionism and its local political nuances, Ratzel posited that human culture could be understood in terms of collective processes dominated by migration and competition originating in the availability of favourable geophysical conditions (Smith 1991: 136, 141–2). Although he refused a rigid environmental determinism, Ratzel nonetheless set forth a nomothetic approach, i.e. a biology-like model ruled by laws regulating state behaviour, based on 'struggle and selection on a national level' (Stoddart 1966: 694), while discarding almost entirely the core Darwinian concepts of randomness, chance and individual specificity, resulting in the justification for both a 'migrationist colonialism' and a policy of expansion (Smith 1991: 91–2, 143; cf. Smith 2008: 184–5). Uniting historical philology and its obsessive focus on invaders with Moritz Wagner's (1813–1887) biological law of migration,[15] Ratzel elaborated the following points:

1. the presence of similar features in material objects produced by distant peoples testifies to previous contacts between them, contacts otherwise documentarily unattested;
2. the larger the number and spatial distribution of said cultural features, the greater the chance to discover a reliable historical route of cultural transmission or pattern of migration;
3. the clustering of such traits into patterns characterizing a specific area denotes what would later be labelled as a proper *kulturkreis*;
4. as a corollary later expanded by Ratzel's followers, the organic unity and interrelation between mental culture and material production would make it possible to infer from non-mathematical but observational trait analysis both the 'formation of whole cultures' and their historical relationships *in toto* (Smith 1991: 142–3, 145).[16]

Notwithstanding Ratzel's apparent acknowledgement of the heuristic inefficiency of the concept of race (replaced by the primacy of culture) and his plea for inclusiveness with regard to other non-Western cultures, his views were unabashedly Pan-Germanic, ethnocentric, teleological and fervently colonialist (Smith 1991: 147–8; Smith 2008: 186–8). Indeed, in the following decades, Ratzel's human geography and ethnography, both focused on historical blocks of unstable ecological adaptation always threatened by human migratory movements in which peoples have to migrate and conquer new territories to survive,[17] became a geopolitical tool sympathetic to imperialist expansion and missiological concerns (that is, reflecting the missionary work of evangelization to convert non-Christian peoples to Christianity). The *Kulturkreislehre*, with its intuitively appealing criteria of form and quantity of material culture identified in the stratification of more or less advanced cultural layers (or *kulturschichte*), developed into a tautological loop easily prone to interested manipulation and confirmation bias, since 'it could be used to disguise blatantly circular reasoning. The *Kulturkreis* – defined as an aggregation of observed traits – came to be employed as the criterion for determining which traits in a people's culture were significant for analysis' (Smith 1991: 158).

Within the nascent movement, two outsiders stood out: Frobenius, who had not completed any official course of study and yet succeeded in being eventually co-opted within German academia (at the University of Frankfurt, since 1932), and Schmidt, who started as seminary professor, and was only later co-opted as lecturer first and then professor in Austria and Switzerland, from 1921 onwards. While Frobenius worked mainly within the framework of an imperialist ethnology, Father (or, in Latin, Pater) Schmidt became the patron of modern missiological anthropology (Smith 1991: 156; cf. Sharpe 1986: 182; Dietrich 1992). Both were able to exploit institutional sources of funds to achieve their goals, that is, private donors and public museums, ultimately gaining the support of German emperor Wilhelm II (for Frobenius) and both the Austrian and Vatican establishments (in the case of Schmidt, who had also been father confessor to the last Hapsburg monarch, Karl I; see Conte 1988; Spöttel 1998; Klein-Arendt 2010).

Although interested concerns had been present in some distinctively different configurations (see Chapter 2, §*A preliminary note on imperialism, postmodernism, and science*), the *Kulturkreislehre* worryingly set a new standard for the outspoken institutional relationship between imperialism, internal racial policy, irrationalism and academic research. Both Frobenius and Schmidt used their theories to support actively the racial politics behind the establishment of the German colonial empire, to foster anti-Semitism and, in the specific case of Schmidt, to advance Caesaropapist politics under the Austrian clerical fascist regime on the basis of racist, ethnocentric and theological theses – as well as advocating the expansion of Italian fascist domination in Africa on the very same racist grounds implied in his formulation of the *Kulturkreislehre* (see Conte 1988; Spöttel 1998; Connelly 2007; Pyrah 2008; Mischek 2008). Finally, swamped by the stationary *kulturkreis*, both Frobenius and Schmidt resorted to explaining cultural processes in terms of forces that acted beyond human rationality and individual cognizance: Frobenius' *Kulturmorphologie* (or 'cultural morphology') was allegedly explained by *paideuma* (Greek for 'education', 'learning'), i.e. the particular, non-universalistic and immaterial soul of every population, while Schmidt's universal history of religions was resolved by and within *Urmonotheismus*, that is, the primordial monotheism that, according to Schmidt, had been revealed by God to the very first human inhabitants of the Earth (Smith 1991: 160; Spöttel 1992).[18]

Austria and Switzerland, 1900s–1950s: Schmidt's apologetic history of religion

How did Schmidt manage to arrive at such a potentially revolutionary discovery? Basically, he identified the potential outcome of his research by eschewing the null hypothesis (e.g. no religion in primeval times) and choosing beforehand the expected result on the basis of a confirmationist position (Zimoń 1986: 248). As early as 1895, in fact, he privately remarked that 'the proper study of [comparative religion] puts in an even stronger light the supernaturalness of our Holy Religion, as well as the excellence of Her essence and outward validity' (Schmidt to Janssen, 5 March 1895; from Dietrich 1992: 113–14; originally in Bornemann 1979: 279). To confirm the 'outward validity' of his own religion, Schmidt successfully established an independent and 'complete research infrastructure' in which ethnological data gathered from fellow missionaries all over the world could be studied, showcased and discussed (Schmidt himself never engaged in field research):

1. a research journal (*Anthropos*, established in 1906);
2. two monograph series on ethnology and linguistics (respectively, 1909 and 1914, then combined in 1951 to create the *Studia Instituti Anthropos*);
3. a museum (the *Pontificio Museo Missionario-Etnologico* in 1927, originally hosted at the Palazzo Laterano, Vatican, and later relocated in the Musei Vaticani complex);
4. an institute (the *Anthropos-Institut*, founded in 1931 near Vienna);
5. a workshop on the comparative history of religion from an ethnological perspective ('Semaine d'Ethnologie religieuse'; for all points see Dietrich 1992: 112).

The creation of such impressive scholarly apparatus might seem a necessary step to foster a scientific environment. Indeed, Schmidt was being gradually co-opted within academia as a Catholic 'scientific expert' in matters such as ethnology and racial policy (Connelly 2007: 821), first at the University of Vienna, Austria (1921–1938), and then, as he had been arrested in 1938 after the *Anschluss* for mere internal dissent and rescued by the intervention of Mussolini and Pope Pius XII, at the University of Fribourg, Switzerland (1939–1951) (Spöttel 1998: 143; cf. Ries 2005: 191–2 for an apologetic account). In scientific research, it is usually held that the assembly bonus effect rewards similar disciplinary organizations by allowing the members of the research group to overcome individual cognitive limitations and provide a mutual and critical peer review. Indeed, despite the public imagery of the scientist as a genius working in splendid isolation, science is – and must be – a collective effort. I will insist on this fundamental point in the final chapter. What I would like to note now, however, is that if a defective epistemology is used to support such organizations, for instance, by promoting confirmatory positions and eschewing negative feedback and continuous revision, the result is but a mere mimicry of scientific research, resulting in the creation of what Noretta Koertge has defined as 'belief buddies', that is 'people who share a firm commitment to the stigmatized knowledge claims and who help collect supporting evidence and arguments but are very reluctant to encourage criticism' (Koertge 2013: 179). In this sense, Schmidt's compartmentalization provides a clear example of the downsides and shortcomings of the *Anthropos* complex. Schmidt's control of dissent, his strict adherence to a theologically dogmatic and pre-established set of goals (to which we will return in a moment), his manipulation of ethnological data, and his unwillingness to confront criticism are all the hallmarks of pseudoscience (Zimoń 1986; cf. Boudry, Blancke and Pigliucci 2015).

Overturning the Humean foundations of academic enquiry on the subject, Schmidt wrote that the recovery and gathering of what he took to be the 'rational proofs of the natural foundations of religion' was a theological necessity (Schmidt 1931: 34; from Dietrich 1992: 114). Indeed, the very adoption of the *Kulturkreislehre* responded to a three-fold necessity, that is, achieving three distinct goals on three intertwined extra-epistemic socio-cultural and geopolitical levels:

1. *European culture*: to strengthen the Catholic faith against the advance of secularism;

2. *Christian apologetics*: to defend Catholicism against those who considered it to be scientifically backward while, at the same time, criticizing godless modernity;

3. *international comparative religion*: to attack strategically from within the discipline those scientific theories sympathetic to Darwinian evolution ('monkey enthusiasm', in Schmidt 1964: 45; from Spöttel 1998: 139) in order to uphold the supernatural core of religion itself (Dietrich 1992: 112, 114).

This three-fold goal coalesced around Schmidt's industrious publication of his 12-volume *The Origin of the Idea of God* (in German, *Der Ursprung der Gottesidee*), a colossal work whose writing spanned four decades (1910s–1950s) and which can be considered as the theological counterpart of Frazer's pro-science, encyclopaedic *Golden Bough*. For obvious reasons of space, I cannot list every modification or update of its whole structure across forty-odd years of continuous reworking. Suffice it to say that the first six books were organized as following Schmidt's organization of degeneration, from *Naturvölker* ('peoples living in a natural state', i.e. hunter-gatherers) to *Kulturvölker* ('peoples living in a cultural state', whose cultural organization included, more or less sequentially, pastoralism, totemism, farming and various mixed forms of religion/culture; Zimoń 1986: 249; for a compelling history of this untenable anthropological distinction, see Smail 2008). Surprisingly discarding any archaeological or palaeoanthropological data or document, Schmidt intended to reconstruct the ultimate origin of religion by studying and comparing certain modern-day hunter-gatherers deemed particularly 'primitive', such as Pygmy peoples in Africa and South-Eastern native Australians. The only key to access the deep past was thus provided by the alleged equivalence between present-day *Naturvölker* and prehistoric ancestors.

The developmental scheme Schmidt devised was built by assuming *a priori* the historical accuracy of the Biblical Genesis. Having ascertained the presence of High Gods of the sky, or Supreme Beings, retaining a principal role even in most polytheistic or henotheistic[19] present-day hunter-gatherer societies, Schmidt argued that, on a theological basis, such primitive peoples were obviously incapable of conceiving of such a 'sublime' idea (Henninger and Ciattini 2005: 8169). Consequently, Schmidt posited that present-day *Naturvölker* had been cognitively prepared for the reception of a perfect, primeval monotheism (*Urmonotheismus*) in the guise of a divine concession. Interestingly, the mechanisms by which, according to Schmidt, monotheism – and thus religiosity *tout court* – was born were two: the primordial revelation itself, obviously as ultimate cause, and causal and teleological thinking, which acted both as proximate means of religious intuition and maintenance (before 1930, personification, that is, anthropomorphism, was also added; Zimoń 1986: 250). Working backwards, the first stage of human society (*Urstufe*) was imagined by Schmidt as a perfect condition announced by a divine revelation (*Uroffenbarung*) which disclosed the flawlessness of monogamy, nuclear family and patriarchal respect and submissiveness. The cradle of this first-ever culture/religion complex (*Urreligion*) was identified in Asia, with subsequent diffusion all across the globe; for instance, Indo-Europeans and Hamito-Semitic tribes civilized respectively Europe and Africa by subduing previous, degraded cultures (Zimoń 1986: 247–8). In any case, diffusion soon entailed degeneration: cultural decay and moral deterioration ensued, resulting in the development of a series of culture areas or stages in which religion *preceded, provided* and *supported* ethics and culture (Dietrich 1992: 115). A possible synthesis might follow this chronological sequence:

0. *Uroffenbarung*: primordial, divine revelation;
1. *Urstufe/Urreligion*: as a consequence of (0), primeval, perfect human religious organization is implemented;
2. remnants of the *Urmonotheismus*: hunter-gatherers as the oldest civilization and living fossils of the oldest religion (1);
3. *Primärkulturen*: degenerative development and diffusion of (2) into three degraded 'primary cultures':

    a. agricultural matriarchy;
    b. hunting and androcentric totemism;
    c. pastoral nomadic patriarchy (Zimoń 1986: 248; Spöttel 1998: 140; Henninger and Ciattini 2005: 8169).

The passage from stages (2) to (3) was possible thanks to orthogenetic *Elementargedanken*, i.e. inborn 'elementary ideas' theorized by anti-Darwinian, anti-Ratzelian (but still nomothetically oriented) ethnologist Adolf Bastian (1826–1905). These hypothetical core ideas were the parallel output of the psychic unity of humankind, and further divergence was due to geo-historical diversification. Following Bastian, these basic concepts were identifiable by the comparison of multiple less-civilized, and thus less-culturally stratified, *Naturvölker* (Brandewie 1982: 155; Sharpe 1986: 180; cf. Smith 1991: 118–19). Stages (3.a) and (3.b), sometimes mixed by Schmidt into a single 'totemistic-matrilinear mixed civilization', were characterized as 'a hotbed for sorcery and secret societies, for a trader mentality, urbanization, materialism and godlessness, for the decline of authority, for hate, violence, decadence and sexual vices' (Spöttel 1998: 140, based on Schmidt and Koppers 1924: 299 and Schmidt 1955). Stage (3.c), finally, was evaluated as the positive opposite, as *Herrschervölker* ('ruling peoples'), bringer of a religion-civilization complex able to revert or contain the ongoing decadence, with Christianity as the teleological apogee able to cure this decay once and for all (Dietrich 1992: 115; Spöttel 1998: 140).

Degeneration did not stop there, as modernity brought about some of the worst cultural developments possible: evolutionism, Communism and psychoanalysis. Schmidt ranted aggressively against a ragbag of topics and authors, including Tylor, materialism, Marxism, '"excessive evolutionism", determinism, promiscuity, animism, magic, mass psychology, and "the destruction of individualism"' (see Schmidt 1919–1920: 548; Schmidt 1926–1955, I: 72; Schmidt 1935a: 134; cit. in Andriolo 1979: 142; on Schmidt's role in the Catholic crusade against psychoanalysis, see Desmazières 2009). In this context, the worldwide, primordial stage of *Urmonotheismus* became the anti-Semitic wedge by which Schmidt tried to renew the Christian accusation of deicide as original guilt (*Urschuld*) and overthrow the theologically sanctioned Jewish uniqueness as the chosen people recipient of the Covenant with God: 'the first Jews were but a population of monotheists among the others' (Schmidt 1914: 348; from Conte 1988:

128). Moreover, Schmidt considered the Jews as guilty of partaking in the modern spread of secularism (Conte 1988: 124; Connelly 2007).

In the end, Schmidt's biased 'ethnotheological' approach was dissociative and accommodationist, in the sense that it worked by isolating items useful for his interpretation, thus breaking the original integrity of the studied cultures to promote a 'positive theological interpretation', with the aim of creating a downward spiral of religious degeneration, a sort of *scala religionum* whose careful organization was thought to provide moral and strategic reasons to support theocratic policy (Dietrich 1992: 119). Interestingly, just as Dutch phenomenology became a sort of evolutionary taxonomy emptied of its epistemological content, by adopting a *Kulturkreislehre* devoid of any evolutionary processes, Schmidt 'was left with a frame of reference which lacked a historical perspective' (Andriolo 1979: 135; cf. Zimoń 1986: 246). Thus, notwithstanding his 'phobia' of evolutionism and his rejection of any nomothetic approach, Schmidt bit the bullet and resorted to specific evolutionary models, i.e. orthogenesis and degeneration, to explain the development of human cultures and religions (Andriolo 1979: 135; cf. Evans-Pritchard 1965: 103–4).

Schmidt's legacy: The strategic rescue of Andrew Lang's High Gods as an early home run for post-truth

The Origin of the Idea of God was a groundbreaking accomplishment in the most literal sense: to reprise our geological metaphor, *The Origin of the Idea of God* shattered what remained of the old comparative science of religion and, in the collision between the Victorian approach and the new phenomenology, something radically different was born. But did this turn really undermine the Victorian evolutionary and positivistic paradigm, as is sometimes stated (e.g. Spineto 2010: 1270–1; Sfameni Gasparro 2011: 83)? Did Schmidt really bring an epistemology of 'moderate realism' to the field (Brandewie 1982: 152)? And is his legacy untarnished, for the questions he tackled have 'now been abandoned by scholars, because it is not possible to provide an adequate scientific response' (Henninger and Ciattini 2005: 8168)?

The short answer to all these questions is *no*. Schmidt's academic work might be best understood as a decoy used to promote an ideologically conservative agenda which had few, if any, epistemic warrants, and which was genealogically nested within the fringe Victorian anti-rational reaction. Let us review and briefly comment on the evidence.

First, the cognitive intuitions which, according to Schmidt, allowed for the understanding of the divine revelation (and religion in general), were not accounted for as biases (i.e. causal and teleological thinking plus anthropomorphism), but fallaciously assumed *ipso facto* as divine, rightful and unbiased mental tools themselves (cf. De Cruz and De Smedt 2015). Epistemologically speaking, around this theological kernel Schmidt built its entire pseudo-historical reconstruction, bypassing Hume's natural history and resurrecting Vico's cultural layers, resorting to Bachofen's penchant for cultural substrates and developmental history towards patriarchy, and interpreting Smith's Semitic totemism as a degenerate cult identifiable in Judaism (Spöttel 1998: 146; Rudolph and Ciattini 2005: 5259). Here is just an example of Schmidt's apologetic theology and multiple endpoints, in which Schmidt explicitly refuted his own concession to an ethno-cognitive interpretation based on causal thinking and teleology so that he could now favour the direct, 'real' revelation from the Christian God himself:

> for one, however, who has had the misfortune of never possessing this belief in God, or who has lost this faith, we have here another powerful proof for the existence of God, a religious-historical proof. The oldest commonly held religions of mankind, taken in their entirety, cannot be understood in their fullness and uniqueness unless we accept the existence and reality of a God who founded this religion inasmuch as He himself personally instructed the people of the earliest time in their faith, in their moral obligations by giving them His commandments and in the ceremonies of worship they were to follow.
>
> Schmidt 1935a: 491–8; from Brandewie 1983: 283

Second, Schmidt elaborated his ideas in the wake of Andrew Lang's (1844–1912) second-hand identification of a Supreme Being who created the world, or the universe, in all the so-called 'primitive' societies. Described as being 'at his best a gifted amateur and at his worst a polemical bore' (Turner 1981: 119), Lang was a Scottish Calvinist by upbringing who studied at St Andrews, Scotland, and Oxford, where he became fellow at Merton College from 1868 to 1874. After a spell as anthropologist there, he left to pursue a career in journalism (Wheeler-Barclay 2010: 110). Author of more than 100 books, which quite inevitably for the most part were 'both trivial and superficial' (Wheeler-Barclay 2010: 110), Lang began as an armchair scholar interested in Tylorian survivals as a lens through which to study folklore scientifically (he was one of the founders and presidents of the *Folklore Society*, established in 1878), and ended as a relentless and tiresome opponent of the Victorian science of religion (specifically focusing on Max Müller's solar mythology and Tylor's evolutionary animism). Interested in the occult since his Oxford years, Lang also became interested in the study of the unconscious 'region of which we know nothing', i.e. the supernatural source of 'miracle, prophecy, [...] vision' and 'supernormal human faculties' which he labelled as 'X region', even serving as president of the Society for Psychical Research in 1911 (Lang 1898: 366, 3, 15; cf. Wheeler-Barclay 2010: 113). Ever the ambiguous outsider, Lang adopted an equivocal stance of epistemological open-mindedness, attacking Hume's definition of miracle as anti-scientific (Lang 1898: 28) while resorting, although reticently, to the argument from design as the ultimate origin of the beliefs in 'High Gods' (Lang to Tylor, 25 November 1898, Box 13, Pitt Rivers Museum, Manuscript Collection, Tylor papers; from Wheeler-Barclay 2010: 130). The ultimate, and quite paradoxical, aim was to dispose of science itself by using the scientific method to prove the existence of paranormal phenomena as scientifically explainable phenomena (Wheeler-Barclay 2010: 130).

Most importantly for the legacy of the HoR and Schmidt's work had been Lang's fideistic degenerationism, within which Lang collocated the beliefs of 'the lowest savages known to entertain ideas of a Supreme Being such as we find among Fuegians,

Australians, Bushmen and Andamanese' as proofs of the absolute precedence of the idea of the High Gods, afterwards seemingly forgotten in favour of ghosts and animistic beliefs: '[…] if, as a result of the ghost theory, the Supreme Being came last in evolution, he ought to be the most fashionable object of worship, the latest developed, the most developed, the most powerful, and most to be propitiated. He is the reverse' (Lang 1898: 230).[20] Anticipating Otto's *ganz Anderes*, Lang also stated that the core of religion might be intended as an 'unanalysable *sensus numinis*', puzzlingly appealing to an understanding of religion as a sort of traditionally validated, tried-and-true science of the supernatural, implying that modern positivistic science itself was 'very far from being exhaustive of the truth' (Lang 1898: 51). Unsuccessful during his time, Lang's devastating plea for a rejection of the Victorian positivistic approach in favour of an emically investigated theological and paranormal degenerationism, was to flourish in Schmidt's *magnum opus*, where Lang's thesis was enthusiastically welcome: Lang 'has caused a revolution in the science of religions, and the rock he threw, while crushing many more hypotheses, will become the corner stone for a new edifice' (Schmidt 1910: 71).

In hindsight, Schmidt's *Kulturkreislehre* provides the first fully fledged case study in the modern history of post-truth HoR, where all the following items, before scattered here and there or contested, were now gathered to form a consistent and coherent anti-Enlightenment worldview:

1. the adoption of a scientific jargon to pretend epistemic and virtuous prestige;
2. a willingness to bend the rules of scientific research, and to manipulate data, within an explicit pro-religious, irrational and actively anti-scientific framework;
3. the assumption of a confirmationist stance, according to which research is conducted by discarding disproving data;
4. the building of a community of 'belief buddies' (Koertge 2013);
5. the control of internal dissent and the neglect of external criticism;
6. an obsessive focus on the hypothetically homological nature of religio-cultural data, that is, descended from a single origin, minimizing the possible role of analogy, namely, parallel but independent development;
7. the construction of a nostalgic pseudo-history based on theological and political *a priori* assumptions;
8. a frankly racist, colonialist and anti-Semitic perspective;
9. the institutional support from conservative, theocratic or fascist governments;
10. the presence of anti-modern ideas supported by logical fallacies, such as:

    a. *ad hominem* attacks via 'black-listing' and 'derogatory word associations' (Andriolo 1979: 142 n. 6);
    b. the use of a domino technique to suggest a chain of *guiltiness by association* in order to discredit opponents;
    c. the *slippery slope* on which such chain of guilty parties was purportedly sliding (e.g. evolutionism leads to moral, social and sexual decadence, via Marxism and psychoanalysis).

In a metaphorical sense, Schmidt had started the construction of an epistemic surrounding wall around a non-negotiable core of religious ideas to rationalize the failure of theological interpretations away from positivistic disconfirmation (see Boudry and Braeckman 2012). The anti-rational dangerousness and misleadingly intuitive appeal of the Schmidtian defence mechanism have been recognized as far back as 1931 by Canadian humanist and Chicago University professor of Comparative Religion A. Eustace Haydon (1880–1975) who, in a review, stated that 'it is doubtful whether even the vast erudition and amazing industry of Roman Catholic scholars can make plausible the theory of primitive monotheism to scientists trained in the field. The uninitiated lay reader, however, will probably agree with Father Schmidt, for this is learned and persuasive apologetics' (Haydon 1931: 611). But what if the 'scientists' in the field were to give way to scholars prone to be fascinated by Schmidt's 'persuasive apologetics'? By mixing intuitive biases mediated by faith and theology with appealing fallacies and an avowed political stance, while at the same time expanding on Lang's ambiguous anti-positivism and rejecting the scientific legacy of the natural history of religion, Schmidt had just sneaked onto the academic High Table.

Italy, 1910s–1950s: Pettazzoni's revolutionary rebuttal

Schmidt's *modus operandi* and works were acknowledged and integrated within the boundaries of a new disciplinary synthesis which attempted to fuse together phenomenology with a more historically oriented approach, while Schmidt's results were at the same time criticized and falsified. During a *tête-à-tête* which spanned almost half a century, 'outstanding Italian scholar' Raffaele Pettazzoni (Sharpe 1986: 184) critically engaged Schmidt's *Urmonotheismus* on the same comparative and ethnological field of the German priest while recognizing the value and breadth of the *Kulturhistorische Methode* (on Pettazzoni's international reception, see Rennie 2013). It could even be stated that the entire academic career of Pettazzoni, a freemason and a socialist who in his youth had abandoned Catholicism, had been dedicated to the falsification of Schmidt's *Urmonotheismus* which, as Pettazzoni recapped, amounted merely 'to a return, by way of science, to the old position of the doctrine of revelation'. The danger posed by Schmidt's thesis was such that, actually, 'the parenthesis which Hume and Rousseau had opened in the eighteenth century was now to be closed, and monotheism brought back to the very fountainhead of religion' (Pettazzoni 1954a: 4; cf. Gandini 2005).

As early as 1911, Pettazzoni delivered a speech at the First Congress of Italian Ethnology entitled *Le superstizioni* ('Superstitions'), in which he tackled the integration of superstitious cults within Catholicism, and praised Schmidt's ethnographical working method (Pettazzoni 1911, from Gandini 1993: 200). One year later, the two scholars started a brief and formal correspondence. However, right at the outset, Pettazzoni's acceptance of Schmidt's *Kulturhistorische* methodological and theoretical systematization stands in stark contrast to his critique of Schmidt's (and Lang's) theological results (see Gandini 1994: 195–8, 240–4). A monograph, entitled *L'essere celeste nelle credenze dei popoli primitivi* ('The Skygod in the Beliefs of the Primitive Peoples') and ready since 1915, was supposed to provide a conclusive rebuttal to Schmidt's *Urmonotheismus*, but the outbreak of World War I prevented its publication until 1922 (Gandini 1996: 117; Pettazzoni 1922). Decades of heated confrontations through a panoply of harsh reviews followed the publication of this and the following works (Gandini 1994: 243–4; Gandini 1999a: 157–8; Gandini 2000: 197–9; Gandini 2001a: 31–3; Gandini 2001b: 106–8; cf. Brandewie 1983: 43–4, 242–6, 250–1). For instance, in a commentary published in 1927, Pettazzoni reported the various critiques already published against the *Urmonotheismus* (among which, most notably, Otto's in his *Das Heilige*) and remarked on the 'extrascientific moments (*momenti extrascientifici*) in Schmidt's thought that invalidate the scientific value of his results, notwithstanding the depth of his ethnological knowledge' (Pettazzoni 1927: 111).

To recap Pettazzoni's rebuttal, monotheism might take place only after the elaboration of polytheism. The formulation and institutionalization of monotheism are inseparable from the action of reformers who establish a new religion, while 'discarding all the previous divinities as demons, affirming the uniqueness of one god', and imposing a new or different moral code (Filoramo and Prandi 1997: 76). Monotheism, according to Pettazzoni, was not the product of evolutionary trends, nor a pre-existent divine truth: it was a *revolution*, a dramatic, sudden, unexpected change. In order to support his thesis, Pettazzoni compared the three Abrahamic religions and Zoroastrianism, their religious texts and new traditions, against the backdrop of each local and previous polytheistic system. The most important methodological point was that only historical data can provide the *sine qua non* framework for an epistemically warranted study of religious development. Ethnology alone was insufficient and misleading (however, palaeoanthropology was almost ignored). This is why Pettazzoni attacked the obsessive focus on the 'uncivilised' (Pettazzoni did not renounce the usual racial jargon) and the lack of knowledge concerning historical documents and data shown by all those scholars 'from Hume to Lang, from [...] de Brosses and Auguste Comte with their fetishist negroes to Father Schmidt's Pygmies', who, in one way or another, approached the 'problem of monotheism' (Pettazzoni 1954a: 4). In Pettazzoni's own words, instead, 'every coming of a monotheistic religion is conditioned by a religious revolution', i.e. a consequence of precise historical factors (Pettazzoni 1954a: 9). Schmidt, in turn, recognized in Pettazzoni's work the 'expression of a former, classic evolutionism', and downplayed his rebuttal as being a merely 'terminological problem' for a Supreme Being is nothing less than a god (Schmidt 1935b: xviii–xxvii; from Brandewie 1983: 243–4).

As to Schmidt's thesis, Pettazzoni stated that there is no such thing as 'primitive monotheism'. Instead, what 'we find among uncivilised peoples is not monotheism in its historically legitimated sense, but the idea of a Supreme Being' (Pettazzoni 1954a: 9), that is, a deity within a pantheon generally provided with omniscience and omnipotence, mostly indifferent or inactive yet sometimes acting as a guarantor of the societal moral order (Sullivan 2005: 8878). Commenting on the research state of the *Urmonotheismus* after the death of Schmidt in 1954, Pettazzoni built on the growing internal dissent within the *Kulturkreislehre* and noted the

vanity of [Schmidt's] attempt to demonstrate the existence of which is nonexistent, that is, of that 'primordial monotheism that existed only in his thought [...]. The concept of the Supreme Being should have never been confused with the idea of monotheism. This confusion has exerted – and still exerts – a bad influence in the field of study.

Pettazzoni's personal notes before Pettazzoni 1956a;
from Gandini 2008: 61

Why did Pettazzoni bother to engage relentlessly Schmidt's works long after having provided a cogent falsification of his theories? Because, interestingly, and notwithstanding his rebuttal of Schmidt's *Urmonotheismus*, Pettazzoni still needed the *Kulturhistorische Methode* heralded by Ratzel, Gräbner, Frobenius and Schmidt to support the autonomy of a new academic study of religions and criticize the apparent lack of historical knowledge exhibited by the various Victorian scholars who established progressive stages of socio-cultural evolution. The aim, as attested by Pettazzoni's earliest writings and personal notes from the 1910s (Gandini 1994: 196–8; Spineto 2012: 12–13, 99), was to unify the study of contemporary 'primitive' peoples with the archaeological and comparative study of past civilizations within a non-evolutionary perspective (Pettazzoni 1913; see Spineto 2012: 95). Thanks to Schmidt's ideas a new 'science of religions', i.e. a multidisciplinary endeavour born from philology and anthropology and pertaining to ethnology and sociology (to which Pettazzoni himself added archaeology, mythology and psychology; Pettazzoni 1912b: x), had the potential to become the true heir to the Victorian science of religion.[21]

Italy, 1920s–1930s: Pettazzoni's two-fold gamble

Despite Pettazzoni's efforts, Schmidt's legacy was not the only tough nut to crack. Any institutional proposal for the reorganization of the discipline had to mediate between phenomenological approaches based on ahistorical uniformity, historiographical approaches based on differential analysis, philological thoroughness in the study of each religion, and the *Kulturkreiselehre*'s historical continuity between ethnology and history. A successful mediation, according to Pettazzoni, could 'resolve' – or 'reduce', as stated in a personal note from 1920 – all the previous 'science of religions' to 'history of religions' (from Gandini 1998: 131; see Pettazzoni 1924: 10; cf. Spineto 2012: 98–9). Pettazzoni's proposal took the name of *historicism*, a bridging branch methodologically engineered to appeal to all those different branches, comparative and functionalist, dedicated to the discovery of the existential sense of religion(s) in a psychological framework to study human beliefs and desires (Spineto 2010: 1273; Figure 11).

In order to achieve this extremely ambitious goal while appeasing and, at the same time, keeping at bay the more unscientific or anti-scientific branches of the field (e.g. like those inspired by Creuzer's meta-historical symbolism and Schmidt's theological approach), Pettazzoni disregarded the Victorian science of religion and resorted to Vico's 'poetical metaphysics' as a sort of unifying psychological framework and ultimate explanation for the birth of religious ideas, specifically referring to Vico's anthropomorphism ('corporeal imagination'), according to which 'the more violent weather-phenomena suggested the first notion of divinity' (Pettazzoni 1956b: 22; see Spineto 2012: 97). And yet, as we have already seen, Vico's primordial history rested on

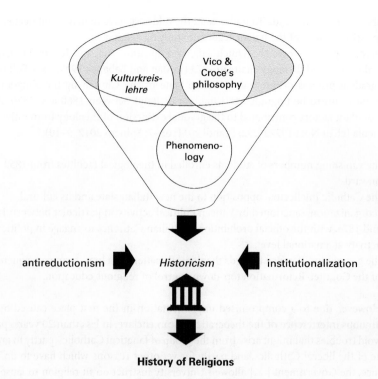

Figure 11 Pettazzoni's history of religions: original features, influences, and main themes

a Providential development of humankind which incredibly sabotaged Pettazzoni's own solution by resorting to the very same supernatural *a priori* adopted by Schmidt – i.e. God (see Chapter 2, §*Rationality as cumulative by-product of comparison*). Another ambiguous statement maintained a distinction between 'true stories' and 'false stories', with 'myth [as] true history because it is sacred history', whose 'origin' is not logical or historical, but magical and religious at once (Pettazzoni 1954a: 15–16). Being an 'absolute truth because a truth of faith', myth is believed by the members of the community 'because it is the charter of the tribe's life' (Pettazzoni 1954a: 21). However, this sociological perspective is merely hinted at (for the autonomy of the HoR prevented any in-depth extra-disciplinary analysis), and a socio-cultural evolutionary perspective on the degradation of myth from truth to falseness adds to the confusion if we really take out 'logic' and 'history' as *explanans* (see Pettazzoni 1954a: 22–3). Such a tautological statement appears as another concession to those unscientific and theologically oriented branches, and its vagueness engineered to provide all the different areas involved academically in the study of religion with a common ground – most of all, as we will see shortly, phenomenology (on Pettazzoni's 'contradictions' and Vico's influence see di Nola 1977a: 292). Since its very beginning, historicism was a two-fold political and disciplinary gamble because the discipline itself was still a faceless *never*discipline without epistemological identity. Would this gamble pay off in the end? To understand the odds, we have to delve a bit deeper into the political history of the Italian academic study of religion.

Preceding by 4 years the Dutch Higher Education Act, the Italian Parliament officially abolished theology chairs in 1873 (Jordan and Labanca 1909: 156). Following the gradual process of annexation of the State of the Church by the Kingdom of Sardinia (soon to be upgraded to Kingdom of Italy) between 1860 and 1870, three concomitant factors contributed to the gradual extinction of theology from university curricula (cf. di Nola 1977a: 292; Prandi 2011: 65–7; Spineto 2012: 9–10):

1. the vanishing numbers of students enrolled in theological faculties from 1859 onwards;
2. the Catholic intellectual opposition to the new Italian state and its cultural reorganization, sanctioned by Vatican political activity in particular between 1868 and 1874, with the official prohibition for Italian Catholics to engage in political activity at a national level;
3. the general indifference towards theological matters as a long-term consequence of the Counter-Reformation top-down control of religious education.

However, due to a compromised implementation in the first place caused by the continuous interference of the theocratic Vatican enclave, in less than 20 years, 'partly to avoid troubles that might arise from the group of fanatical Catholics, partly to gratify some of the liberal Catholics, and partly for yet other reasons which have to do with finance, the Government [...] allowed University instruction in religion to lapse and [...] proceeded to wash its hands of the whole matter' (Jordan and Labanca 1909: 185). A faulty distinction between *professional* and *confessional* theology and its impact on the academic teaching of religion(s), the persisting focus on Christianity in spite of the institutional advocacy for the establishment of a comparative discipline, the debated presence of Catholic priests and scholars as teachers in such a politically heated context, and other socio-political considerations that could not be severed from the very presence of the Vatican state and its influence (e.g. how should 'religion' itself and its study be operationalized in such a new national context?), all these problems contributed to hinder any proposal and prevent any major steps forward towards the establishment of the academic and comparative study of religions (Jordan and Labanca 1909: 287–8; Spineto 2012: 8, 13–14).

At last, after 'a much prolonged apathy', something started slowly to change (Pettazzoni 1912a: 110; in Gandini 1994: 232). Uberto Pestalozza (1872–1966) became lecturer at the University of Milan in 1911, where he led an interdisciplinary team whose aim was the allegedly pre-Indo-European religious substrate and the reconstruction of a primordial, pre-Neolithic Indo-Mediterranean Great Goddess under racial assumptions and sexist wishful thinking (Di Donato 2015; cf. Casadio 1993, and Carozzi 1994). Although Pestalozza became full professor in 1939 at the same University, Pettazzoni preceded him by being appointed as the first Italian tenured professor in the History of Religions in 1923, just one year after the fascist coup, at the University of Rome, where he previously served as lecturer in 1913–1914 (Gandini 2005; Spineto 2010: 1272–3). Despite 'Pettazzoni's reputation as an anti-fascist', political involvement and compromise in the form of 'consent and opportunism' drove Pettazzoni's academic career under the fascist regime (Junginger 2008: 62–3), in a mutually 'fruitful working relationship' where the regime extended his academic legitimacy and Pettazzoni, publicly subservient to fascist policies and eager to recognize the institutional role of the new HoR as auxiliary to state and colonial policy, obtained funds and honours (Stausberg 2008: 375, 381–8; cf. Gandini 2001a: 40, 131).[22]

Upon his appointment, Pettazzoni immediately started working on a virtual common ground for a new international HoR (hence the confrontation with Schmidt's works) while struggling with a complicated national context. As summarized by Marcello Massenzio, the Italian branch of the discipline was 'the product of a subtle intellectual ability to correlate two apparently irreconcilable schools of thought', that is, 'religious phenomenology' and neo-idealism, in particular that of Benedetto Croce (1866–1952), a most influential anti-positivistic yet agnostic philosopher, conservative yet (mostly) anti-fascist, literary critic, and politician who was a firm critic of both the comparative method and the autonomy of religion as a subject (Massenzio 2005: 213). Religion, according to Croce's view, should have been included in a unique, historically grounded, neo-Hegelian philosophy: 'religious history resolves itself, firstly, into the history of ideas [*storia del pensiero*], that is, of philosophy' (Croce 1947: 227; cf. Spineto 2012: 102–3). Another neo-idealist philosopher, Adolfo Omodeo (1889–1946), stated that 'the comparative method is the absolute contradiction of history', which is based upon absolute individuality of human beings and facts (Omodeo 1929: 85; from Spineto 2012: 103). Before such mighty institutional opposition, Pettazzoni entertained a *double or nothing* gamble, compromising all the way down to the very label of his HoR, for *historicism* – quite confusingly – was also the label used by Croce himself, a label which, interestingly enough, was already designed as a complicated, conservative and moderate middle ground between philosophical, social, moral and political opposites (Roberts 1987: 3, 145). On the other hand, Pettazzoni was to enjoy the institutional support of another neo-idealistic philosopher and Croce's frenemy, fascist Minister of Public Education Giovanni Gentile (1875–1944), who issued a competition for the new Roman chair of History of Religions aimed at Pettazzoni less than two months after he was appointed in October 1922 (Gandini 1998: 163–6; Stausberg 2008: 367–70).[23]

Woefully for the field, and for Italian culture in general, science had been long depreciated and obliterated by neo-idealism, with Croce deriding it as a collection of 'pseudo-concepts', i.e. hypothetical abstractions useful for practical purposes but historically and philosophically useless – such, for instance, was the comparative method for Croce (Collingwood 1946: 198; cf. also Ambasciano 2013: 157–8). Therefore, conceptually embedded within a neo-idealist culture, and deprived of scientific references (apart from some psychology which he never seemed to have fully mastered; see Spineto 2012: 105), Pettazzoni nevertheless needed something more to justify and ground methodologically the reference to the existential and experiential dimensions of religion, which the Italian scholar, 'in the wake of Vico, conceived as the product of human "fantasy"' (Filoramo and Prandi 1997: 76). From the 1930s onwards, this psychological 'something' (or 'psychologism') was to be found in van der Leeuw's *Religion in Essence and Manifestation*, hailed as a 'classification of religious phenomena by structures and types; yet, this typology is not purely descriptive and empirical, but rather, and conditioned by, the psychological comprehension of religious phenomena' (Pettazzoni 1933: 243). Notwithstanding the many criticisms Pettazzoni reserved for van der Leeuw's ahistorical, theological and 'static phenomenology', to which he added a complementary 'dynamic phenomenology' (that is, historically oriented and focused on the internal development of each single religious tradition), the final judgement is more than positive, for now Pettazzoni has found a 'precious working tool for any scholar of the religious sciences, no matter their orientation' (Pettazzoni 1933: 243). Van der Leeuw reciprocated by stating that Pettazzoni was 'one of the most outstanding scholars of religion today' (van der Leeuw 1933: 478; from Gandini 2001a: 157).

1950: The foundation of the IAHR and the defeat of science

Pettazzoni was aware of the potential risks posed by van der Leeuw's phenomenology, 'a compromise between theology and history of religions' (Pettazzoni's personal note, *ca.* 1956; in Gandini 2006: 134). However, eager to strengthen the disciplinary position on an international level and to promote the international acceptance of the HoR, Pettazzoni gradually welcomed hermeneutics and phenomenological approaches on a par with his historicism, paradoxically undermining everything he did to counter Schmidt's influence. In 1959, for instance, Pettazzoni acknowledged that 'religious phenomenology and history are not two sciences but are *two complementary aspects* of the integral science of religion, and the science of religion as such has a well-defined character given to it by its unique and proper subject matter' (Pettazzoni 1959: 66; my emphasis).[24]

The institutional success of such an unexpected coalition was unprecedented. Having found a sort of unstable and ambiguous middle ground between the pro-theological phenomenology and the more historically based historicism, Italian and Dutch scholars were instrumental in the establishment of the International Association for the History of Religions in Amsterdam in 1950 (IAHR; initially International Association for the Study of History of Religions), with Pettazzoni appointed as second President (1950–1959) after the untimely death of van der Leeuw, President elect for just a couple of months. The first issue of the official disciplinary review of the association was published in 1954, and the review was named *Numen: International Review for the History of Religions* (Gandini 2006: 179–87, 222; see Jensen and Geertz 2015).

However, in the long run, Pettazzoni's two-fold gamble was to backfire spectacularly. From an institutional point of view, Pettazzoni's ambiguous and conciliatory approach towards phenomenology was to have the most egregious consequences within twentieth-century academia. Donald Wiebe once remarked that Pettazzoni's decision to include phenomenology as *the* privileged psychological reference to complement the historiographical study of religion(s), thus bargaining the existence of the HoR with the estrangement of the new discipline from science-informed research, proved to be *the* point of no return for the discipline: 'what Pettazzoni [saw] as a most important innovation to revitalize the academic study of religion [...] is essentially its subversion' (Wiebe 1999: 175). Quite expectedly, the accords between historicism and phenomenology to create a new HoR implicitly safeguarded and gave substantial historiographical and institutional acceptability to the *homo religiosus* concept.

Eventually, Pettazzoni felt that the two strands of the study of religions could not actually be united under the umbrella term of the HoR, so he came up with another term, 'sacrology', which he was about to officially present at the 10th Congress of the IAHR (held in Marburg in 1960), as a new moniker for the science of religions as a whole (Severino 2015). Pettazzoni's death in 1959 prevented him from doing so and, ironically, the Marburg Congress became subsequently famous for the following frankly fideistic statement by then IAHR Secretary, Dutch Reformed church minister and professor of phenomenology of religion at Amsterdam University, Claas Jouco Bleeker (1898–1983), whose approach stemmed from that of his teacher Kristensen (see Chapter 4): 'the value of religious phenomena can be understood only if we keep in mind that religion is ultimately a realization of a transcendent truth'. This much feared 'theological drift', in turn, prompted a vigorous response in the form of a pentalogue of 'basic minimum presuppositions for the pursuit of our studies' to prevent the re-emergence of a *theologia naturalis* within the field, and which was signed by seventeen renowned scholars, including the drafter, Hebrew University of Jerusalem professor Raphael J. Zwi Werblowsky (1924–2015) (Schimmel 1960: 236–7). Unfortunately, following the disciplinary loop of scientific advancement and fideistic neutralization I have briefly sketched at the end of the first chapter (§*The arms race of natural theology*), such efforts were to be soon disregarded. In any case, Bleeker's blunt statement and related response exemplify why Pettazzoni's naïve ideas that the sheer change of the discipline's name could resolve any epistemological issue and that historicism could oppose a bulwark against the 'irrationalistic directions' of phenomenology amounted to mere wishful thinking (Severino 2015: 9; from Lanternari 1960: 54–5). Indeed, Pettazzoni's decision to create a common ground between phenomenology and historicism started a definitional infinite regress within the IAHR about the name and the scientific statute of the discipline (Severino 2015: 9; Martin and Wiebe 2016: 9–13) and, eventually, located the field on the edge of what is known in philosophy of science as a degenerating research programme, a topic which we will tackle in the following chapters.

Similarly, on a national level, the neo-idealistic roots, the ambiguous methodological charter of the discipline, the presence of a persistent socio-cultural anti-evolutionism, the neglect of scientific literacy, and the lack of any solid epistemological support, all contributed to the gradual disintegration of the original Pettazzonian project. In an undated manuscript from the Pettazzoni archive, possibly written in the late 1950s, the Italian scholar listed the 'enemies of the History of Religions' as follows: '1. Theology 2. Psychologism 3. Phenomenology 4. Philology' (from Severino 2015: 12 n. 22, with 'psychologism' used to describe psychoanalytical trends based on supposedly universal archetypes). According to such a list, with a discipline left in epistemological disarray, what remained to approach the HoR was just historiography, which could barely offer comfort to sustain the autonomy of the discipline. Indeed, with the partial exception of philology, Pettazzoni's successors within the so-called 'Roman (or Italian) school of the HoR' tried to escape the cul-de-sac in which the discipline was stuck by resorting to the help of those very 'enemies', in one way or another re-fuelling the endemic and emic fascination for pseudoscience.

In 1958, three outstanding candidates were selected by a panel chaired by Pettazzoni himself to expand the national roster of historians of religions (Spineto 2012: 109–19). These scholars were:

1. Hungarian-born Angelo Brelich (1913–1977), who directly succeeded Pettazzoni in Rome and was one of the signatories to the Marburg document, started his career in the 1930s from the standpoint of the universalizing, phenomenologically friendly 'psychologism' of Hungarian classical philologist Károly Kérenyi (1897–1973), aimed at identifying uniform, universal and experiential archetypes within ancient cultures. However, from the 1950s onwards, Brelich gradually rejected this framework to embrace a historicist approach which moved away from Pettazzoni's autonomy of religion and, eventually, resulted in the abandonment of any dialogue with phenomenologists (most importantly, Kérenyi), the development of a differential comparativism based on the analysis of peculiar differences between different cultures and the keen analysis of the cultural matrix of each religion, and the methodological critique to strict philological principles and outdated historical interpretations (Brelich 1979; Massenzio 2005: 217–19). Because of the gradual voluntary estrangement from national and international disciplinary milieus due to an increasing discomfort with the HoR as such (e.g. the resignation from the office of president of the *Società Italiana di Storia delle Religioni* in 1967), Brelich's later proposals, which were nonetheless stuck in the usual disciplinary antireductionism, failed to have any significant resonance;

2. Ernesto de Martino (1908–1965) landed a professorship at the University of Cagliari (Sardinia) and advocated an eclectic mix of Crocian historicism, Marxist and Gramscian philosophy, existentialism and, most of all, phenomenology, to approach lower classes' folklore in Southern Italy, resulting in the emphasis on the constant precariousness of human existence. Key concepts elaborated by de Martino were *critical ethnocentrism*, that is, the use of Eurocentric conceptual tools with the caveat that concepts may have a burdensome ethnocentric historiography which scholars should be aware of, and *dehistorification*, i.e. the ritual suspension of daily routine to start everyday, profane life with renewed strength. Like Lang, the main driver of de Martino's folkloric interests was constituted by the emic study of the supernatural on ethnological grounds and the firm belief in the reality of paranormal powers (De Matteis 1997; Ferrari 2014; Di Donato 2013; see Gandini 2009: 84–6 for Pettazzoni's private critique of de Martino's approach);

3. Ugo Bianchi (1922–1995) was appointed professor first at the University of Messina (Sicily) in 1960, then Bologna (1970), and finally Rome (1974–1995), while also holding teaching positions at Catholic institutions and serving briefly as consultant of the Vatican Secretariat for the Non-Christians (later renamed Pontifical Council for Interreligious Dialogue; Casadio 2005b: 862). Bianchi, who did not sign the Marburg response but penned a softer response of his own (Bianchi 1961), was elected President of the IAHR in 1990, after having served 10 years as vice-president of the same organization. With Bianchi, the formal

confrontation with Crocian historicism came to an end, thus allowing the official re-entry of frankly pro-theological and/or hermeneutical stances. Working from an avowed Catholic stance, Bianchi sponsored the continuing dialogue between theology and HoR while supporting the autonomy of the latter (e.g. Bianchi 1989), and developed a post-historicist approach which accounted for the typological study of 'concrete universals' while re-elaborating phenomenological tenets, e.g. the transcendental and primordial *rupture du niveau* (i.e. the 'break of [existential] level' towards otherworldly experiences/realities), to approach reputedly universal or innate religious ideas (cf. Casadio 2005b; Sfameni Gasparro 2016).

Regrettably, for reasons of space, I cannot provide a comprehensive list of all the nationally and internationally relevant Italian scholars who tried to advance a more restrained, non-phenomenological and critical approach to the application of the comparative method, such as Cristiano Grottanelli (1946–2010), Alfonso Maria di Nola (1926–1997), and Vittorio Lanternari, another signatory of the Marburg pentalogue (1918–201) (see di Nola 1977a: 292–7; Filoramo and Prandi 1997: 79–96; Spineto 2012: 1289–92). Suffice it to recall here that what held together Pettazzoni's ideal, and unstable, discipline was the opposition to Crocian philosophy and the autonomy of religion (i.e. religious facts and ideas) *per se*. As soon as philosophical opposition began to slowly wane with the death of Croce, any justification for the very autonomous existence of the discipline as envisaged by Pettazzoni himself seemed to fail (Spineto 2012: 155). Interestingly, opposition to the autonomous charter of the HoR *within* the Italian field paralleled the never-ending discussions on the name, and implicitly the scope and extent, *within* the IAHR (e.g. Spineto 2010: 1273, 1291). As Grottanelli and Bruce Lincoln have cogently argued,

> in order to justify a disciplinary autonomy which he considered politically and institutionally advantageous, indeed indispensable, Pettazzoni felt constrained to speak of the unique ontological status of religion – its autonomy or irreducibility – something which had (and among many still has) a certain rhetorical appeal, but was (and is) difficult if not impossible to justify in strictly logical terms.
>
> Grottanelli and Lincoln 1998: 317

Crucially, Brelich's professional career attests to the impossibility of pursuing any relevant non-phenomenological approach within the echo chamber of the HoR, while also showing that an epistemologically warranted historiographical study of religion(s) simply develops, in fact, into cultural historiography. Instead, Pettazzoni's heirs de Martino and Bianchi turned to other non-historicist options to justify the very existence and epistemology of the HoR, fuelling the disciplinary confrontational stance against science and reinforcing fideistic approaches, and engaging consistently and empathetically the neo-phenomenological works of another historian of religions, Pettazzoni's friend and colleague Mircea Eliade, whose works and ideas will be the theme of the next chapter.

5

Eliadology

Eliade is an enemy of 'history', he stands for 'proto-history' [...]. I find it hard to believe that academics will be such idiots to let the Trojan horse of 'permanent prehistory' inside their citadel, i.e. those who attack the very foundations of 'civilization' as such [...]. Still, academics do not get it.

Eugène Ionesco

Neither with you nor without you: Mircea Eliade

By the mid–1950s, HoR had finally become the theory of everything imagined in Victorian times – although with a twist. Thanks to the *Kulturkreislehre*, which provided HoR with a historico-ethnological method, and phenomenology, which supplied scholars with a quasi-psychological and theology-friendly taxonomy, academic institutionalization was finally achieved *at the very expense of science*. However, science could not be easily shrugged off, for science *was* and still *is* – or should be – at the foundation of academia. Therefore, a cognitive dissonance, if not an outright schizophrenic attitude towards science, began to swamp the field, and this became sorely evident in the aftermath of the publication of the Marburg manifesto.

On the surface, a scientific ethos shone through the manifesto, whose five basic presuppositions can be summarized as follows:

1. a constant 'alert to the possibility of *scientifically legitimate* generalizations concerning the nature and function of religion' (original emphasis);
2. the anthropological nature of the HoR, in the general sense of a worldly, human endeavour and the exclusion of emic theological definitions;
3. the rejection of '*theologia naturalis* or any other philosophical or [*sui generis*] religious system' as 'terms of reference of *Religionswissenschaft*';
4. the affirmation of epistemological antireductionism, but within a '*culture* pattern that allows for every quest of *historical truth*' (my emphasis);
5. the rejection of religious propaganda, ideological commitments and fideistic advocacy within the IAHR (Schimmel 1960: 236–7; cf. Sharpe 1986: 276–8, and Wiebe 1999: 145).

However, the manifesto made a concession to different 'nuances of methodological questions' (Schimmel 1960: 237), which basically opened the backdoor to extra-epistemic, crypto-religious phenomenology. The fact that institutionalized HoR did strive to differentiate itself from ecclesiastical and institutional theology did not mean that science was to be automatically embraced, or that an emic and appreciative study of religion was to be abandoned. Within the same institution, obviously, there were scholars seriously concerned by the abandonment of a truly scientific commitment, but they had been deceived in that, as we have seen, the very foundations of the discipline itself justified a cultural environment sympathetic to conservative, reactionary, fideistic, apologetic and esoteric scholarship from an emic viewpoint (Wiebe 1999: 146).

In order to further the religiously appeasing and crypto-theological agendas of the IAHR, as remarked by Werblowsky (1960: 218), the powers that be were tolerant enough, or eager, to expand their ranks by letting in 'dilettanti' and non-ecclesiastically oriented theological scholars. These were all breaches in the structure of the epistemological edifice, which could withstand only so much before becoming unstable. And yet, the first clues of such schizophrenic developments have always been there, right under the scholars' noses. Ironically, among the signatories of the Marburg manifesto there was Mircea Eliade (1907–1986), a Romanian researcher who previously taught at the École des Hautes Études, La Sorbonne, Paris, and who had been recently appointed professor in the United States, a scholar whose best-selling works were to unabashedly break the Marburg statements (bar, quite conveniently, the fourth point on antireductionism). The astonishing fact is that, despite his hard-core phenomenological approach, Eliade was a self-avowed disciple of Pettazzoni.

In 1947, Eliade wrote to Pettazzoni: 'you were my first master in this exciting [in French, *passionante*] but elusive [*insaissisable*] 'science' of religions, and your approval makes me proud' (Eliade to Pettazzoni, 2 November 1947; letter XLV in Spineto 1994: 166). As painstakingly documented by Natale Spineto (1994), Pettazzoni and Eliade engaged in a 33-year-long correspondence, the first known letter of which is dated 1926. In the same year, a 20-year-old Eliade wrote a newspaper article in which he cited Croce and Gentile for the first time (Eliade 1998: 207–9). Indeed, the relationship of Eliade with Italian culture was far-reaching, and Croce, Gentile and Omodeo's writings were all part of his background: 'in 1925, [Eliade] took exams on the logic of Croce and Gentile; in 1927 he took a class held by Gentile in Rome and, the year after, he interviewed the philosopher; he will [also] correspond with Croce' as well as with many other Italian authors (Spineto 2006: 103). In 1928, Eliade graduated from the University of Bucharest with a thesis on philosophy of religion in the Italian Renaissance (Mincu and Scagno 1986: 125–52). From that point onwards, although sloppily conflated with different and often contradictory trends, Italian culture and historicism were to become a central aspect of Eliade's thought and harsh criticism – for Eliade ultimately rejected Pettazzoni's agnostic historicism in favour of an emic, enthusiastically positive view on X-claims and psi phenomena (e.g. Eliade 1960; Kripal 2007: 153; Law 2018; see Spineto 2006: 102–6; Vanhaelemeersch 2007). In addition to Italian Renaissance religious philosophy, this stance resulted from the intellectual interaction between Interwar Romanian ultranationalism, a fascination for occultism and the paranormal, and Eliade's experience in India during his PhD on yoga techniques (Idel 2014: 280). In any case, Eliade remained always well acquainted with the ongoing disciplinary debates and publications in Italy (e.g. Carozzi 1994; Grottanelli 2000; Casadio 2002; Spineto 2006; cf. Ambasciano 2014: 242 n. 360). When, for instance, Eliade reviewed the 1948 updated French edition of van der Leeuw's 'excellent' *Religion in Essence and Manifestation*, judged 'the best introduction to the general history of religions', he began by recalling Croce's own critical review from 15 years earlier, to highlight the resistance of some philosophers before the irreducible and autonomous character of religion *per se* (Eliade 1950: 108, 110).

After the death of van der Leeuw, Eliade's works became *the* morphological-phenomenological reference *par excellence*. The most important Italian historians of religions, in particular de Martino and Bianchi, were to engage in a constant, albeit complicated, dialogue with Eliade. Pettazzoni himself, while arguing that only phenomenology could provide a much needed 'deeper understanding' to historical documents (Pettazzoni 1954a: 217), remained ambiguously undecided between public warm acceptance and private firm refusal of Eliade's method and theory (Severino 2015).[1] Such 'deeper understanding' stretched beyond the usual scholarly topics. For instance, in 1956, during a conference held in Royaumont, France, and organized by the Parapsychology Foundation, de Martino and Eliade were involved in a lively discussion in which the former, resorting to Lang, exposed his ideas on the possible field collaboration between parapsychology and ethnology to ascertain the reality of paranormal powers, and the latter speculated that Palaeolithic or Neolithic shamans could really fly (Angelini 2005: 126–39).[2]

Whether or not scholars in the field were sympathetic to such themes, Eliade's growing fame forced historians of religions and other scholars to confront the supernatural and the paranormal. In a sense, Eliade's frankly emic approach towards unscientific or fideistic topics exacerbated the many epistemological contradictions of the discipline. When, in the 1970s, Ugo Bianchi wrote about the 'methodological disagreement' and the 'high regard' (Spineto 2012: 151) in which he held Eliade's work, he described this methodological relationship with a Latin motto: *nec tecum nec sine te*, 'neither with you nor without you' (Bianchi 1975b: 171; cf. Martial, *Epigrams* xii, 46, and Ovid, *Amores* 3.ii.39; from Spineto 2012: 151). And, because of the intuitive appeal exerted by the X-claims entertained by Eliade, the fame of his works gradually obliterated any kind of effective, etic epistemological resistance (Brelich 1979: 9; Idel 2014: 282). This peculiar situation was not limited to Italian historicism – by the 1960s it had become an *international* issue.

Indeed, the outstanding success of the Chicago School of HoR led by Eliade (Wedemeyer and Doniger 2010) constrained *volens nolens* the diffusion of any non-phenomenological approach. Writing in 1967 on the official journal of the IAHR (*Numen*), Swedish historian of religions Geo Widengren (1907–1996) noted that

some of [Eliade's] books have been real best-sellers, also in English and German translation [at that time Eliade wrote primarily in French], for Prof. Eliade is possessed of an easy style and a suggestive and persuasive manner of arguing. It is quite obvious that these books, very often concerned with the study of symbolism, have met a real need among our time's educated public.

Widengren 1967: 165

Twenty years later, noting the Eliadean use of 'nonphilosophical and nontheological terms in an elegant literary style', a posthumous encyclopaedic entry dedicated to Eliade and written by University of Chicago colleague and professor Joseph M. Kitagawa (1915–1992), confirmed that, 'during the latter part of [Eliade's] stay in Chicago, fame and honor came his way from various parts of the world. By that time, many of his books, including his literary works, had been translated into several languages' (Kitagawa 2005: 2756–7). Kitagawa also recalled that Eliade 'had his share of critics', but also warned that Eliade 'held a consistent viewpoint that penetrated all aspects of his scholarly and literary works, so that it is difficult to be for or against any part of his writings without having to judge the whole framework' (Kitagawa 2005: 2757). Which were the bases of such a 'consistent viewpoint', anyway?

Eliade, 1920s–1980s: From Romanian post-truth to American New Age

It is not easy to summarize Eliade's ideas. A precocious writer who started publishing regularly when he was 12 years old, Eliade produced more than 2,500 works (counting literary, academic and journalistic contributions, translations included; Handoca 1997), and wrote an astonishing number of unpublished diaries, private notes and letters (Ambasciano 2014: 23). However, thanks to recent, critical systematizations of Eliade's life and works, it is anyway possible to briefly outline the most important biobibliographical events in Eliade's career and reconstruct in broad strokes his intellectual and professional networks. This is all the more important because Eliade's works are so relevant for the contemporary HoR that a case could be easily made for the division of the disciplinary history into 'before Eliade' and 'after Eliade'. Considering that Eliade's own approach did not change much in more than five decades, that he accumulated over the years some minor and contradictory readjustments which we can skip here, that his pivotal ideas recur in all his works, and that he mostly accommodated data chosen expressly to fit into his own research programme, I will focus on a couple of major issues and themes (i.e. diffusionism, prehistory and shamanism), preferring to give a more comprehensive epistemological overview and referring interested readers to the much more detailed biobibliographical analyses included in Țurcanu (2007), Ambasciano (2014), and Idel (2014).

Greatly interested in biological sciences and evolution, fascinated by entomology, and fiercely anti-Darwinist since his youth, Eliade promoted a peculiar blend of social and biological evolution typical of the anti-modernist vogue of Romanian Interwar culture. For instance, in 1927 a 20-year-old Eliade eulogized Vasile Conta (1845–1882), 'the greatest Romanian philosopher', and his political blend of racism, anti-Semitism and anti-Darwinism in which natural selection was discarded in favour of a focus on (and control over) 'external influences' such as 'emigration' and 'cross-breeding' (Eliade

1996: 48, 50; Ambasciano 2014: 82). The Romanian cultural setting was characterized by the typical features of aggressive European nationalism and involved the consistent exploitation of a top-down reinvention of racial, archaic and prestigious tradition(s) through media coverage, tabloid journalism and right-wing politics increasingly disengaged from truth values, in a way not dissimilar to the current post-truth era (cf. Idel 2014: 34). As described in painstaking detail elsewhere (Ambasciano 2014), Eliade's viewpoint was initially shaped by the following intellectual and cultural trends:

1. *Orthodox creationism* and *Biblical literalism* as held by his mentor, extreme right-wing pundit and University of Bucharest professor of Logics and Metaphysics Nae Ionescu (1890–1940). Ionescu mixed these beliefs with his own philosophical system called *trăirism*, a form of vitalism which rejected rational authorities and privileged mystical experiences. As University lecturer, Eliade also served briefly as Ionescu's assistant.
2. Lucian Blaga's (1899–1962) *metaphysical saltationism*, namely, the metaphysical and discontinuist character of human culture in evolution as inherently theological and mystery-driven.
3. Constantin Rădulescu-Motru's (1868–1954) *energetic personalism*, which was a cosmic, teleological evolution guided by a vitalistic force.
4. Bogdan Petriceicu Hasdeu's (1838–1907) *providential evolution*, i.e. evolution as guided by divine and transcendent Providence (Eliade 1967a; Eliade 1967b; Eliade 1967c; Eliade 1987; Eliade 1993; Eliade 1991: 255–60).[3]

A huge and precocious interest in occult, folkloric and fantastic literature,[4] in the paranormal as a way to experience otherworldy realities, the orgiastic-erotic and narcotic experiences which were to define his approach to the study of religion (see Oişteanu 2010: 374–413), and the fascination for cultural trends that emphasized the primeval revelation of God's will and knowledge to make sense of the similarities between different cultures and religions (for instance, various forms of modern and contemporary esotericism, diffusionism and, later, Schmidt's *Kulturkreislehre* and *Urmonotheismus*), completed the list of Eliadean extra-epistemic fundamentals.

These motifs were all combined in a single, eclectic vision, one that remained incredibily stable for 50 years (Figure 12). Eliade seemed to believe in the possibility of retrieving the vestiges (or 'living fossils') of antique religious truth, as well as techniques apt to experience again allegedly paranormal and prehistoric powers (especially for some gifted individuals), from the *sui generis*, historico-religious 'metapsychoanalysis' of European and Asian myths and folklore (Eliade 1961: 35; cf. Ambasciano 2014, and Idel 2014). This mental framework was accompanied by Eliade's deep commitment to the revanchist and extremist programme of a far-right revolutionary social order insofar as the prestigious Eurasian past recovered by his HoR would have brought about the sheer uniqueness of the Romanian nation. In other words, since Romania lacked the ancient, classical or medieval documents or monuments found in other European nations, HoR was supposed to provide the academically recognized support for the creation of a powerful, xenophobic and

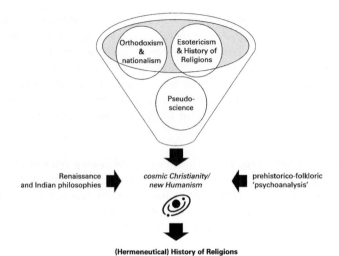

Figure 12 Eliade's (hermeneutical) history of religions: original features, influences, and main themes

totalitarian national state. As a result, Eliade developed a resentful approach to history *per se*, focusing instead on phenomenological escapism, wishful thinking and colonial fantasy, taking advantage of his role as a public intellectual, using his novels, his academic works, and his newspaper articles to justify extremist violence, and getting directly involved in the construction of both a nationalistic mythology and a 'new man' at the service of ultranationalist, fascist, anti-Semitic and fundamentalist political organizations born from the orthodoxist *Legiune Arhanghelul Mihail* ('Legion of the Archangel Michael'; also *Mişcarea Legionară*, 'Legionary Movement', and *Garda de Fier*, 'Iron Guard') and the related party *Totul Pentru Ţară*[5] ('Everything for the Fatherland'; see Laignel-Lavastine 2002; Ioanid 2005; Heinen 2006; Junginger 2008; Ambasciano 2014; Idel 2014; R. Clark 2015; Dumitru 2016; Ambasciano 2016a; Eliade's propaganda articles from this period have been collected in Handoca 2001; cf. Oişteanu 2012 for a history of anti-Semitism in Romania, and Griffin 1993 for palingenetic ultranationalism).

After his Romanian career as a political commentator in far-right tabloid journalism, and having been arrested in 1938 for his political involvement in the legionary party and its institutional organization,[6] Eliade was granted a diplomatic appointment in London and Lisbon that allowed him to leave Romania and carry out institutional duties and propaganda activities for a new, monarchy-supported but still far-right, authoritarian, anti-Semitic, Romanian government during World War II (Ţurcanu 2006: 387–436; Handoca 2008: 379–83; Junginger 2008; Ricketts 2008; Ambasciano 2014: 356–7).[7] However, as we will see later, the knowledge of Eliade's political activity remained quite limited in academic circles before the mid–1970s.

With the war coming to an end, Eliade went to France, which provided him with a remarkable network of scholars. He lived for slightly more than a decade in Paris (1945–1956), where, invited by classicist and right-wing intellectual Georges Dumézil

(1898–1986), he taught a 10-lesson course at La Sorbonne in 1946, completed some of his most important academic and literary works, and took part in the *Eranos* workshops held at Ascona, Switzerland, from 1949 to 1961, where he was introduced to mostly phenomenologically oriented colleagues interested in symbolism and depth psychology, among whom was Swiss psychoanalyst Carl Gustav Jung (Spineto 2006: 57–70; cf. Țurcanu 2013). Between 1949 and 1954, Eliade completed his most important academic works, most notably the anti-historicist *Le mythe de l'éternel retour. Archétypes et répétition* (later translated into English as *Cosmos and History: The Myth of the Eternal Return*, 1954), the phenomenological handbook *Traité d'histoire des religions* (*Patterns in Comparative Religion*, 1958a), a re-elaboration of his old PhD thesis on the history and philosophy of yoga (*Yoga: Immortality and Freedom*, 1958b), and what is probably his most relevant academic contribution, *Le chamanisme et les techniques archaïques de l'extase* (*Shamanism: Archaic Techniques of Ecstasy*; Eliade 1964). Meanwhile, Eliade kept on promoting the folkloric supernatural and the prehistoric paranormal with a slightly softened understanding (e.g. Eliade 1948a; Eliade 1960; Angelini 2005: 126–39), and acknowledged as foundational the speculative works of Pierre Lecomte du Noüy (1883–1947) and especially Pierre Teilhard de Chardin (1881–1955) who, respectively, justified the beginning of life as ignited by a divine will and saw the culmination of the orthogenetic process of evolution as reached on Earth by the 'Omega point', that is the Christian *Logos* (Christ himself).[8]

Invited to hold the Haskell Lectures at the University of Chicago in the mid–1950s by German phenomenologist Joachim Wach (1898–1955), and shortly after appointed as Wach's heir in the Divinity school of the same University, Eliade consolidated his fame in the field and rapidly became the most renowned morphologist-phenomenologist and influential historian of religions of the past century (Casadio 2005a: 4046–7). Historian of religions Jeffrey J. Kripal has recently suggested that the Eliadean use of the label 'cultural fashions' during his Chicago years (e.g. Eliade 1967d) was an academic umbrella term which served to camouflage an emic, appreciative treatment of the paranormal (Kripal 2011a: 202). In the wake of such interpretation, Eliade's academic work during his Chicago years also could be considered as part of the pseudoscientific movement that in the 1960s and 1970s tried to 'combine multiple unorthodoxies into a single volume' (Thurs and Numbers 2013: 139), from esotericism to archetypical psychoanalysis, from the paranormal to Intelligent Design (Ambasciano 2015a; cf. Pigliucci and Boudry 2013). As to the old, loose link between historicism and phenomenology, in the late 1970s Eliade concluded that Pettazzoni had taught him '*what* to do, not *how* to do' (Eliade to I. P. Culianu, 3 May 1977; from Culianu 1978: 6). Additionally, since the 1960s Eliade had been advocating the active use of the HoR as a tool for Western or worldwide spiritual renewal, continuing his old advocacy of the orthodoxist 'new man' in different terms (Ambasciano 2014: 301, 357, n. 617).

During his entire post-war career, Eliade became the proponent of what David Hackett Fischer called 'the fallacy of archetypes', that is a critical stance against history and change as degradation (whenever historical facts were not seen as the vehicle for spiritual renovation), coupled by Romantic anti-modernism. As Fischer argued, when this idea 'is used by a historian to conceptualize his subject, then it becomes a fallacy, for the myth implies that what is real does not change. His time series is bent back upon itself in a sterile series of cyclical enfoldments' (Fischer 1970: 151; cf. Spineto 2006). This interpretation extended from periodic mythical patterns to recurrent theories of myths themselves, to embrace not just Eliade's 'subject' but the whole of history (Eliade 2000; Eliade 2010). As epitomized by Ivan Strenski, Eliade was simply 'no historian' (Strenski 2015: 143; cf. Dudley 1976). The Marburg manifesto remained a dead letter, and a new, post-truth natural theology conquered the HoR.

The sacred from the Stone Age to the present and back again

With such a cultural background, it is no wonder that Eliade was a staunch advocate of human discontinuity and human exceptionalism in evolution and, due to the influence of Ionescu's philosophical works, he apparently struggled to find a balance between his fascination with natural history and the outright rejection of evolution for all things concerning humankind, a conflict that ultimately led to his ambiguous and quasi-theological view of an orthogenetic, non-Darwinian evolutionary path. As a result, he developed what is one of the most striking combinations of pithecophobia and Asiatism in the HoR (see Chapter 3, §*Ladders, progress and pithecophobia*). This stance is evident in the first pages of Eliade's celebrated volume on the *History of Religious Ideas: From the Stone Age to the Eleusinian Mysteries* (Eliade 1978a, published 2 years earlier in French), a book that prompted English historian of religions Ninian Smart (1927–2001) to remark the 'questionable, possibly confused, or even perchance false, premises' of Eliade's 'greatly fruitful work' (Smart 1980: 68). Let us delve deeper into these premises, starting from evolution.

First of all, in such an academic work dating from the late 1970s and whose title indicates the Palaeolithic date of religious beginnings, Eliade's treatment of evolution is nothing short of perplexing. Except for a couple of introductory pages where evolution's role is downplayed, where monkeys are pithecophobically depicted as very cognitively limited animals, nonhuman apes are never cited, and human beings as 'prehominians' or 'Paleanthropians' appear on the scene fully formed and practically limited to *H. sapiens* itself, no updated scientific explanation is provided for the readers.[9] After an anti-evolutionary rant about the attribution of 'the theory of the nonreligiosity of the Paleanthropians' to the 'heyday of evolutionism, when similarities to the primates had just been discovered' (Eliade 1978a: 5) – a theory which is thought by the author to be just a 'misconception' – Eliade admits that 'it is difficult, if not impossible, to determine what [the primordial religious content] was'. Quite puzzlingly, he explains that the palaeoanthropological documents available are insufficient, yet they do contain a language which can be identified through the unconscious via a hermeneutical effort aimed at discovering the underlying archetypes. Once that is done, those documents will reveal 'a universe of mythico-religious value' (Eliade 1978a: 6). In order to sustain his own version of human exceptionalism, Eliade argues that

> if the Paleanthropians are regarded as complete men, it follows that they also possessed a certain number of beliefs and practiced certain rites. For, as we stated

before, *the experience of the sacred constitutes an element in the structure of consciousness.* In other words, if the question of the religiosity or nonreligiosity of prehistoric men is raised, it falls to the defenders of nonreligiosity to adduce proofs in support of their hypothesis.

Eliade 1978a: 5; my emphasis

While Eliade's justification here amounts to a reversal of the burden of proof, and a rejection of the null hypothesis, and thus falls within the range of immunizing logical fallacies (cf. Shermer 2013: 218; Pigliucci and Boudry 2014), his peculiar idea of the sacred deserves a brief explanation. Since the 1930s, Eliade envisaged the sacred as the result of deep, symbolic contents settled and deposited in the human subconscious. This content is encased within archetypes, in turn transmitted through a 'transconscious', which is a transcendent vertical link that connects the human conscience with an unspecified otherworldy and divine reality through the virtual mediation of the subconscious as receptacle and 'apish' imitator.[10] Therefore, every activity is *ipso facto* religious and derived from a divine source nested in human conscience, and archetypes contain the instructions, so to say, for every folkloric, fantastic, religious and non-religious action, belief or symbol. All hark back to prehistory, and archetypes are thought to be (also) the remnants of long-lost truths now encoded in myths, dreams, visions, imagination and desires.[11] Finally, such archetypical contents are expressed differently in 'traditional' and modern Western societies, for the latter merely experience a degraded, degenerated, secularized and profane version of such themes. The fideistic hermeneutics promoted by Eliade's comparative study of 'living fossils' and contemporary Western vestiges is thought to help in recovering and re-establishing a new *transconsciously* reinvigorated religiosity (e.g. Eliade 1958a: 450, 454; Eliade 1958b: 226–7; Eliade 1961: 17, 37, 120; for the differences between 'traditional' and modern societies see Eliade 1959; cf. Spineto 2006; Ambasciano 2014: 66, 105, 124–5, 133, 281, 451–61). Eliade envisioned a metatheological agenda for his HoR in that he created a theology of primeval nostalgia centred on the human longing for the now lost, divine contact (or coexistence) with(in) an otherworldly reality (Olson 1989). *Religion is the result of such nostalgia*; but since the sacred is a 'an element in the structure of consciousness', and thus cannot be obliterated in any way whatsoever, it lives on (as a more or less degraded expression) in arts, leisure activities, rituals, holidays, dreams, visions, love, sexual activity and the occasional superpowers showed by some particularly gifted individuals who attained freedom from physical constraints and 'victory over death', such as the 'levitating' Italian Franciscan St Joseph of Cupertino (Giuseppe da Copertino, 1603–1663) – as testified by 'a witness' (Eliade 1964: 481–2; see Figure 13).[12]

Now, according to Eliade, the scientific underpinnings of these subconscious structures are to be found in the works of Romanian, anti-Darwinist biospeleologist Emil Racoviță (1868–1947), and they represent the link between psychoanalytical and palaeoanthropological explanations.[13] Eliade had been trying to translate Racoviță's ideas from cave biology to racial ethnohistory since the 1930s, justifying each nation's own eschatology and mythical arc, the historical resilience of certain archaic practices (such as yoga), and the persistence of some populations apparently stuck in time,

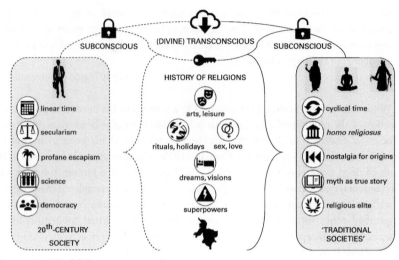

Figure 13 Eliade's 'transconscious' system

within an evolutionary, but not Darwinian, process (Eliade 1991: 113–16; cf. Ellwood 2001: 682). All of these were reputed 'living fossils', and Eliade applied this definition indiscriminately to both cultural and social items. As cave biology discovered organisms which were supposed to be living fossils of long-extinct faunas, Jungian archetypal psychoanalysis allegedly uncovered the depth of otherwise inaccessible human knowledge buried within both the individual and collective subconscious (see Ambasciano 2014: 89–97, 124–5; Strenski 2015: 144). Thus, in *From the Stone Age to the Eleusinian Mysteries*, Eliade cites Racoviță as an authority and recalls the equivalence between *troglobites*, i.e. animals that, quoting Eliade, 'inhabit caves [and] belong to a fauna that has long since been transcended', and current 'archaic civilizations' that have 'survived until recent times on the margin of the ecumene (in Tierra del Fuego, in Africa among the Hottentots and the Bushmen, in the Arctic, in Australia, etc.) or in the great tropical forests (the Bambuti Pygmies, etc.)'. All of these 'civilizations, arrested at a stage similar to the Upper Paleolithic, thus constitut[e] a sort of living fossils' (Eliade 1978a: 24 and n. 36).

And yet, Racoviță's ideas have long been shown to be false. Troglobites are not living fossils, the fauna they were a part of has not been 'transcended' and, moreover, the concept itself of 'living fossils' has been disproved (Ambasciano 2014: 89–98; cf. Gould 2002: 937; Switek 2012; on cave biology and troglobites, see Romero 2009 and Culver and Pipan 2009). The level of technological knowledge exhibited by the cited populations is different and does not reflect a single historical cause.[14] And this is without even mentioning the fact that no human population has ever remained stuck in an ahistorical void capable of preserving it as a social, cultural, religious and technological 'living fossil' – whatever that may mean.[15] More worryingly still, Eliade kept on using such dated terms in an academic publication in the late 1970s to combine the ideas of a *scala naturae* with a *scala religionum*, to showcase 'primitive' human populations as an open-air museum for Westerners' past,[16] in a way not dissimilar to Schmidt's own approach (Schmidt's works are actually referenced in *From the Stone Age to the Eleusinian Mysteries*).

Schmidt's 'stupendous learning and industry'

Interestingly, those Eliadean living fossils scattered all around the globe were the same 'present-day primitives' Schmidt identified as the unmistakable proof of an Asian origin of humankind.[17] In addition to the degree of geographical and environmental remoteness of certain *refugia*, the regions listed by Eliade answered neatly to Schmidt's following rule: 'those who wandered farthest from the entrance place [i.e. central Asia], or were forced by successive waves of people farthest from the point of entry, represent the oldest cultures, while those groups who are closest to the point of entry have to be more recent in time' (Schmidt 1926: n.p.; from Brandewie 1983: 145–6).

Indeed, the similarities between Eliade's and Schmidt's works have already been noted, as has the Eliadean debt towards the *Kulturkreislehre* (e.g. Saliba 1976: 109; Ambasciano 2014: 144 for a list of commentators). In this sense, one remarkable source was constituted by the diffusionist works of Italian archaeologist and palaethnologist Pia Laviosa Zambotti (1898–1965), who tried to fit more palaeoanthropological data and theories into the *Kulturkreislehre* (Ambasciano 2014: 179–242). In the 'Introduction' written for Laviosa Zambotti's French translation of *Origini e diffusione della civiltà* (*Les origines et la diffusion de la civilisation*, 1949), Eliade adopted her modified *Kulturkreise* model (particularly evident in his *Shamanism*; cf. Ambasciano 2014: 222), which retained Asia as the cradle of humanity, and elaborated further on her progressionist history of human populations. In Laviosa Zambotti's model, and in line with coeval palaeoanthropology, prehistoric Neandertals were thought of as beastly brutes, whose modern-day descendants were Schmidtian human vestiges immobilized in a perpetual past (see Ambasciano 2014: 212–13; on the falsification of Neandertal's brutal primitiveness, see Ambasciano 2014: 126–7). As recapped and expanded by Eliade,

> Neandertal man survives today, even somatically (especially concerning the face, the part of the body whose evolution is the lowest) in Australia, where we find the Mousterian industry typical of prehistoric Neandertals, dating back as far as 100,000 years ago. But true progress, as well as the almost integral transmission of the cultural heritage, began with Upper Palaeolithic men, that is 40,000 years ago.
>
> Eliade 1949: iv

Again, this reconstruction is unsupported by scientific evidence. Just to point out a few considerations: by the late 1940s, the proofs adduced to justify the Asian origin of humankind, already doubted on sound comparative and primatological grounds by Darwin himself in favour of Africa (Darwin 1871, 1: 199), were under critical scrutiny (and later proved false; Ambasciano 2014: 137–8, 212–13), while the alleged 'brutal' similarities between Native Australians and Neandertals had been already disproved at the beginning of the twentieth century (Boule 1913; cit. in Sommer 2006: 214).[18] Also, even though the differences between the two aforementioned lithic productions were noted in 1971 (Gould, Koster and Sontz 1971),[19] Eliade managed to eschew every serious confrontation with (palaeo)anthropological research and never updated his viewpoints basically inherited from Schmidt's *Kulturkreislehre*, reproposing them unabashedly in the decades to come.

In the early 1960s, Eliade acknowledged the publication of the first volume of Schmidt's *Ursprung der Gottesidee* in 1912 as a landmark moment for the HoR, along with Freud, Jung and Durkheim's coeval works (Eliade 1984: 12–13). Later in the same venue, Eliade praised Schmidt's brave refusal to accept 'Tylor's animism as well as preanimism, totemism, and vegetation gods' (Eliade 1984: 14), stating that

> despite its polemical excesses (chiefly in the first volume) and apologetic tendencies, *Ursprung der Gottesidee* is a great work. Whatever one may think of Schmidt's theories on the origin and growth of religion, one must admire his stupendous learning and industry. Wilhelm Schmidt was certainly one of the greatest linguists and ethnologists of this century.
>
> > Eliade 1984: 24

Apparently, Eliade was displeased by the criticism reserved for Schmidt's works, and both Australia and Schmidt kept on occupying a position of utmost importance in Eliade's paradigm. The reason is simple: as seen in the previous paragraph, remoteness from Asia and geographical inaccessibility were reputed to be two of the most basic factors to determine the degree of primitiveness of a religion, and Australia ticked these and other boxes in the diffusionist form. Moreover, because of cultural, ethnological, theological, missiological, and political reasons and preconceptions, Australia was thought of as the classical hotspot in the history of HoR's living fossils (Lucas 2005; for Schmidt, see Brandewie 1990: 135). Eliade dedicated to Australia a series of articles published between 1966 and 1968, later collected in a book, in which he reported and commented upon past ethnographical works about Native Australian religions. What better occasion to come back to Schmidt's ideas?

In the first article in the series, Eliade recalled Schmidt's 'prestige [as] a great linguist and ethnologist', he rehabilitated Schmidt's theses, commended Schmidt's working ability to 'substantiat[e], correct, and systematiz[e] the ideas of Andrew Lang' (Eliade 1966: 118), remarked on Schmidt's critiques of Pettazzoni's rebuttal and highlighted Pettazzoni's late 'partial agreement' with Schmidt on some specific points (Eliade 1966: 122–3). The only real weakness in Schmidt's interpretation was, according to Eliade, his idea of the 'Supreme Being belong[ing] to a religious stage preceding any mythological formulation', which, according to Eliade, ran counter to 'everything that we know of *homo religiosus* in general and of primitive man in particular' (Eliade 1966: 119). Thus, from Eliade's viewpoint, Lang's and Schmidt's interpretation based on (divinely granted) cognitive and logical faculties allowing for the primeval intuition of a god was not entirely wrong, but was only too 'rational and even elevated' (Eliade 1966: 117; cf. Brandewie 1990: 120–40). In line with an interpretation of religion that favoured the Romantic and anti-rational exaltation of myth as a true story replete with crude and savage beliefs and behaviours, as James Cox recapped, Eliade 'was not challenging the theory promoted by Lang and Schmidt that a belief in a Supreme Being stood at the foundation of religion, but [he] rejected what they called their "rationalist" interpretation of myth as somehow representing a degenerated form for expressing the original idea of a High God' (Cox 2014a: 29–30). In other words, Schmidt's *Uroffenbarung* was safe.

If this re-evaluation was not clear enough, Eliade also stated that Lang, Schmidt and the scholars that adopted their perspective,

> have the merit of having studied an important aspect of primitive religions in general and of Australian religions in particular. The conception of a High Being – *no matter how different* this High Being might have been from the Supreme Beings attested in other, more complex cultures – was at least something which a great number of religions could be said to have in common.
>
> Eliade 1966: 121; my emphasis

Downplaying the whole *Urmonotheismus* diatribe, the paradigmatic shift which led to the rebuttal of Schmidt's theses was merely due to an unspecified 'change' in 'the Western *Zeitgeist*' that caused the interest in the problem of the High Gods to 'fade out', merely suggesting a series of cyclical vogues (Eliade 1966: 121; Eliade 2000: xiii; see also Ambasciano 2014: 129–34). Leaving basically untouched Schmidt's hard core, Eliade focused on some trivial matters pertaining to the *Kulturkreislehre* external belt of additional ideas, such as the role of myth and irrationality, thus rescuing Schmidt's central and anachronistic ideas concerning diffusionism and divine revelation.

A few years earlier, in a commentary dedicated to Eliade's *The Sacred and the Profane* (1959), anthropologist William A. Lessa (1908–1997) noted something that could easily be extrapolated to describe Eliade's entire academic production:

> there is no history in this work, except some dubious assumptions regarding the sequences through which man and religion have passed (the author at one point invokes Bachofen and Pater Schmidt as authorities for the matriarchate). What passes for history is a series of imaginative reconstructions on highly selected data and contradictory ethnographic and historical materials omitted.
>
> Lessa 1959: 1147[20]

Given the complicated history of cross-breeding between the *Kulturkreislehre* and historicism that we have seen in the previous chapter, Eliade's *fallacy of archetypes* was disciplinarily legitimated and, in some sense, justified.

An epistemic twilight: Pseudoscience and esotericism

Eliade had always entertained the possibility that major religious substrates from human protohistory might have outlived their primeval cultures, even before reading Schmidt. Eliade's entire career was obsessively focused on the uncovering of unconscious and religious survivals of religious elements from protohistoric or prehistoric 'autochthonous substrata' in spite of successive cultural conquest, colonization or assimilation by foreign powers. For the existence of such substrata would explain worldwide mythical and symbolic similarities whose veridicity was beyond question (thanks to the HoR axiom of myth as true story; e.g. Eliade 2000). According to Eliade, one of the main cultural substrata which gave rise to alchemy, yoga, folkloric beliefs and other religious or esoteric knowledge came from some unknown proto-historical Australasian or Melanesian culture, and he nurtured this belief by accumulating throughout his career increasingly contradictory examples of hypothetical colonization (concerning, for instance, ancient Phoenicia, ancient Romania, Europe, North America, Japan, Southern India and some African regions; Eliade 1936: 293–4; Eliade 1939 in Ciurtin 2000: 308; Eliade 1935 in Eliade 2001: 63). Once more, as shown elsewhere (Ambasciano 2014), such ideas were unsupported by any evidence, as Eliade got the entire process backwards. Human colonization of the world proceeded from Africa outwards, in multiple and complex migratory waves, potentially involving interbreeding, with Oceania being the latest area ever to be reached by human beings in historical times (Ambasciano 2014: 193; Bae, Douka and Petraglia 2017; Rabett 2018; cf. Cavalli-Sforza, Menozzi and Piazza 1994 for an overview). One exception could be Australia. But even Australian colonization itself, one of the most ancient and successful intercontinental migrations, dates back to *ca.* 65,000 years ago, which is geologically quite recent if we think that the most ancient fossil remains of *Homo sapiens*, and related palaeogenetic analyses, currently push back the African origins of the taxon to more than 300,000 years ago (Clarkson *et al.* 2017; Stringer and Galway-Witham 2017). However, all of this does not take into account the extra-epistemic bases of Eliade's thesis. Also, Eliade's ideas showed a persistence of racial and cultural biases and a disinterest towards both evolutionary anthropology and Africa; consequently, his religious analyses of the African continent were characterized by what anthropologist Pascal Boyer called a 'caricatural comparativism' (Boyer 1983: 44; cf. Ambasciano 2014: 136–7).

Although Eliade found in Schmidt's proposal a more academically acceptable presentation, the ultimate origins of such a mindset came from Eliade's eclectic youth readings; indeed, his interest in prehistory and Asia was a direct consequence of his acquaintance with both historico-religious works and esoteric, occult literature (cf. Eliade 1991: 208; Spineto 2006). Disparate trends within esotericism had always been fascinated by lost, pseudo-mythical lands and kingdoms (such as Atlantis or Shambhala) to explain the origins, diffusion or degradation of human cultures by resorting to sci-fi-like mythical narratives, omnipotent invisible agents, paranormal abilities and supernatural causes. One of the most famous modern attempts in this sense was that devised by the Theosophical Society, an esoteric organization founded in 1875 by Russian spiritualist Helena Petrovna Blavatsky (1831–1891). Claiming to have received spiritual revelations from unknown masters eager to counteract modern spiritual depauperation via selected spokespersons, and able to perform magic tricks such as materializing objects, Blavatsky came up with a mix of Western esotericism, HoR, Eastern religions and, most of all, pseudoscience. In the wake of the cultural diffusionism that tried to explain every major religion's ancestry via a unique cradle, and eager to exploit the appeal of scientific jargon, Blavatsky chose the biogeographic theory of lost land-bridges, which preceded plate tectonics, and focused on the hypothetical and submersed Pacific continent called Lemuria, recently hailed by German evolutionary biologist Ernst Haeckel (1834–1919) as a potential cradle of humanity (van Wyhe and Kjærgaard 2015: 61). In Blavatsky's storytelling, Lemuria became one of the veritable crucibles of cosmic existence, and the pivot for a new eschatology. As writer Lyon Sprague de Camp recapped, according to the Theosophical mythography (later further embellished),

> life evolves through seven cycles or 'Rounds', during which humankind develops through seven Root Races, each comprising seven sub-races. The first Root Race, a kind of astral jellyfish, lived on an Imperishable Sacred Land. The Second, a little more substantial, dwelt in the former arctic continent of Hyperborea. The Third were the apelike hermaphroditic egg-laying Lemurians, some with four arms and some with an eye in the back of their heads, whose downfall was caused by their discovery of sex [. . .]. The Fourth Root Race were the quite human Atlanteans. We are the Fifth, and the Sixth will soon appear.
>
> de Camp 1970: 56; cf. Nunn 2009: 118–29

Personal spiritual evolution was thought necessary to reverse this degrading process of incarnation into physical forms (Kripal 2014: 64). In 1927, Eliade rebuked Blavatsky *not* for such ideas as grotesquely fanciful, but for being *too open* to evolution and positivistic materialism, an opinion he reiterated in 1962 in his personal diaries (resp., Eliade 2003: 329, and Eliade 1989a: 176–7; see Ambasciano 2014: 191–8). When Eliade came back to this argument in the early 1970s, he claimed that 'the most erudite and devastating critique' of Blavatsky's ideas and all the various more or less faithful followers came 'not by a rationalist "outside" observer', nor 'from a skeptical or positivistic perspective, but from what [is known as] "traditional esotericism"', that is French 'learned and intransingent' esotericist René Guénon's (1886–1951) perennialist reinterpretation (Eliade 1976: 51). If that was a critique at all, it came emically from within the same ideological spectrum.

Perennialism, also known as Traditionalism, is a meta-religious esoteric doctrine which posits that ancient divine knowledge (also called *philosophia perennis*) and supernatural powers were granted to a privileged human race in primeval times and are now scattered or lost, substituted by a symbolic web of hidden meanings to be retrieved and decoded in sacred texts by privileged, enlightened individuals, heirs of that primeval race. Perennialism also asserts the unavoidable crisis of Western society and culture and is characterized by recurrent historico-cosmological cycles and renewals which purify the moral decadence of the previous, final cyclical phase. In practice, Perennialism took to the extreme HoR's fideistic parallelomania, disregarding science, historiography, philology and every Western academic endeavour, and resorting instead to anti-modernism, conspiracy thinking, appeal to a reinvented tradition, clustering illusion and confirmation bias (Eco 1989; Eco 2008: 301–6; cf. Spineto 2006: 133–63; Ambasciano 2014). These ideas, already tied to a reactionary and conservative cultural milieu, were to be explicitly elaborated within an extreme far-right worldview by Eliade's acquaintance, Italian fascist intellectual and racist ideologue Julius Evola (1898–1974), who accentuated the political clash between modern and traditional worldviews. As a dialogue between HoR and perennialist esotericism began to take place in the early decades of the twentieth century, the works of Evola, Guénon and other perennialists were to leave a mark on Eliade's works and strengthen his conservative, pseudoscientific and diffusionist ideas, although within a distinctively eclectic worldview (Spineto 2006: 133–63; Ṭurcanu 2007: 491; Ambasciano 2014: 330–68; cf. also Eliade's distaste for philology in Smith 2004: 367–8).[21]

The Eliadean fusion of Interwar Romanian provincialism, anti-Enlightenment Romanticism, reactionary politics, esotericism and phenomenological HoR led to an anti-establishment stance: Eliade disregarded the scientific method, despised the reference system of citation, and constantly favoured authors and themes that were anti-scientific or critical towards mainstream science (Ambasciano 2014: 268–71; cf. Pernet 2012: 25 n. 21). In a sense, Eliade could not care less about epistemology: science meant almost nothing to him if it was not to conform to his *a priori* ideas and confirm his own ideas, just a bit more restrained than those of actual esotericists (Ambasciano 2014). Eliade himself called his own academic strategy a 'Trojan Horse'. On 2 February 1944, he wrote in one of his journals: 'I consider myself a Trojan Horse in the scientific camp, and that my mission is to put an end once and for all to the "Trojan War" that has lasted too long between science and philosophy [i.e. metaphysics]. I want to validate scientifically the metaphysical meaning of archaic life' (Eliade 2010: 104). To achieve this goal, a road map was implicitly followed, whose general points, with the benefit of hindsight, might be summarized as follows:

1. Western science in its current form was to be delegitimized, and any pseudoscientific contender sufficiently imbued with religion or spirituality advertised as superior;
2. Western science in its current form was to be shown as incomplete and dogmatic;
3. research activity was to be tied to ethnic and spiritual roots to show its current limits and boost nationalistic fideism – thus, HoR was reputed able to explain the ultimate mechanism of scientific discovery itself;
4. HoR was to be demonstrated to be able to go beyond the limits of science *within academia*, as science was slowly discovering 'truths' already known in esoteric, mystical and folkloric notions;
5. finally, HoR was to supplement and help *changing* science in the quest for spiritual truth (e.g. Eliade 1985; Eliade 2000; Eliade 2006; Eliade 2008: 65–8; Ambasciano 2014: 105–9, 271–4, 291–6).

Paranormal abilities, the existence of supernatural powers and entities, the folk knowledge of the afterlife, the possibility to reach otherworldly dimensions; these and other sensational notions were chosen by Eliade to promote the superiority of his HoR. Considering the outstanding success of his proposal, and how Eliade's HoR preceded current post-truth attacks on science as well as specific contemporary fields such as 'Religion and Science' (e.g. Kripal 2014; McGrath 2015; De Cruz 2017; cf. Coyne 2015), he did succeed admirably, and nowhere else was this success more tangible than in the study of shamanism.

Shamanism, 1200s–1800s: Heretics, noble savages, (super)heroes

In order to fully understand Eliade's contribution to the sub-field, we need a brief historiographical introduction. First of all, shamanism is a modern label with no general emic equivalent, a definition that has been historiographically used to describe a specific set of ritual practices from central Asia, characterized by a specialized local figure well-versed in artful performances that combine music, frenzied dance, role-playing, healing practices, ecstatic trances, divinations, foretelling and contact with culturally postulated superhuman beings, usually for specific individual or social purposes. Two of the most ancient accounts concerning Asian shamanism came from Italian travellers from the thirteenth and fourteenth centuries, i.e. diplomat and Catholic archbishop Giovanni da Pian del Carpine (*ca.* 1185–1252) and Venetian merchant Marco Polo (1254–1324), whose descriptions vividly portrayed what is known today as a *séance*, i.e. a ritualized, highly emotional session featuring the aforementioned elements (see DuBois 2009: 14–15; precedent reports in DuBois 2009: 12–13).[22] Closing a spell of curiosity and fascination, the subsequent history of shamanism follows closely the socio-political and cultural vicissitudes of Europe: from the European wars of religion and witchcraft trials to Czarist imperialistic expansion in Asia, shamanism was used to project heresy and delegitimize, eradicate and Christianize local 'primitive' beliefs and, more or less at the same time, justify national, imperial or colonial domination over those populations. With the Enlightenment, increased ethnographic knowledge spurred rational denunciation of such superstitious behaviours, although accompanied by 'a good deal of condescension from worldly men of science' (DuBois 2009: 21; see also Hamayon 1993, and Znamenski 2007).

The unresolved tension between fascination, outraged consternation and desired subjugation for such alien behaviours was somehow resolved in the following century, when the process of colonization gave way to full institutional control over those communities that featured such rituals. With shamanism falling out of practice during the nineteenth century,

> Western scholars began to seek out shamans to interview, observe, and analyze, so confident at last in the inexorable triumph of Judeo-Christianity over all trappings of barbarism that they almost regretted the fact. It was in this context that shamanism could at last be viewed with a semblance of scholarly neutrality or even sympathy, the object of a new 'science' of the history of religion.
>
> DuBois 2009: 23

Subsequent positivistic and psychosocial explanations of shamanism between the nineteenth and early twentieth centuries identified the interpretive key of shamanism in the extreme weather conditions of the circumpolar region. As a result, dietary deficiencies, lack of vitamins and insufficient sunlight, with all their grave neurocognitive consequences resulting in psychiatric and behavioural disorders, were the suspected culprits. All these causes could effectively explain the manic performances enacted by shamans, which involved 'trembling and the compulsive imitation of words and gestures', and the resulting syndrome was recapped with the label 'arctic hysteria'

(Lewis 1978: 43; Znamenski 2007: 83, 86–100). Such was the main thesis advanced by British–Polish anthropologist Marie Antoinette Czaplicka (1884–1921), who also noted ritualized similarities in other challenging, Tropical climates (e.g. the Malay peninsula in South-East Asia), and thus tied hysteria to 'climatic extremes' in general, although she warned about the necessity of more field research (for such 'environmental explanation' was apparently disconfirmed by its absence elsewhere) while not excluding a racial component (Czaplicka 1914: 323–4). However, it is important to note that the institutionalization of shamanism in those societies implied a standardization of the beliefs and practices that far exceeded any kind of health issues: some shamans might have been perfectly healthy and fine in all regards, and faked their behaviours. Indeed, pointing out these very social aspects of shamanism, Russian anthropologist Sergei M. Shirokogoroff (1887–1939) interpreted instead hysterical acts encoded in the séances as a (mostly) non-verbal means to soften in-group conflict, to make people laugh, and to mock someone else's behaviour or ideas in order to relieve the community from stress. Shirokogoroff's shamans were funny social heroes (Shirokogoroff 1935; on both see Znamenski 2007: resp., 86, 87, 98, 112).

Two features that intervened further to revise the general approach to shamanism were tied to the diffusion of European Romanticism as a conscious reaction against the Enlightenment. *First*, as we have seen, early HoR was obsessed by the role of Asia in the birth of whatever was considered religiously prestigious. Asia, and India in particular, were reputed to represent the earliest stages of modern human civilization, a belief boosted by the discovery of the linguistic affinities among extinct and extant Eurasian languages. Shamanism rose to the utmost relevance, and quickly became the focus of new research. *Second*, the post-Romantic pantheism of the German *Naturphilosophie* posited that shamanism was not a clever invention by impostors or charlatans (as sometimes held in the previous centuries), but a true and vivid manifestation of worship inspired or even caused by the natural landscapes in which shamans lived. The indigenous genius at work – reputed to be divinely inspired – was thus recovered as the religious manifestation of the 'noble savage', that is, the idea that primitive peoples were not tarnished or corrupted by modern civilization, and the subsequent admiration was inscribed into a degenerative dialectic. From this perspective, mainstream Western civilization at the dawn of the Industrial Age was considered regressive (Znamenski 2007: 17–28). As such, the study of shamans as noble savages might have helped to recover a more pristine form of natural devotion, and a quest for the purer forms of shamanism was thus started. Such enquiry was tied to specifically conservative, ideological and political propaganda (Znamenski 2009). As we shall see shortly, Eliade contributed in an appealing way to the popularization of these two ideas (i.e. the pristine origin and the Asian cradle of human origins). The resulting image was that of the shaman as a superpowered superhero.

Shamanism, 1937–1946: Eliadean superpowers

In 1946, Eliade published an article that was going to have an unprecedented impact on the field. Entitled *Le problème du chamanisme* (*The Problem of Shamanism*, Eliade 1946), that article had two main aims:

1. Eliade wished to use the prestigious academic journal in which he published his paper, the French *Revue de l'histoire des religions*, as a springboard and a Trojan horse to push some esoteric and pseudoscientific ideas (i.e. the existence of supernatural powers and the exploration of the afterlife), as he confessed in a letter written in 1947 and sent to Perennialist thinker and historian of Indian art Ananda K. Coomaraswamy (1877–1947; Eliade to Coomaraswamy, 26 August 1947, Eliade Papers, box 1, SC, RL, Regenstein Library, University of Chicago; from Țurcanu 2007: 489–90);
2. Eliade wanted to offer a rebuttal to Swedish scholar Åke Ohlmarks' (1911–1984) work published in 1939 and entitled *Studien zum Problem des Schamanismus* (*Studies on the Problem of Shamanism*; see Eliade's note dated 10 June 1946, in Eliade 1990a: 18).

According to Ohlmarks, who worked from both a nationalistic and racial perspective within a *Kulturkreislehre* framework, the most extreme, and almost pathological, form of shamanism coincided with the most psychologically and physiologically challenging experiences and was found in the area surrounding the Arctic Circle (North America included), which he labelled the *Urheimat des Schamanismus*, the 'original fatherland of shamanism' (Ohlmarks 1939: 58; see Znamesnki 2007: 98–100). Southbound, with milder weather, shamans were forced to use psychoactive substances (fly agaric mushroom, tobacco and alcohol) or prolonged drumming and masked dancing in order to elicit the same ritual dramatization. Ohlmarks' ultimate aim was to dissociate the contemporary shamanic practices of the Sámi, the discriminated autochthonous, non-Germanic, Uralic speaking population of Northern Scandinavia, from an ancient Norse ritual known as *seiðr*, as both featured shamanic-like rituals. In particular, Ohlmarks identified in the Norse practices the patriarchal divergence from the (Asian, sub-arctic) *Urkultur*, i.e. the original, primeval human civilization. Exploiting an *ad hoc* clause to bypass the potential precedence of Sámi shamanism and make it more degenerated, matriarchal (in Schmidtian terms) and primitive despite the similarities, and in order to prove that the *seiðr* was instead a noble and advanced Indo-European form of ecstatic ritual tied to the Greek Delphic oracle and Tibetan shamanism, Ohlmarks argued that 'it would [have] be[en] impossible for a primitive culture such as the Saami culture to have had an impact on the higher developed Norse one as this would be against something he call[ed] the "rules of cultural impact"' (Åkerlund 2006: 215, citing Ohlmarks 1939: 347–9; cf. von Schnurbein 2003; on Sámi's ancient rituals, cf. DuBois 2009: 12).

Eliade's rebuttal was atypical: he was not interested in criticizing or disconfirming such *Kulturkreislehre* understanding, nor was he really concerned by the distinction between shamanism and more or less similar Indo-European practices.[23] Eliade just wanted to dissociate shamanism and psychopathology and expand the main features of Arctic shamanism to the *entire world* within the same, degenerative Schmidtian frame (for the resulting conceptual and racial ladder of shamanic degeneration, see Ambasciano 2014: 53; cf. also Pharo 2011 for Eliade's network of influences). Reprising some motifs from an older article from 1937 entitled *Folclorul ca instrument de cunoaștere* (*Folklore as an Instrument of Knowledge*, Eliade 2006), Eliade set out to equate modern shamanism with a reimagined prehistoric religion, when more people could easily access *supernatural powers* and enter into divine realms. Resorting to the usual 'living fossils' trope, some 'primitive' peoples alive today were thought of as stuck in time, while others were merely considered degraded versions of the primeval humankind: therefore, always according to Eliade, certain 'living-fossil' populations could have exhibited fantastic superpowers today not experienced any more by Western populations (quite confusingly, these powers were labelled as 'living fossils' too). Thus, the study of these superpowers (which Eliade wanted to recover also in local European folklore) might have sparked and occasioned a contemporary and revolutionary Renaissance in magical knowledge (Eliade 2006). Now, the two superpowers that Eliade identified as universal patterns of shamanism were *ecstasy*, i.e. a 'euphoric altered state of consciousness which shamans used worldwide to interact with the sacred', and 'the shamanic ascent (*flight*) to the heavenly world' (Znamenski 2007: 172; my emphasis).

As Andrei A. Znamenski has noted, Eliade 'saw his own task as a comparative historian of religion as one of uncovering common ancient patterns hidden under the thick layer of "civilization"' (Znamenski 2007: 172). 'Traditional societies', an Eliadean motley crew that included societies characterized by the presence of shamanism as well as 'Stone Age people, classical civilizations, and modern "primitives"', provided a convenient case study to ground Eliade's diffusionist ideas and locate in ethnological terms his decipherment of allegedly transconscious 'universal archaic patterns' while looking incessantly for a 'primordial wisdom lost by modern civilization' (Znamenski 2007: 172; see Figure 12 and Figure 13). Therefore, Eliade proceeded to project modern Asian shamanism back in time, making it the most ancient religion of humanity further identified in prehistoric rock art, an idea that has exerted a certain disciplinary appeal, and subsequently has attracted many palaeoanthropologists and archaeologists, but which lacks a sufficient epistemic warrant (Bahn 2010; Ambasciano 2014; Ambasciano 2016a).[24] Asian shamanism was chosen as the absolute starting point because Eliade thought it was the 'most complete form' of such technique, as he described it in the 1951 French edition of his monograph (Eliade 1951: 23). This idea had long been reinforced by Eliade's reading of some works by Turkish historian and politician Mehmed Fuad (or Mehmet Fuat) Köprülü (or Köprülüzade; 1890–1966) and Laviosa Zambotti. These two scholars shared the view of a predominant or exclusive role of central Asia, respectively, in the ethnological, Indo-European spread of shamanism and in the evolution of humankind and its religions (Ambasciano 2014: 179–246, 373–422). Finally, Eliade was possibly influenced by esoteric speculations, based on Indian mythology (reputed true), which asserted the existence of a mythical race of people coming from the North (the so-called 'hyperboreans') as the custodians of the most primeval and divine revelations (Znamenski 2009).

Shamanism, 1951–1970s: the Eliadean synthesis

Eliade's *magnum opus* on shamanism, updated and translated into English in the early 1960s as *Shamanism: Archaic Techniques of Ecstasy* (Eliade 1964), made clear from the very title the equation he initially recovered in 1947 between contemporary 'primitives' and prehistoric 'primitives'. The *trait d'union* became ecstasy as an 'archaic technique', the shaman being its 'great master'. The minimum definition of the phenomenon was synthesized as 'shamanism = technique of ecstasy' (Eliade 1964: 4). In order to show accurately the continuity between past and present shamanism, Eliade set out to compile an outstanding, 600-page account of previous ethnographic works. As anthropologist Homayun Sidky has noted, Eliade's book

> contained a vast amount of information gleaned from an immense assortment of materials in multiple languages, ranging from the well-known to the most recondite sources. The book thus appeared to be a comprehensively investigated and painstakingly documented scholarly masterpiece. It quickly became the definitive study of shamanism published during the last half of the 20th century and had an enormous intellectual impact upon scholars and popular audiences alike.
>
> Sidky 2010: 72; cf. DuBois 2009: 24

However, three issues affected Eliade's qualitative meta-analysis of most of the then-known ethnographic works:

1. he relied heavily on 'German translations, reviews, and digests' of the main Russian ethnographic accounts (Znamenski 2003: 2);
2. he filtered the results of those studies through his preconceived biases in order to provide an interpretation that fitted his reconstruction (Sidky 2010: 86);
3. he did not provide anything new in terms of field research, but merely borrowed ideas following a confirmationist perspective and avoiding disconfirming data (cf. Ambasciano 2014; Ambasciano 2018a).

If we reverse-engineer Eliade's *Shamanism*, we can better discern what he borrowed and from whom, and consequently identify his peculiar reinterpretations:

1. *Shamans as superheroes*: Eliade borrowed Shirokogoroff's idea of shamans as heroes (Znamenski 2007: 113). However, Eliade took this idea to the extreme, idealizing shamans as an elite of saintly heroes. As such they were needed by defenceless, vulnerable communities: they were super-powered spiritual warriors that fought against the spirits of the netherworld and were capable of wandering in every region of the mythological realms (Znamenski 2009: 198) – an image not far from what Eliade envisaged for historians of religions as the Interwar vanguard of Romanian spiritual and palingenetic revolution (Ambasciano 2014: 276–7; see Handoca 2008: 330).
2. *Indocentrism*: Eliade endorsed the view of a continuous co-dependence of (proto-) historical shamanism and Indian/Tibetan religions; thus, he adopted Shirokogoroff's ideas about the influence of Indian religions on shamanism (Znamenski 2007: 173). Building on his previous works, Eliade created a dichotomy between internalized and externalized archaic techniques to achieve

mystical states, i.e. shamanic ecstasy and yogic 'enstasis'. Both yoga and shamanism were considered as the specialized mastery of supernatural powers that existed in modern times as 'living fossils', although Eliade considered yoga a purer and more elevated form (Ambasciano 2014: 53, 56; Eliade change his idea though; see below).

3. *Universal (perennialist) symbolism*: as a result of the coeval and constant dialogue between perennialists and historians of religions, Eliade took the idea of pan-Eurasian religious symbolism embedded within shamanic complexes (such as the so-called *axis mundi*, the tree of the world that connects the various mythological dimensions of the cosmos) from Finnish scholar Uno Harva (also Harva-Holmberg; 1882–1949), who, in turn, was influenced by Guénon (Harva 1922; see Ambasciano 2014: 363). Eliade, however, took these comparative patterns and applied them virtually everywhere, past and present, from contemporary native American myths and Australian initiations to ancient Mithraic rituals and worldwide heroic epic, with shamans idealized as forefathers of dramatics (cf. Ambasciano 2014: 55). Consequently, according to Eliade, the alleged universality and coincidence of dream patterns, visions and mythological motifs (such as the hero's ascent to heaven or the descent to hell) prompted a meta-psychoanalytical review that was based on the recovery of such archetypical symbolism – the subconscious remnants of primordial true stories (e.g. Eliade 1948b; see Ambasciano 2014: 364).

4. *Degeneration and psychoactive substances*: Eliade re-elaborated Ohlmarks' distinction between arctic and sub-arctic shamanism and Czaplicka's psychopathological frame in Schmidtian terms. Subarctic shamans were reputed to be failed and aberrant, abnormal, degraded practitioners who relied on drugs to achieve the required ecstatic trance. True ecstasy was available only to circumpolar shamans – heirs of the primordial revelations by the god(s), while the other shamans from Southern regions were forced to consume drugs and hallucinatory substances to reach a similar, and yet degraded, ecstasy. However, late in the 1970s, Eliade seemed to have changed his mind, considering hallucinogen usage as important as the 'true' circumpolar ecstasy (Ambasciano 2014: 53, 364; Znamenski 2007: 141).

More generally, how does this reconstruction fit in Eliade's overall comparative project? In the 1960s, old racial classifications and provincial narrow-mindedness were used to describe the academic tenets of his restructured HoR as a 'new humanism' and a 'saving discipline' (resp., Eliade 1984: 1–11; note dated 2 March 1967, in Eliade 1989a: 296; cf. Ţurcanu 2007: 561). With regard to non-Western religions, in this period Eliade's HoR became a sort of hermeneutical missiology aimed at investigating the resilience of pre- or (allegedly) proto-Christian tenets in non-Western cultures and religions for the greater good of Western spiritual renovation. In turn, the non-Western world had the duty or, according to Eliade's terms, the 'responsibility' to come out of the shade and embrace their HoR-designed 'destiny' identified thanks to a patronizing teleology (Eliade 1984: 57). Notwithstanding the fact that, according to Eliade, the presence of similar archetypical religious elements is due to a unifying transconscious, every religion is implicitly interpreted through the eschatological lenses of 'cosmic Christianity', i.e. the folkloric survival of the proto-historical, idyllic union with the cosmos (i.e. the divinity itself) under a Christianized valorization. The key to understand this explanation, which apparently contradicts the Eliadean HoR's claim for a global and egalitarian study, lies in Eliade's theology of nostalgia: given that Christ is nonetheless reputed as the ultimate religious manifestation (or, according to the Eliadean vocabulary, 'theophany'), the most accomplished transconscious expression of archetypes, elaborated through universal rituals and religious mysteries, is to be found in the spiritual cradle of Eliade's reimagined South-Eastern Europe, in general, and Romania, in particular. Shamanism, intended as the survival of Asian archaic techniques to master superhuman powers granted in ancient times, became the keystone to approach this eclectic, Christianized, ethnohistorically reinvented *philosophia perennis* (e.g. Eliade 1992a; Eliade 1980: 25; cf. the bibliographical discussions in Ambasciano 2014: 255–6, 290, 301–4).[25] With this interpretive tool, elaborated all over the course of Eliade's career, the reinvention of both national and disciplinary traditions came full circle.

Resolving 'The problem of shamanism': An unwarranted answer to a non-existent question

Eliade's works on shamanism tried to resolve a non-existent question by advancing an unwarranted answer: 'the problem of shamanism', from the Eliadean viewpoint, concerns the supernatural origin of religion *tout court*, which should be reframed in an emic way in order to extrapolate the necessary general features common to *homo religiosus*. This study was supposed to lead to the enrichment of Western civilizations, in that it purported to retrieve ancient powers currently psychically or subconsciously unavailable. Eliade was mostly interested in the 'experimental knowledge' about the afterlife and the supernatural, and in the resulting 'comparative history of mysticism'; therefore, he thought that the investigation of shamanism could provide researchers with valuable information about the existence of supernatural dimensions and paranormal powers (Eliade's undated note, 1952; Eliade 1990a: 180–1).

As Homayun Sidky has remarked, Eliade's works on shamanism were destined to great success, both for the discipline and for the scholar himself: Eliade modernized the field by providing the first overarching synthesis, 'situating the problem "in the context of the history of religions", as Eliade (1981: 117) explained', and re-establishing 'a new sense of significance, legitimacy, and relevance' for this field of study (Sidky 2010: 73; cf. Znamenski 2003: 34; DuBois 2009: 24). In the end, 'Eliade's efforts were enormously successful and propelled him to the category of an academic superstar in his field of expertise and beyond' (Sidky 2010: 73). Even though Eliade's justifications fall outside the scientific paradigm to such an extent that archaeologist Paul G. Bahn labelled his synthesis as a 'fraud' (Bahn 2010: 80–2), Eliade's *Shamanism* and his prehistorico-ecstatic model are still used or recalled as a pivotal reference in many contemporary analyses (e.g. Anati 1999; Lewis-Williams 2010; Pharo 2011; Clottes 2011; Noiret 2017; Singh 2018; cf. Ambasciano 2016a; see also DuBois 2011, and Currie

2016; Sanderson 2018: 187–8 refers to the Eliadean understanding of traditional beliefs regarding agrarian societies to explain the shift from shamanic belief systems).

Two additional reasons that need to be factored in when considering the spread of the Eliadean paradigm from the 1960s onwards were the explosion of the New Age and counterculture youth movements and the fact that many former students and colleagues of Eliade became affiliated to other US academic institutions, spreading further Eliade's ideas and granting academic citizenship and popular stardom to Eliadean shamanism – notwithstanding the insufficient evidence for such a reconstruction (the cultural milieu also prompted Eliade to change his attitude with regard to drugs, as we have seen above). As Sidky concluded, in the second half of the twentieth century, both the US cultural environment and Eliade's works were tied in a loop which led to the 'uncritical acceptance of the fantasy that Eliade and his followers have built upon this historical/ethnographic complex' (Sidky 2010: 86).

In a sense, Eliade brought to a close the metamorphosis of the HoR from incipient science to fully fledged pseudoscience. Although his ideas can be collocated and studied in their original Interwar context, the remarkable fact is that they sustained and supported well into the incipient twenty-first century and inside academia what philosopher of science Imre Lakatos (1922–1974) called a *degenerating research programme*, that is, an ever-expanding complex of *ad hoc* explanations influenced by extra-epistemic factors (e.g. socio-political) intended to immunize the hard core (here, the *homo religiosus* and myth as true story) and bypass criticism and falsification, creating a prosperous and thriving 'Eliadological' niche within the HoR faithful to the authoritative *ipse dixit* of Eliade (Ambasciano 2018b). As recalled in the opening epigraph, the fear of Eliade's frenemy, Franco-Romanian playwright Eugène Ionesco (1909–1994), did eventually materialize: Western academics let the Trojan horse of pseudoscience in (Vianu and Alexandrescu 1994: 233; cf. Lakatos 1989: 34, and Dubuisson 2005). To be sure, as soon as Eliade's works became best-sellers in Europe and in the USA, some scholars began to react against the explosion of irrational, emic and fideistic trends within the HoR. And yet, there was no sudden overthrow, no immediate rejection of the *status quo*. What followed was a revolutionary schism in slow motion, punctuated by localized outbursts of guerrilla warfare.

6

The Demolition of the *Status Quo*

It helps to know the tradition if you want to subvert it.

Daniel C. Dennett

Point of (k)no(w) return: The politics of the Eliadean HoR

Preceded by some methodological skirmishes during the 1960s, mostly ignored by Eliade himself, a heated debate broke out in the 1970s when the Eliadean research programme became the object of vehement international attacks that questioned its epistemic consistency, the fideistic *a priori*, and the implicit socio-political tenets (e.g. Leach 2006; Pernet 2011; cf. Țurcanu 2007: 620–2; Ambasciano 2014: 166 n. 768). At the same time, morphological systems of classification and the phenomenological approach as a whole were being challenged as incoherent, incomplete or fallacious methods of social-scientific analysis (e.g. Penner and Yonan 1972; Penner 1989; cf. McCutcheon 1997). Paradoxically, Eliade's works were becoming extremely popular among the youth counterculture, as they contributed to the shaping of the New Age's syncretic spirituality. Rather reluctantly, Eliade became a 'guru' for a new, US post-war generation eager for non-mainstream figures, an academic persona whose ideas on the liberating power of myth were able to captivate the youth aspirations for spiritual and social renovation (Eliade 1982a: 109–11; see Țurcanu 2007: 598). Carlo Ginzburg defined this paradox as the result of an Eliadean 'ambivalent legacy', strategically supported by the historian of religions himself during his lifetime (Ginzburg 2010). In order to achieve such a goal, Eliade kept on fictionalizing his biography, exploiting the serendipitously convenient difficult access to Interwar documents located beyond the Iron Curtain, taking advantage of the new Romanian People's Republic regime to rebuild implicitly his identity as a victim of (a different) totalitarianism, rewriting his memoirs, and pursuing his idea of the HoR as a metatheological, saving discipline.

A turning point in this pseudo-bibliographical narrative production was the publication of the so-called 'Toladot dossier' ('Dosărul Mircea Eliade'), written in Romanian and published in early 1972 in the bilingual Israeli journal *Toladot. Buletinul Institutului Dr. J. Niemirower*. Edited by *Yad Vashem* historian Theodor Lavi (Theodor Löwenstein; 1905–1983), the Toladot dossier was a partial collection of Interwar Romanian testimonies extracted from the diary of Eliade's former acquaintance, Jewish writer and playwright Mihail Sebastian (Iosif Hechter;

1907–1945), which revealed the historian of religions' intellectual involvement and active participation in extreme-right politics (Scagno 2000: 281–9 for the dossier itself; see Ambasciano 2014: 299–300 n. 333; Idel 2014: 193–212). A copy of the Toladot dossier was to reach Eliade along with a letter sent to him on 25 June 1972, by his colleague, Hebrew University of Jerusalem professor of Jewish Mysticism, Gershom Scholem (1897–1982). Prompted by Scholem's polite but seriously concerned letter, Eliade produced a long, and ultimately unconvincing, apology in which – as was discovered later – he lied about his own extreme-right past, denying any allegation concerning his Interwar political participation (an admission would have jeopardized Eliade's life and work in the USA; see Dubuisson 2014: 277–88; Junginger 2008: 38–9 for the context of the letters; cf. Ambasciano 2014: 357 n. 618, 482 n. 35; Scholem's letter is in Handoca 2006: 315–16).[1] However, the knowledge of Eliade's political past circulated within European cultural and diplomatic entourages at least after his Parisian years, and got in the way of Eliade's potential academic affiliation in France (Spineto 2006: 34; Țurcanu 2007: 441–2). Unsurprisingly, considering the shared disciplinary history, Italy found itself in the eye of the Toladot hurricane. The dossier achieved national resonance thanks to historians of religions Alfonso Maria di Nola and Furio Jesi. In 1977, di Nola took the opportunity of the concomitant Italian edition of Eliade's journal (1945–1969) and wrote a powerful *j'accuse* against the Eliadean web of reactionary politics and academic production, specifically underscoring that his critical reflections about the dossier wanted

> to invite the youth – and many among them, as a matter of fact – captivated by the spell cast by the Eliadean discourse, to the duty of remembering, to render *memory and history as forms of knowledge*. [These notes] want to be of help in the discovery of the concealed roots of the evils that circulate amongst us and which might bring back – without even noticing it – a renewed *sleep of reason, producer of monsters*.
>
> di Nola 1977b: 15, my emphasis; see also di Nola 1977c

Two years later, Jesi, in his book entitled *Cultura di destra* (*Right-Wing Culture*), delved deeper into the Eliadean worldview in the wake of the Toladot dossier, bringing to light previously omitted connections between the Eliadean socio-political, right-wing mindset and his academic *mythological machine* (Jesi 2011; cf. Rowland 2014; see Chapter 1, §*Maps, compasses and bricks*). Notwithstanding some factual inaccuracies and wild guesses due to the incomplete knowledge of Eliade's Interwar production, the volume was a cultural watershed and had a significant impact on the reception of Eliade's works (Manera 2012). Eliade, once made aware of such a book (which he did not read), believed that Jesi's 'campaign' and 'perfidious attack' was aimed at 'eliminat[ing]' him from the preliminary evaluation for the Nobel Prize in Literature during the late 1970s (personal notes, resp. dated 23 July 1979 and 6 June 1979; from Eliade 1990b: 22, 16; cf. Ricketts 2000b: 375 and Țurcanu 2007: 606–11, 627–9; on the late-1970s Nobel Prize nomination, see Handoca 1998: 122–3).[2] In 1989, another article by di Nola summarized the growing interpretive literature on the Toladot dossier, noting a polarizing increase in both the use of logically fallacious immunizing strategies and critical scholarship, respectively for and against Eliade's research programme (di Nola 1989). Meanwhile, the international academic environment continued to support a generally sympathetic view with regard to the Eliadean paradigm (e.g. Bianchi was elected President of the IAHR in 1990).

Thanks to a disciplinary environment built in a way not dissimilar to Schmidt's research complex, Eliade kept on wilfully ignoring every criticism, favouring instead confirmatory scholarship (e.g. Eliade 1978b). Indeed, there was no reason to do otherwise. Since the 1960s, Eliade had received awards from prestigious academic institutions on a regular basis, accumulating academic honours such as the election to the American Academy of Arts and Science (1966) and the appointment as Corresponding Fellow at the British Academy (1970). Eliade was also awarded honorary degrees from Yale University (1966); Universidad de la Plata, Argentina (1969); Ripon College, Wisconsin (1969); Loyola University, Chicago (1970); Boston College (1971); La Salle College, Philadelphia (1972); Oberlin College, Ohio (1972); University of Lancaster (1975); La Sorbonne, Paris (1976) (Burgess 1989: 248; Spineto 2006: 72). On top of that, in 1985, an anonymous donation of one million dollars allowed the trustees of the University of Chicago to establish the 'Mircea Eliade Chair in the History of Religions' in his honour at the local Divinity School (Țurcanu 2007: 646).

However, in the same year, Eliade finally acknowledged in a private note that his non-confrontational strategy was backfiring (15 September 1985; in Eliade 1990b: 143). The strategic promotion of Eliade's students to occupy academic positions all across the US also turned out to rebound on him (Țurcanu 2007: 575). In the wake of the Toladot dossier, critical scholarship interested in uncovering the Romanian Interwar roots and continuities in Eliade's thoughts gained momentum between the 1980s and the 1990s thanks to the increasing success of Eliade's works, the growing group of students and scholars interested in reconstructing his life, and the ongoing biobibliographical systematizations of his production (the most impressive products of the period probably being the 'annotated bibliography' by Douglas Allen and Dennis Doeing, 1980, and the 2-volume Romanian-years biography in Ricketts 2004, originally published in 1988; see Țurcanu 2007: 623–4, 629–39, 643–4; cf. Ambasciano 2014: 298). Regardless of the defence strategies deployed by Eliade's supporters (recently reviewed in Bordaș 2012), and even if some of the original allegations were demonstrated to be wide of the mark (e.g. Strenski 2004; Spineto 2006: 43–4), the long-awaited republication of some among Eliade's most remarkable Interwar works has confirmed the picture resulting from the Toladot dossier, demonstrating beyond any reasonable doubt that the Eliadean research programme had its roots in the reactionary socio-political ideology in which he vehemently and fervently believed (Handoca 2001; cf. Ambasciano 2014).

As a result, to summarize, Eliade's original idea of what the morphological and phenomenological HoR should have been is confirmed as that of an *instrumentum regni for* the elite and *by* the elite, where historians of religions were actively engaged *ex cathedra* as the academic *longa manus* of ultranationalism to study and classify subordinate peoples and their supernatural knowledge, whether folk peasants at home or 'primitives' abroad, for the greater good and excellence of a divinely inspired, new Romanian nation (e.g. Handoca 2008: 330–2, 335). Eliade never disavowed his extreme-right political belief in the creation of a 'new man' within the spiritual renovation of humankind; he only became bitterly disillusioned after the defeat of the Axis powers and began rethinking in religious terms the aim and scope of the discipline (e.g. Eliade 2010), while continuing to collaborate with extreme right-wing scholars and journals, camouflaging his creed to rebuild an acceptable reputation in the mutated socio-political context, and pursuing opportunistically the same conservative agenda (Țurcanu 2007: 576–9; Gardaz 2012; Strenski 2015: 152; cf. Ambasciano 2014).[3]

Sexist biases and gender issues: Eliade's tunnel vision

Eliade's right-wing tunnel vision for the discipline relegated to the background many important themes and topics, the most dramatically evident of which were perhaps tied to his sexism. We can briefly tackle this point by approaching, *first*, a case study of local shamanic practices in classical Japanese myth and ritual (Strenski 2015: 161–3), and, *second*, the more general disciplinary reaction to such sexism.

Arguing for a locative reinvestigation of the mythological 'realm beyond ordinary experience that is also thought to be the abode of the *kami*' (i.e. Shinto superhuman agents), and pushing for an analysis able to retrace socio-political intents in both religious and non-religious power networks and spatial hierarchies (Grapard 1991: 3; cf. Brown 1993: 9), Japanese historian Allan Grapard notes that male shamans are mostly absent while female shamans are the norm.[4] The classical, phenomenological interpretation is simply that those women are more apt to channel certain spiritual powers which, in turn, allow them to contact spirits and access some otherwise inaccessible sacred worlds described in the local mythic topography. Thus, according to the standard interpretation, the scholarly exploration of the religious role of those female shamans would accept *ipso facto* the emic descriptions of the local worldview *once* filtered through the phenomenological strainer. What would be fatally lost in this scholarly analysis is the fact that the domain of religious, shamanic performance represents *what has remained to those local women to express themselves* in a way suitable to be accepted by the dominant, androcentric social order. Grapard, who studied the gendered mythologies and ritual performance related to female priesthood in Japan, writes that

> ritual power is divided between the sexes: purification is essentially a male prerogative and [. . .] is directly related to the creation of cultural emblems and, therefore, of domains of knowledge. The only field of speech-knowledge that is left to women is that of communication with the introverted negativities of the worlds beyond.
>
> Grapard 1991: 17

In a sense, those shamans made a virtue out of necessity, for this is the only way to 'achieve ends which they cannot readily obtain more directly' (Lewis 1978: 85) in *androcentric and/or patriarchal social systems*, that is (a) communities dominated by a set of masculine, and potentially sexist or misogynist, schemata which shape beliefs, customs, policies, institutions and individual expectations, in (b) societal settings whose power roles from family to politics are under the more or less exclusive control of men (see Ambasciano 2016c: 118). Therefore, female shamanic abilities in the aforementioned Japanese setting were located in a mythological setting which justified the limited role assigned to women: 'males separated themselves from females, in those myths, whenever they treated them, not as human beings, but as objects of curiosity' (Grapard 1991: 17). The religious involvement of women as shamans, then, is the result of the androcentric social network which confines those women to this subordinate religious performance, while their alleged ability to travel to other supernatural worlds and communicate with imagined superhuman agents (such as spirits) is what has been left to reclaim a social space otherwise denied – but always under male strict supervision (Strenski 2015: 162). From this new perspective, the critical study of local religions reveals the existence of the underlying, constraining power dynamics, something almost unthinkable within classical HoR.[5] One of the most striking features of shamanism, i.e. the spiritual voyage to other mythical dimensions during the séance, betrays such a gendered construction. As Grapard concludes, the 'cosmology of [shamanic] travel chambers is the *mental map of a social code,* the genealogical ground of relations of power between the sexes' (Grapard 1991: 21; my emphasis).

All these themes are something which phenomenologists, for all their focus on emic religious grandeur, could never aspire to identify: it is fair to say that, blinded as they were by the recovery and the reinvention of prestigious spiritual, religious, social and national traditions, they were not interested in uncovering the network of power relationships behind whatever was their particular interest. Nor were they interested in understanding the social and religious roles generally reserved for subordinate groups and individuals. In the wake of the phenomenologically pretentious observance of historicist precision (heritage, as we have seen, of the HoR's constitutive paradigm), those scholars accepted as a fact the masculine exercise of power and moral prestige revealed by androcentric mythological complexes and, consequently, took for granted the top-down imposition of legitimacy on subordinates. In particular, they accepted the *sociodicy* about the female social role as narrated in ancient religious storytelling and practices, i.e. the mythological justification for social subordination as sanctioned, or provided, by the dominant system (Bourdieu 1971: 312). In so doing, past historians recognized as legitimate the dominant social order emblazoned in the foundational mythical charter of the societies they studied, conveniently immunized from change because they were naturalized as eternal (Bamberger 1974; Juschka 2005; C. Martin 2014: 49–69).

Why did those scholars approach their field with such a bias? Because most historians usually endorsed the very state of affairs behind androcentrism. As a consequence of both the androcentric history of the discipline itself, and Eliade's own chauvinistic, jingoistic, macho, sexist views (cf. Eco 1995; Casadio 2011), *homo religiosus* was etymologically restructured as *vir religiosus*: not the history of religious humankind, but a history of religions *for* men and *by* men. Women were not considered part of the religious equation, and most of the time they were considered a mere complement to men's *own* HoR, willing adjutants or companions who helped their men to achieve a higher spirituality alternatively through chaste marital love or extraordinary sexual techniques (e.g. the Eliadean analysis of yoga techniques; Eliade

1958b). Women were academically studied as objects and not subjects because they were mostly considered unworthy of historiographical attention; indeed, their religious roles in certain societies were downright minimized and 'effeminate' behaviours labelled as degradation. For instance, with regard to shamanism and sex and gender issues, one can note the retrograde psychological framework that encompassed Eliade's analysis of homosexuality in shamanism. In the index of Eliade's *Shamanism*, under the voice 'homosexuals', the reader is referred to two other entries, 'inverts' [*sic*] and 'pederasty' (Eliade 1964: 587), thus harking back to 'the appearance in 19th-century psychiatry, jurisprudence, and literature of a whole series of discourses on the species and subspecies of homosexuality, inversion, pederasty, and "psychic hermaphrodism" [which] made possible a strong advance of social controls into this area of "perversity"' (Foucault 1978: 101; cf. Taylor 2017: 39, and 106–7 on psychoanalytical interpretation). In line with this discursive system aimed at classifying perversions and creating an intellectual scaffolding further politically exploited to exercise institutional and normative power, Eliade used 'inversion' to conflate homosexuality, paraphilia, and aberrant or psychopathological development of sexual desire within a theory of sexual and religious degenerescence tainted by a bad company fallacy and a slippery slope argument. Eliade's *Shamanism* posited homosexuality, 'transvestitism' and 'pederasty' as middle terms (or missing links) between *normal*, male shamanism and *abnormal*, female or 'effeminate' shamanism, indexing a regressive deviation or a minor survival from a Schmidtian 'archaic matriarchy' (Eliade 1964: 125 n. 36, 258).[6] This obsolete and groundless reading is at odds with subsequent ethnographic studies that have gone beyond the record of 'gossip and jest' reserved for cross-gendered practitioners to assess the empowering social function of shamanism for 'community members which avoid[s] binary oppositions that had become normative in Western notions of sexuality' (DuBois 2009: 79; cf. DuBois 2011: 106).

Dismantling *homo religiosus*: Rita M. Gross

To sum up, the dangerous intellectual short-circuit was caused by the correspondence between androcentric objects of study (i.e. the cultures and civilizations studied) and the androcentric subjects who studied them (i.e. the scholars who investigated those cultures and civilizations). With regard to the HoR, the field was obsessed with the religiosity of men because the majority of scholars involved were chauvinistic males writing in a chauvinistically androcentric context (Kinsley 2002; Gross 2005; Juschka 2005; Mikaelson 2005; Korte 2011). This most despicable sexist viewpoint was widespread in academic historiography. In 2009, Roman historian Amy Richlin recalled that 'a noted Roman historian, when approached in the early 1970s by a group of women students requesting a course in women's history, is said to have replied, "Why not dogs' history?"' (Richlin 2009: 146). Rita M. Gross (1943–2015) remarked that in the early 1970s

> the 'women and religion' movement existed mainly in the frustrations of some graduate students and a few young professors of religious studies. No institutional

format existed at that time and neither the American Academy of Religion nor the International Association for the History of Religions paid much attention to the topic of women and religion. If someone brought up the idea of specifically focusing on women and religion, the response was a combination of indifference, hilarity and hostility.

Gross 1980: 579

When Gross was about to work on her PhD and she approached her mentors to talk about her intention to tackle such issues, her proposal 'upset' them, while one unnamed professor 'actually told [her] that an intelligent person like [herself] should realize that the generic masculine "covered and included the feminine", making it unnecessary to say anything about women specifically' (Gross 2009: 4–5). Thanks to young scholars like Richlin and Gross herself, things were about to change drastically. As far as the HoR was concerned, in 1977 Gross denounced this blatant sexist bias by attacking the hard core that supported the old HoR, i.e. *homo religiosus* as such, disentangling the fallacious scholarship behind it. In her article, entitled *Androcentrism and Androgyny in the Methodology of History of Religions*, Gross wrote that

> *homo religiosus* as constructed by the history of religions does not include women as religious subjects, as constructors of religious symbol systems and as participants in a religious universe of discourse. History of religions really only deals with women and feminine imagery as they are thought about by the males being investigated, whether specific males in a specific religious situation or the abstract model *homo religiosus* are the subject of inquiry. Since the discipline of history of religions is basically concerned with discovering and understanding humans as religious beings, the androcentric limitations of the construct *homo religiosus*, religious humankind, constitute a very severe liability indeed.

Gross 1977: 10

Those words, in the late-1970s, marked the very beginning of a revolution in the field. While Gross' specific target was disciplinary sexism, her article was just one of many contributions that in the 1970s started to raise all sorts of legitimate questions about the methodology and epistemology of the classical HoR. All those papers, one way or another, converged on what was their common target: the Eliadean research programme. As noted previously, Eliade's works and teaching were instrumental in the birth of new and critically informed ways of doing research in the field – as a negative example not to follow. Gross attended Eliade's 1967 class at the University of Chicago about 'Primitive Religions', based on the previously recalled series of articles on Australian religions (see previous chapter; Gross and Ruether 2016: 41–4). In the wake of Eliade's interest in Australian religions, and disinterest towards women's religious involvement, Gross produced the first ever PhD dissertation in women's studies in religion, expanding a term paper on *The Role of Women in Aboriginal Australia* (1975). Her mentors judged such choice as something 'redundant and unnecessary' because of the 'high level of sexual segregation' of Australian religions, and thus tried to discourage her from doing so (Gross 2009: 5). Eliade showed some interest in the literary data collected by Gross but he was resolutely unwilling to concede that the methodology of the discipline was in need of a thorough revision. The very subject was deemed unapproachable and labelled as not rigorously academic. As Gross recalled later, 'the Divinity School faculty was [. . .] fighting about the appropriateness of studying women and religion' (Gross and Ruether 2016: 43).

Eventually, in 1974, after a 6-year-long ordeal, the powers that be at the University conferred the degree on her, but only after having bullied her and discredited her work. In the same year, Gross, unflinching in her resolve, headed a recently established section of the American Academy of Religion called 'Women and Religion'; 6 years later, she co-edited with Nancy Auer Falk, another former student who attended Eliade's class on 'Primitive Religions', a trail-blazing volume on women and religion from a global, cross-cultural and comparative perspective (Falk and Gross 1980) and, finally, she became Professor of Comparative Studies in Religion at the University of Wisconsin (Eau Claire; Gross 1994; Love 2006: 190; cf. Pernet 2012: 12–13; for a fuller account of Gross' work, cf. Gross 2009). Gross was just one among many young scholars, students and colleagues of Eliade who started questioning the works of the Chicago Professor. They were dissatisfied by his teaching and disappointed by the phenomenological approach. Although most of them were not eager to abandon the HoR, and even less interested in demolishing their own field, their studies represented the first steps towards the creation of the new field of Religious Studies (henceforth, RS). As usual, a caveat: the following choice of scholars, topics, and works represents a personal and minimum selection for the sake of simplification.

Dismantling the Eliadean research programme: Henry Pernet

Eliade's treatment of Australian religions as a blueprint for the primitive, primeval religion of *homo religiosus* proved to be a turning point in the critical re-evaluation of the entire field. Four years before Gross' experience, Henry Pernet (1940–), a Swiss freshman who decided to enrol in the Chicago Divinity School because he was fascinated by Eliade's personal charisma, attended Eliade's 1963 class on 'Primitive Religion', the first part of which was focused on Australian religions (Pernet 2011: 29; Pernet 2012). Already disillusioned with Jungian psychoanalysis and esoteric interpretations, Pernet was looking for a more empirically based and epistemically warranted methodology to approach the study of religion. With such a critical mindset, for the term paper of the class Pernet welcomed a suggestion made by Eliade and approached the 'instances of the Master of Animals collected in European and Eurasian legends by Leopold Schmidt, an Austrian folklorist' (1952). In his paper, Pernet expressed some 'methodological reservations' about the comparative system of analysis. This reading ignited a life-long interest in folklore, which eventually led Pernet to study the ritual use of masks in a comparative light (Pernet 2012: 13).

The following year, in preparation for the class on 'Prehistoric Religions', Pernet read a recently published book on *Les religions de la préhistoire* (1964) by French palaeoanthropologist André Leroi-Gourhan (1911–1986) and, at the first opportunity, he presented his thoughts along with the book itself to Eliade. Pernet pointed out the epistemological inconsistencies of the ethnographic analogy to interpret prehistoric artefacts as religious, but Eliade, dodging the problem, 'remained unconvinced' and answered to those criticisms by giving Pernet a signed copy of his recently translated *Shamanism*, simply telling the young student that 'Leroi-Gourhan was wrong and that one should not in any case give up ethnological parallels' (Pernet 2012: 15). Pernet analysed the new English edition and compared it with the French one, and was struck by the following new sentence: 'recent researches have clearly brought out the "shamanic" elements in the religion of the Paleolithic hunters' (Eliade 1964: 503). This apodictic, and aprioristic, shamanic interpretation of prehistoric religion, based on scanty artefacts and questionable ethnographic parallels inspired Pernet to delve deeper into possible alternative interpretations. Therefore, he prepared a critical assessment of Eliade's main reading for his class, a book on prehistoric religions (Maringer 1960), noting a discrepancy between excessive interpretation and insufficient critical analysis of the data at hand. Pernet focused on Palaeolithic female statuettes, and he went 'back to the sources that the author cited [to show] that he had wrongly interpreted them' (Pernet 2012: 16). Pernet also started questioning 'the use of terms like "structure" and "archetype"'. Anyway, Eliade accepted the paper 'without problem' (Pernet 2012: 16).

By early 1967, Pernet was becoming increasingly troubled by the severe criticisms that disciplines potentially considered allies of the HoR reserved to the Eliadean research programme (i.e. anthropology, palaeoanthropology, history). In April, he drafted an extensive methodological discussion for a seminar, entitled 'From History of Religions to Comparative Science of Religions', which he offered as a contribution to the reassessment of the status of the field. In that paper, Pernet presented an exit strategy and a scientific road map to escape the confusion surrounding 'a discipline totally lacking unity and, to a large extent, still struggling to find its self-identity' (Pernet 2012: 23; cf. Pernet 2011: 27). His recommendations to abandon the lack of a 'common method' and 'goal', the crypto-theological approach, and the recourse to depth psychology, are worth recalling here in their entirety:

1. drop unnecessary presuppositions, such as the *sui generis* nature of the religious experience and its expression;
2. renounce the presuppositions of the existence of a 'trans-conscious' (Eliade 1964: 454) and, in general, refrain from relying on psychological conceptions until they had received some empirical verification;
3. abandon the contention that the religious man is the total man, i.e. that total man is religious; to me, this postulate was unacceptable as an onto-theological position (Eliade 1989a: 164–5; note by Pernet);
4. agree on a method of empirical verification that fits both the requirements of a more scientific approach and those of the particularity of our field, granted that this would be found only after a certain amount of trial and error (Pernet 2012: 24).

This paper was accepted in partial fulfilment of Pernet's Master's Degree (AM 1967, Chicago), and, although Eliade never talked to Pernet about it, Eliade reported in his diary Pernet's 'revolt':

like my other former students Kees [W. Bolle] – more or less 'unconsciously' –
believes that he too can construct an 'original position' in opposition to my ideas
[K. W. Bolle, 1927–2012, had been the first doctoral student of Eliade back in the
mid–1950s]. I must add, not *all* my former students but only those who were closely
associated with me, with whom I worked seriously, whom I helped. I'm not angry. It's
natural. It's the revolt against the 'father' (the most characteristic example: Pernet).

> 15 March 1969; unpublished note from
> Eliade's journal; Pernet 2012: 9–10

Unable to understand how science works, Eliade misinterpreted the progress of
research, and misunderstood psychoanalytically and religiously any criticism. Indeed,
Pernet's suggestions were doomed to be ignored by Eliade, with whom he discussed in
vain. At last, after 2 years of enforced idleness in Switzerland caused by a delay in
receiving his visa to go to Chicago to engage in PhD research, Pernet proposed a
feasible structure for his dissertation on *'Primitive' Ritual Masks in the History of
Religions: A Methodological Assessment*, and Eliade gave his consent. However, worn
out, Pernet gave up his ambition to produce an entire PhD dissertation on the scientific
revision of the HoR (Pernet 2012: 25–6).

Pernet eventually obtained his PhD degree from the Chicago Divinity School in
1979, and his critical take on the subject was later re-elaborated and reintegrated in a
much later volume about ritual masks (Pernet 2006, originally published in 1988). He
also kept on corresponding regularly and amicably with Eliade from 1961 to 1986,
among other things, helping the Chicago professor to correct the proofs of the 1968
English republication of *Shamanism* (Pernet 2012: 97–101). Nonetheless, despite the
friendly relationship, Pernet had clearly identified the problems with the unfalsifiable
method heralded by Eliade: 'I had slowly reached the conclusion that Eliade was
fundamentally a philosopher [and not a historian]; as such the history of religions he
defended was not the science I had hoped to find with him, but an unfalsifiable system
in which the main notions were held as postulates' (Pernet 2012: 25). Even though both
Pernet and Gross were forced to withhold criticism while experiencing a temporary
academic limbo, for different reasons, there was no turning back: an entire new
generation of scholars was slowly discovering the shortcomings of the HoR.

Dismantling the primacy of shamanism: Mac Linscott Ricketts

Mac Linscott Ricketts (1930–), former Methodist pastor, professor first at the Department
of Religion at Duke University (1964–1969), and later at Louisburg College, North
Carolina (1971–1994), is the author of what is generally considered the most complete
biography concerning the Romanian years of Eliade (Ricketts 2004). Ricketts also
translated many non-scholarly works of the historian of religions into English (i.e.
Eliade's autobiographies, diaries, novels, short stories). Ricketts completed his PhD
research at the University of Chicago in 1964, under the formal supervision of Eliade,
although, not unlike his fellow students, he probably worked most with Eliade's colleagues
Charles H. Long (1926–) and Kitagawa (Ricketts, pers. comm., e-mail, 25 May 2015).

The subject of Ricketts' dissertation was the mythical figure of the trickster, chosen in the wake of an interest spurred after attending Eliade's 1961 class about 'The Problem of the High Gods', focused on Schmidt and Pettazzoni's works (Ricketts 2007: 212). As defined by Ricketts, the trickster is the 'creative transformer of the world and heroic bringer of culture' who combines and embodies in just one character (a 'trickster-fixer') a plethora of different, and sometimes opposite, social roles and moral codes (Ricketts 1966: 327). More specifically, the Native North American trickster, a zoomorphic figure with anthropomorphic behavioural features (e.g. coyote, raven, rabbit, etc.), is someone who cunningly steals and brings knowledge or technology to humankind as the unintended by-product of ludicrous, tragicomic tricks and swashbuckling adventures (cf. Grottanelli 1983). Now, following the phenomenological and morphological classification of the HoR, Ricketts claims that the trickster is also a '"type" with a recognizable form wherever he is found, regardless of innumerable variations in history' (Ricketts 1970: 3). Examples of tricksters as recovered by the HoR are the German god Loki, the Latin Mercury, the Greek god Hermes and the titan Prometheus, and the anthropomorphic Coyote in California (e.g. Bianchi 1958; Ricketts 1970: Chapter 20). In his PhD dissertation, Ricketts posited the 'trickster and the shaman as rivals', with the shamans explicitly downplaying the religious role of the trickster and the storytellers that narrate trickster tales joyfully desecrating shamanic ideology (Ricketts 1970: Chapters 15, 17). By pointing out this dichotomy, Ricketts undermined the archaic, absolute value of *homo religiosus*, deconstructing the Eliadean link between the archetypal religiosity and current 'primitive' religions. Ricketts achieved this result by interpreting the trickster as the

> embodiment of a certain mythic apprehension of the nature of man and his place in the cosmos. In [the] myths of theft we see the trickster as man fighting alone against a universe of hostile, spiritual powers and winning by virtue of his cleverness. The trickster is man, according to an archaic intuition, struggling by himself to become what he feels he must become – master of his universe.
>
> Ricketts 1966: 336

Resorting to a phenomenological morphology, Ricketts reframed the trickster narrative as an agnostic, cunningly camouflaged religious critique embedded in a religious discourse, locating it at the dawn of human culture and religiosity. He associated the religious philosophy of the trickster with two of the things that Eliade disliked, ignored or minimized throughout his entire scholarly production, i.e. *fun and laughter* (as a way to live fully and desecrate ignorant precepts which impede a fuller, free life) and *science* (as a way to obtain knowledge about the natural world instead of explaining it away with obscurantist theological dogma; Ricketts 1966: 336; Ambasciano 2014: 324–7). Finally, as anticipated above, Ricketts explicitly contrasted the mythologies built around the trickster as the opposite pole of shamanic mysticism, futile reverence and arrogant superstition (Ricketts 1966: 338). The myths of the North American trickster, Ricketts concluded, were about human beings celebrating themselves as clever conquerors of knowledge, while parodying shamanism as backward.

The binary reconstruction of *dogmatic religion vs. progressive agnosticism* discredited the Eliadean autonomy and primacy of religiosity over other domains of human knowledge. By doing this, Ricketts also single-handedly deconstructed the Eliadean interpretation of shamanism, which represented, as we saw in the previous chapter, the keystone of the Eliadean research programme. Eliade dealt with Rickett's thesis in a chapter included in his volume *The Quest: History and Meaning in Religion*, published in 1969 (Eliade 1984; originally published in French with the title *La nostalgie des origines*). Drawing heavily on Rickett's own work, Eliade attempted with difficulty to undo Ricketts' criticism, refusing to engage in any discussion, stating explicitly that 'it is not a question of an image of man in a humanistic, rationalistic or voluntaristic sense', and simply reframing the trickster myths as religious expression of a hubristic 'mythology of the human condition' which resulted in a degraded religious condition (Eliade 1984: 157). Unfortunately, apart from some scattered contributions, Ricketts' entire dissertation circulated only as an unpublished manuscript among historians of religions, and Ricketts continued nonetheless to work as editor and translator of Eliade's books refraining from criticizing further Eliade's methods and defending his persona and work (e.g. Culianu 2003; see Ricketts' letter to Culianu, 21 May 1986, in Bordaş 2012: 102–5).

Dismantling phenomenological morphology: Ioan P. Culianu

In the winter of 1988, Ricketts received a telephone call from a young Romanian professor, and he wrote down this note: 'Culianu called Nov[ember] 6 about plans for a book in defense of Eliade, to include all allegedly incriminating articles of 1937–38. I would translate these and write a commentary article' (memo note by M. L. Ricketts, 6 November 1988; in Bordaş 2012: 151). Even though such a project was never to be (and the articles published much later; Handoca 2001; Rennie 2006: 412–22), that note testifies to the intellectual relationships that cemented the network of Eliade's former students and colleagues, of which Culianu was certainly one of the liveliest members.

Ioan Petru Culianu (1950–1991) entered the Faculty of Romanian Language and Literature at the University of Bucharest in 1967 (Anton 2005: 65). In June 1972, after a change of heart and faculty, Culianu graduated almost *magna cum laude* with a dissertation on Italian Renaissance philosophy of religion entitled 'Marsilio Ficino and Platonism during the Renaissance' under Italian literature scholar Nina Façon (1909–1974) (Anton 2005: 86–7). Culianu's refusal to comply with the plan devised by the Romanian secret security police, the *Securitate*, which wished to recruit him as an informant, began to affect his career, leading to editorial opportunities inexplicably refused and scholarship applications denied. In the same year, Culianu, already acquainted with Eliade's works, started a life-long correspondence with Eliade in which the Chicago Professor suggested that the young student should leave Romania and ask for political asylum in Italy or France (Anton 2005: 75–91; see the correspondence between Eliade and Culianu in Culianu-Petrescu and Petrescu 2004).

A scholarship from the Italian Foreign Ministry allowed him to bypass temporarily the many institutional hindrances in Romania and study ancient culture at the Università per Stranieri of Perugia. Eventually, Culianu decided not to go back to Romania and registered to be recognized as a political refugee (Anton 2005: 104). Once granted refugee status, in 1973 Culianu applied for and won with the highest grade an assistant professorship with Ugo Bianchi at the Università Cattolica del Sacro Cuore, Milan, while studying for a PhD in the HoR (Anton 2005: 104–5). In 1976, Culianu landed an assistant professorship at the University of Groningen, Netherlands. Two years later, he published the first biography dedicated to Eliade (Culianu 1978). Thanks to Eliade's support and professional network, in 1980, Culianu earned a second doctorate about cross-cultural analysis of religious ecstasy at La Sorbonne, and, finally, fulfilling his professional dream, in 1986 he became a visiting professor at the University of Chicago, where he stayed until 1991 (Anton 2005: 119, 151; see also Culianu-Petrescu 2003). Increasingly engaged with politics, and slowly switching from an apologetic stance with regard to Eliade's Interwar past to a lucidly critical attitude which led him to plan several editorial projects with Ricketts and other scholars about Eliade's political engagement, to confront vocally the neo-fascist legacy of the Iron Guard, and to denounce publicly the anti-democratic, authoritarian continuities in post-Communist Romania, Culianu was shot dead at the University of Chicago on 21 May 1991. Despite an FBI enquiry (Culianu just got his green card), Culianu's murder is still unsolved (Anton 2005; see Culianu 2005a).

Coming from a family steeped in science and technology (Anton 2005), and exasperated by HoR's indeterminacy (e.g. Culianu 1978: 17–18), Culianu tried to justify the autonomy of the morphological and phenomenological HoR while expanding its scientific borders by appealing, in various forms and different times, to the following dimensions:

1. *Ethology and primatology*: Inspired by Konrad Lorenz's *On Aggression* (1966), and a plethora of other sources ranging from mass psychology and Edgar Morin's essay on *homo demens* (1973) to the study of social hierarchies in chimpanzees, Culianu suggests a biological relationship between aggressivity, dominance, and religious power, and argues that such 'phylogenetic roots even predate the evolution of hominins' (Culianu 1981: 182). In his attempt to update Eliade's morphological phenomenology and naturalize the human sciences, Culianu investigates magic as mental manipulation and religious rituals as human ethology in a way somewhat similar to sexual selection theory (Culianu 1987; cf. Slone 2008). While approaching the deep roots of magic and religion, fashion, and sexuality, Culianu notes what today would be called *supernormal stimuli* (i.e. the artificial exaggeration of a certain sensorial stimulus to elicit an equally exaggerated response, such as make-up, 'candy sweeter than any fruit, stuffed animals with eyes wider than any baby, pornography, propaganda about menacing enemies'; Barrett 2010: 6), reflects on the evolution and development of human imagination, includes Desmond Morris' *The Naked Ape* (Morris 1984) as well as other ethologists' ideas, and concludes that 'the human species is highly *neophilic* (fond of novelty), and it is *neotenic* (fixated on youth and youthful traits; from the Greek *neotes*, "youthful") to such a point that we may owe our lack of body hair to this latter characteristic. Accordingly, neoteny is fundamental as both a process and in its influence on fashion' (Culianu 1991a: 79).

2. *Morphology and cognition*: Eager to systematize further his interdisciplinary synthesis, Culianu translates Scottish mathematical biologist D'Arcy W. Thompson's

(1860–1948) *morphometrics* and Polish-born, French-educated, American mathematician Benoît Mandelbrot's (1924–2010) fractals into the study of culture and religion. Morphometrics was the study of functional constraints on the shape of organisms, according to which 'various types of biological "form" can potentially be explained by various types of physical forces' identified through the application of geometrical grids superimposed on physiological structures, such as bones, within a saltationist mindset, a thesis later disconfirmed by evolutionary developmental biology (Horder 2009: 768; see Gould 2002: 1181). A fractal, instead, is a 'curve or a surface generated by a process involving successive subdivision'; within a mathematical pattern known as the Mandelbrot set, a fractal becomes able to 'produce complex self-similar patterns' (Daintith and Martin 2010: 332, 503). Culianu thus arrived at the following *morphodynamic* concept of religion: a self-replicating, fractal system which consists of diachronic actualizations of ever-existing synchronic possibilities, and which develops following an ever-expanding digital ramification (Culianu 2002; Culianu 2005b). Combining this cultural mathematization with anthropologist Claude Lévi-Strauss' (1908–2009) proto-cognitive and structuralist view on myth as the primal and most important expression of underlying cognitive processes (L. H. Martin 2014: 241) and Eliade's phenomenological primacy of myth, Culianu reframed myth as a fractal 'will to repeat a piece of narrative submitted to continuous reinterpretation', without core nor variants, myth being instead the 'repetition of a hollow plot conveying different messages', a 'mechanism of make-believe intended to establish some perfectly arbitrary and illusory continuity in the otherwise tricky and everchanging world' expressed through a computational mental 'process' – which, in a sequential procedure, would feed into the self-replicating system, and so forth (Culianu 1990a: 280, 282, 287–8).

The ultimate aim was to update previous phenomenological classifications, while maintaining a non-historical, morphological approach to systemic analysis (Culianu 1990b). For instance, the 'real-because-mythical' explorations of otherworldly dimensions that characterized Eliade's descriptions of shamanism were reframed as explorations of inward, infinite, mental spaces, elaborated historically by local cultures according to some contextual and cognitive rules (such as mind–body dualism) and reflecting cognitive limitations to describe what Culianu referred to as a four-dimensional space. Also, Culianu stressed the role of panhuman mental processes as the source of cultural similarities while discarding historical descent and development (Culianu 1991b). Despite this quite exceptional open-mindedness, Culianu's model suffered from several shortcomings, which I will come back to later.

Dismantling classification: Jonathan Z. Smith

Culianu's initial turn to science to bypass phenomenological sterility was certainly remarkable, but not unprecedented. A similar interest in natural classification prompted Jonathan Z. Smith (1938–2017) to advocate the use of numerical taxonomy to overcome phenomenological and morphological limitations and to redefine in more epistemically warranted terms the historical relationships within and between religions.

Inspired by an ongoing interdisciplinary Marxist reconciliation with Kantian philosophy and Freudian psychoanalysis, impressed by German philosopher Ernst Cassirer's (1874–1945) article 'Structuralism in Modern Linguistics' (1945), and critical of the phenomenology of religion since his structuralist BA thesis entitled 'A Prolegomenon to a General Phenomenology of Myth' (Haverford College, 1960), Smith earned his PhD at Yale Divinity School in 1969, discussing a dissertation on 'The Glory, Jest and Riddle: James George Frazer and *The Golden Bough*' (Smith 2004: 3, 7, 29 n. 38). While working on his PhD project, Smith was employed first at Dartmouth College (1965–1966) and then at the University of Santa Barbara, California (1966–1968), where he met Eliade for the first time when Eliade was serving as Visiting Professor, while Smith had just returned from an interview at the University of Chicago. In 1968, Smith became assistant professor at the Divinity School of Chicago, and later full professor from 1975 until his retirement in 2013, resigning from his Divinity school affiliation in 1977 (Smith 2004: 4–12).

Depite these humanistic interests, the methodological roots of Smith's ideas were deeply embedded in the fertile humus of natural history. From his youth, Smith had a keen interest in biology. His first scholarly paper publication was a quantitative study about the 'average number of milkweed plants (*Asclepias* sp.) per acre, undertaken in support of the Royal Ontario Museum's monarch butterfly migration project' (Smith 2004: 2). However, agrostology, the botanical study of grasses, was his first and foremost passion. 'That', he said in an interview in 2008, 'was what I wanted to do with my life' (Sinhababu 2008). Intending to study agrostology at Cornell Agricultural School, Smith worked on a farm as a practical preparation for his future career. However, he eventually decided against enrolling at Cornell because of the impossibility of including liberal art courses in his curriculum (Sinhababu 2008). Anyway, as Smith himself recalled in 2004, 'agrostology led [...] to a deep interest in natural history [which] remains today; taxonomic journals are the only biological field I still regularly read in' (Smith 2004: 2).

Comparison was the *trait d'union* between his botanical and taxonomic interests and religions: 'I think that's what got me interested in grass – how many kinds of grass there are. I'm fascinated by how many kinds of religions there are, how many kinds of Bibles there are. Linnaeus gave us a way of talking about the diversity of grasses' (Sinhababu 2008; cf. Smith 2004: 19). However, the most important taxonomic influence for Smith's subsequent theoretical works on religion was not Linnaeus's *Systema Naturae* but modern quantitative taxonomic classifications which, starting from the 1960s, marked a paradigm shift in the biological sciences (Smith 2004: 22).

The two major innovations in the field of biological classification, at that time still stuck in an unresolved tension between progressive, evolutionary, Darwinian and static, non-evolutionary, Linnean concepts of taxa, were phenetics and cladistics, promptly acknowledged by Smith (1990: 47–8 n. 25; Smith 2000). Phenetics, also known as numerical taxonomy, was a 'phenomenological approach to systematics' in which classification renounced phylogeny (i.e. historical relationships) and focused on natural similarity in order to create a theory-free model and avoid preconceived ideas about ranking and clustering (Rieppel 2008: 296). In a word, phenetics avoided *interpretation*, which had been instead reputed the key to order organisms by previous evolutionary systematists (Stuessy 2009: 71). In particular, evolutionary systematists weighed differently specific physiological features in order to correlate characters to heredity and change over time. However, lacking precise quantitative methods, those empirical classifications were prone to various biases, the most important of which was the limited amount of character coding. Phenetics represented a direct response to such a state of affairs and took advantage of cutting-edge developments in computational power and computer programming to measure, code, calculate and cluster a previously unmanageable amount of data (Hamilton and Wheeler 2008: 335). Once analysed, the resulting data matrices were clustered in 'phenons', i.e. new taxonomical groups, and graphically represented in 'phenograms', tree-like representations of similarities among taxa which put aside diachronic development.

Betraying, like Culianu, the influence of ahistorical morphology, Smith focused on synchronic phenetics for the analysis of a historical case study (in Smith's case, circumcision in early Judaism) (Smith 1982: 4, a paper originally delivered in 1978). In his article, Smith pleaded for the adoption of numerical taxonomy in the HoR to cluster religious taxa within the same tradition according to a polythetic classification. However, as Benson Saler has noted, there was no attempt to use phenetics *practically* in Smith's essay, it was merely exploited as a tool with which to think (Saler 2000: 180). Limiting his foray into taxonomy as a preliminary theoretical effort, Smith recognized that it was 'premature to suppose a proper polythetic classification of Judaism [although] it is possible to be clear about what it would entail' (Smith 1982: 8). And yet, subsequent research by Smith did indeed retreat from such a practical attempt to an increasingly theoretical corner, possibly because disillusioned by the increasing criticism faced by phenetics, especially by its main competitor and eventual defeater, cladistics, or phylogenetic systematics (Saler 2000: 180–96).

Notwithstanding this setback, Smith's interest in theoretical taxonomy, fuelled by his passion in natural sciences, was to lead him to outshine his peers and colleagues. It is impossible to provide here a comprehensive account of Smith's career (e.g. McCutcheon 2008), but I reproduce below a single excerpt, published in 1982 as an introduction to the same volume that included his reflections on numerical taxonomy:

> while there is a staggering amount of data, phenomena, of human experiences and expressions that might be characterized in one culture or another, by one criterion or another, as religious – *there is no data for religion*. Religion is solely the creation of the scholar's study. It is created for the scholar's analytic purposes by his imaginative acts of comparison and generalization. Religion has no existence apart from the academy. For this reason, the student of religion, [. . .] must be relentlessly self-conscious. Indeed, this self-consciousness constitutes his primary expertise, his foremost object of study.
>
> Smith 1982: 11; original emphasis

This is, by far, the most ground-breaking statement in the whole modern HoR, the one that enjoyed the most outstanding popularity, the one which attracted the most visceral attacks, and the first to endorse a transition from HoR to RS, creating from the inside a disciplinary niche to pursue such a postmodern, critical study. We will come back to Smith's statements and his relationships with phenetics later in this chapter.

Dismantling right-wing ideology: Bruce Lincoln

Smith's interest in taxonomy and phenetics fuelled his virtuoso approach to theory and method in the HoR. As another scholar has recently remarked, Smith's break with the traditional emic, fideistic approach of the HoR was nothing short of remarkable: Smith 'insisted on considering all religious phenomena in their social, historic and cultural context, whereas his predecessors tended to treat them as a-temporal expressions of eternal truths. [...] In this way, he re-theorized religion as something human and thereby opened it up to critical investigation' (Shimron 2018). The scholar who has delivered this fitting summary is Bruce Lincoln (1948–), Professor Emeritus of the History of Religions at the Divinity School of the University of Chicago.

For all the importance of Smith's theoretical works, Lincoln's breakthrough in the field proved to be no less important. While Smith was trying to overcome the many methodological problems of morphology and phenomenology by appealing to phenetics, Lincoln set about the most extraordinary operation of reverse-engineering that the HoR has ever seen.

After a BA in Religion at Haverford College, PA (1970), Lincoln enrolled at the University of Chicago, where he focused on comparative religion and Indo-European studies. Smith was assistant professor there, and Lincoln remembers Smith's introductory class to the HoR in the autumn of 1971, at a time when Smith was 'locked in a personal and intellectual struggle with Mircea Eliade, then the dominant figure in the discipline' (Lincoln 2012: 117). In 1976, Lincoln completed his PhD dissertation on 'Priests, Warriors, and Cattle: A Comparative Study of East African and Indo-Iranian Religious Systems' under the supervision of Eliade himself.[7] Lincoln passed his PhD viva with distinction, and in that same year he became professor at the Comparative Studies in Discourse and Society Program, University of Minnesota, until 1994, when he became Professor of the History of Religions at the University of Chicago (The University of Chicago Divinity School 2012).

Lincoln's interests coincided with those upheld by the new generation of scholars: he tackled women's ritual initiations as a sort of religious smoke and mirrors to relegate girls to a life of subordination inscribed in the hierarchically androcentric social organization. Although still within a classical HoR framework, Lincoln suggested poignantly that such rituals offered a 'religious compensation for a socio-political deprivation. Or, to put it differently still, [they were] an opiate for an oppressed class' (Lincoln 1981: 105). Constantly committed to the unrelenting, critical scrutiny and updating of his own positions and stances and heading towards a more historically based enquiry thanks to his friendship with Cristiano Grottanelli (Clark 2005: 11, 15; see Lincoln 2015), Lincoln set out to study Georges Dumézil's Indo-European trifunctionalist hypothesis. This thesis had been proposed back in the 1930s and had since been the implicit backbone of the comparative HoR of old. This hypothesis was built on the implicit equivalence of Eurasian populations speaking Indo-European languages (i.e. Italic, Germanic, Slavic, Baltic, Celtic, Hellenic, Indo-Iranian, etc.), and the ethnic groups speaking them, and it posited the reconstruction of the primordial Indo-European society and culture via the analysis of their ancient mythologies (i.e. ancient Greek, Latin, Celtic, Norse, Slavic, Hittite, Vedic, etc.). According to Dumézil, 'the Proto-Indo-Europeans had an ideology that structured existence into three functions (those of sovereignty, war and production), where the first function additionally was divided into a dark, magical aspect and a light, legal aspect' (Arvidsson 2006: 307). Basically, according to such a view, Indo-European life revolved around a rigid social structure made of kings, warriors or soldiers, and peasants, inherited and conserved in mythological themes by all of the subsequent historical societies.

Dumézil was a close friend and colleague of Eliade, who adopted Dumézil's reconstructions (e.g. Eliade 1982b). They helped each other professionally and they were both in various degrees involved in Interwar extremist politics (e.g. Dubuisson 2014). And, just like Eliade, Dumézil's works were tainted by his political beliefs. Lincoln, as a young scholar encouraged by Eliade, initially tried to expand and build upon Dumézil's theories, but he rapidly became disillusioned about the scientific status of the tripartite trifunctionalism (accounts in Clark 2005 and Arvidsson 2006). Finally, starting from the mid-1980s, Lincoln became a staunch, critical reviewer of both Dumézil's theory and the current consensus in the HoR, because he noticed the close relationship between the reactionary ideology of Dumézil and his subject of study, idealized in a racial and anthropological way as something pure, admired and worth retrieving today. Thanks to Antonio Gramsci's (1891–1937) notion of cultural hegemony and Roland Barthes' (1915–1980) semiotics of myth, Lincoln's interest shifted gradually towards questions of implicit ideology and power dynamics (cf. Lincoln 2000), and he came to understand that the comparative HoR of old was nothing more than an emic, shared, and admired repetition of what the sacred texts reported: Lincoln's 'studies of Indo-European mythology have now made him question the very belief in an objective historiography, and he sees the scientific search for knowledge as a site for political power struggles. The work of cultural studies is, according to Lincoln, "myth plus footnotes"' (Arvidsson 2006: 306; cit. from Lincoln 1999: 209; cf. Clark 2005: 16, and Schilbrack 2005). Basically, as Stefan Arvidsson remarked, Lincoln approached Indo-European mythographical reconstructions as normative devices that engender right-wing societal constraints of dominance and subordination, and the original myths themselves as the legitimization of social injustice (Arvidsson 2006: 303). Therefore, historico-religious taxonomy, whether in original ancient documents or in contemporary analysis, (re)creates hierarchies of power and social discrimination (Lincoln 1989: 7; Lincoln 1999: 147; from Clark 2005: 12).

Mythology naturalizes the social hierarchy, that is, it makes acceptable the idea that things have always been like *this* and that their being like *this* is not only good for the society as a whole, but even necessary, because of their religious prestige and charismatic appeal. In other words, dominant groups exert and extend their power by suggesting and promoting an internalization of religious precepts and behaviours in the subordinated classes themselves. Religion might still be reclaimed as an empowering discursive tool to subvert oppression. However, the study of ancient mythologies, and the religious taxonomy the discipline promotes while looking for pristine origins and prestigious relationships, has served purposes other than just the advancement of academic knowledge, as it had been exploited to advance and support reactionary politics (see Clark 2005).

In order to break this pattern, Lincoln has promoted a self-critical perspective which asks incessantly 'to the benefit of whom this religious discourse?' and, consequently, has redefined myth as 'an authoritative mode of narrative discourse that may be instrumental in the ongoing construction of social borders and hierarchies, which is to say, in the construction of society itself', for myth acts as 'the bearer of ideology' (cit. resp. from Lincoln 1991: 123, and Clark 2005: 14; cf. Lincoln 1986: 4–5). With regard to the HoR itself, Lincoln has recognized with disturbing lucidity the disintegration of the old HoR. In an article published with Cristiano Grottanelli in 1985, the two authors acknowledged the 'radical decontextualization and deprocessualization of religious data' operated by phenomenological HoR, and wrote that 'despite its initial promise, this field never inherited the critical and methodological legacy of the figures we have treated [i.e. Marx, Engels, Weber, Durkheim, Malinowski; see Chapter 2, §*Whodunnit?*] and consequently remained relatively fruitless and isolated' (Grottanelli and Lincoln 2005: 316). The very institutionalization of the HoR as an autonomous discipline, as we have seen, ignored the relationships between ideology and religion to the benefit of those scholars that embraced and shared the same ideology, and created 'a field which wandered into a dangerous blind alley while pursuing autonomy' (Grottanelli and Lincoln 2005: 320, cf. 318). Maybe, the most concise and brilliant synthesis Lincoln has ever produced, and certainly one of the most read manifestos in the field, is his thought-provoking *Theses on Method*, originally conceived in 1995 and first published in 1996, in which he deconstructed – and rebuilt – the entire HoR as a critical meta-discourse able to denounce the disciplinary misrecognition of power structures and dismantle the emic, fideistic sympathies which conceal a defence of ideological, regressive, conservative, religious or spiritual beliefs. 'Destabilizing and irreverent questions' should be employed to counteract any effort to silence those criticisms, for in the very moment 'one permits those whom one studies to define the terms in which they will be understood', failing to 'distinguish between "truths", "truth-claims", and "regimes of truth", one has ceased to function as historian or scholar', and has already started working with other non-scientific and opportunistic agendas in mind (Lincoln 1996: 227; see Lincoln 2012: 135–6).

New generations, poststructuralism and Religious Studies

The previous list of scholars is woefully incomplete and could be obviously expanded and perfected. My only aim, however, is to highlight as succinctly as possible the fact that many young historians of religions, during the decades of Eliade's professorship at the University of Chicago, tried hard to update the field according to epistemically warranted perspectives, providing renewed paradigms for their subjects of study. Moreover, while anthropologists, sociologists, and other historians had already criticized Eliade's approach, some of the most constructive critiques came from the inside of what could be called the Eliadean Chicago school of HoR (*contra* Wedemeyer

2010). Confused by the ideological and apodictic methodology of the HoR, and variously troubled by Eliade's past and academic legacy (hence the focus on critical attention and ideological roots), second- and third-generation historians from this school ignited the revolution HoR needed to reform itself (see Table 3; please note that the table is not exhaustive and does not reflect all the areas of research in which the mentioned scholars were involved within the chosen timeframe).

Indeed, as Lincoln argued, a significant part of classical HoR could be understood as a way to engender and/or support a *politics of nostalgia* by providing prestigious, rhetorical storytelling to reactionary, and sometimes extremist and racist, political discourses, emically endorsing the dynamics of social power implied in the study of ancient documents. As historians Jeppe S. Jensen and Armin W. Geertz have remarked in 1991,

> some historians of religion have advocated a personal and existentially relevant attitude to the world's religious traditions. Foremost among these is Mircea Eliade who presented modern man's estrangement from tradition as fundamentally detrimental to individual and social balance, hence the politics of nostalgia which seeks, on the basis of a universalist interpretation of religions, to restore Man as a complete and inherently spiritual being.
>
> Jensen and Geertz 1991: 13; from McCutcheon 1997: 32

And yet, all those Chicago scholars were still trained to be *historians of religions*, thus most of them were unwilling to renounce completely the HoR research programme: if what they did was a deconstruction, the final goal was the reconstruction of the field. But was a reconstruction possible at all after such annihilation of the previous paradigm? Once the power dynamics in the form of ideological biases and preconceived ideas underlying the classical HoR had been revealed, was going back to that field really an option?

Starting from the late 1980s to early 1990s, a new generation of scholars well-read in these critical works, and ready to rescind the umbilical cord that still connected the previous Chicago scholars from the morphological and phenomenological HoR, migrated, so to say, to the field of RS, a vast network of interrelated approaches informed

Table 3 Falsification of the Eliadean Research Programme: Reaction from within the Chicago School

Disciplinary tenets questioned or falsified	Chicago scholars, *ca.* 1960s–1990s					
	Gross	Pernet	Ricketts	Culianu	Smith	Lincoln
Sexism, androcentrism	✓					✓
Epistemology		✓				
Undisputed sacred ontology		✓	✓			
Morphology (classification)				✓	✓	
Power, politics, ideology				(✓)	✓	✓

by *poststructuralism*. While in some cases the aforementioned Chicago scholars were merely acquainted with this philosophy, RS scholars adopted it as their user manual to approach religion. Poststructuralism is a loose label which gathers together a vast array of critical approaches and discourses born as a reaction against the previous attempts to study human cultures as epiphenomena of rigorous underlying linguistic, social, economic, psychological and anthropological *structures*. A poststructuralist approach ideally entails the use of some of the following tools (Figure 14):

1. *linguistic turn*: language is an imprecise and unreliable instrument to describe ontological reality;
2. *crisis of representation*: the author is an entity not entirely consistent nor coherent;
3. *decentring of the subject*: everything can be analysed as a textual item;
4. *critique of essentialism* (not to be confused with cognitive essentialism): ultimate values, whatever they might be, are questionable;
5. *cultural turn*: a focus on subordinate groups or subaltern social classes and their cultural representations, and a move away from dominant narratives;
6. *social constructionism* (not to be confused with social constructivism): reality is what culture makes of it (Angermuller 2015: 15; cf. Jameson 1998, and Sokal 2010: 94–5).

As far as the HoR was concerned, the knot made up by the entangled strands of disciplinary method and theory and Eliade's persona was to be cut with a *poststructuralist* axe, dispensing once and for all with the heavy baggage of the discipline. However, the usefulness of the term 'poststructuralism' has been disputed mainly for two reasons. *First*, poststructuralism is a North American conventional label used to group together many different and cross-disciplinary European trends, in particular from France

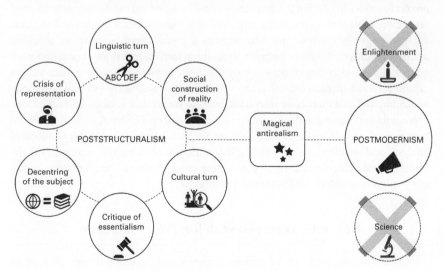

Figure 14 Poststructuralism and postmodernism: a tentative chart

(Angermuller 2015: 16). In the 'Anglo-American debate, poststructuralism is sometimes understood in close connection with "postmodernism" and seen as an antipodal project to modernity' (Angermuller 2015: 18). When French philosopher Jacques Derrida (born Jackie Élie; 1930–2004) commented on the US appropriation and re-elaboration of the postmodern turn and the proliferation of cultural studies, he acknowledged them as being somewhat radically different from their European roots (Creech, Kamuf and Todd 1985: 2, 4, 29). Therefore, differentiating between poststructuralism and postmodernism might be a useful historical strategy to tackle this argument. *Second,* many allegedly 'poststructural' scholars saw themselves as successors and heirs, even if critical, of the allegedly 'structural' predecessors (cf. Wolfart 2000). Just like most Chicago scholars that we have seen did not sever their ties with the old HoR, poststructuralists recovered, updated and expanded some previous trends from the social sciences, especially those focused on the social ways in which social dynamics produce knowledge (e.g. Marx and Engels' historical materialism, Weber's sociology, Durkheim's institutionalization processes, etc.). In so doing, as Johannes Angermuller has pointed out, poststructuralists

> have problematized older philosophical traditions conceptualizing knowledge as a set of pure ideas, abstract concepts and universal truths. If knowledge is constructed by the community which recognizes certain ideas as relevant, legitimate and true, theory too, is involved in social struggles over what counts as legitimate knowledge in which the participants mobilize their non-theoretical resources, such as time, relationships or money.
>
> Angermuller 2015: 84

But what happens when such legitimate critical tools are *really* applied with the intention of deconstructing the whole modernity package and implement an 'antipodal project to modernity'? It is not difficult to see how, in the wrong hands, the poststructural set of tools, most of them originally conceived of as instruments of literary and semiotic analysis, might be carelessly used to promote a pseudoscientific stance of rebellion against the academy itself. For instance, the cultural turn is a milestone in contemporary academia: new inter-disciplinary trends like postcolonial studies or gender studies assumed a much-needed critical, if not vitriolic, stance against the cultural hegemony which implied, consciously or unconsciously, the survival of various degrees of top-down political domination and androcentric paternalism. However, when the cultural turn was mixed with social constructionism, forgetting the original literary context of analysis and resulting in the unbridled criticism of scientific research, or the Enlightenment *tout court*, the result put to sleep modern reason. Paraphrasing Goya, it has produced monsters. And postmodernism as well.

Be careful what you wish for: Postmodernism

The most concise definition of *postmodernism* could be 'cultural turn plus social constructionism amplified to the power of ten' (cf. Figure 14). In brief, postmodernism entails the rejection of all previous modernist tenets and assumptions. If Modernism was a product of the Enlightenment, which posited a rigid demarcation between (a) scientific, innovative and rational approaches to obtain reliable knowledge and (b) non-scientific, traditional, irrational trends that favoured the institutional, socio-political and cultural *status quo*, postmodernism regards the very distinction as ungrounded, that is, lacking sufficient justification: 'postmodernists have shown that neither religion nor science is exempt from socio-cultural influence' (Wiebe 2000: 353). Instead, power dynamics between dominants and subordinates are assumed as the new methodological engine and epistemology is substituted by sociology (cf. Segal 2006: 169). The modern appeal to reason is thus radically challenged as a bourgeois, elitist power play to exert social, political and economic domination. Expanding what I have previously defined as the main tenets of poststructuralism, postmodernism holds that knowledge is bound by space and time, that no reliable scientific knowledge exists outside the socio-political network of power relationships, and that science, as a whole, is an enterprise tainted by its association with capitalism, an enterprise that colluded and conspired to exploit the prestigious label of 'scientific research' in order to impose dogmatic control over modern populations (Cavalli-Sforza 2010: 44). The reprise of the old Romantic dissent against reason, as Nikolai G. Wenzel has recently summarized, implied that

> where some saw aberrations or challenges within the modern project, postmodernism saw unavoidable and logical consequences: colonialism; fascism/ communism and industrially planned genocide; the destruction of the natural environment in the name of unfettered progress and technology; the North's exploitation of the South; the horrors of modern warfare, compounded by methodical application of the very science and technology initially meant to liberate humanity; and the spiritual poverty and alienation of mass consumerism.
>
> Wenzel 2009: 174

From this perspective, for instance, it becomes possible to assert that Darwin and evolutionary biology have been active partners in crime with Victorian capitalism to create the basis for imperialism and Nazi genocide – which is, in Harry Frankfurt's terms, *bullshit* (Richards 2013; cf. Desmond and Moore 2009). Consequently, postmodernism 'rejects any claim of absolute truth', as truth is seen as a top-down coercion aimed at 'impos[ing] a "master voice" or "meta-narrative"' (Wenzel 2009: 175). As advanced by French philosopher Jean-François Lyotard (1924–1998), reality is indistinguishable from fiction and, thanks to the current mass media production of cultural contents, fiction provides a more compelling reality than reality itself. Science, as a way to obtain reliable knowledge about the social and the natural world, loses its prestigious status and becomes just another way to manage and impose social power in hierarchical settings – if not the *longa manus* of political domination itself.

With regard to religion itself, postmodernism entailed an apparent reinforcement of secularism, while fostering a decline of traditional religions and a sharp rise of spirituality and regressive fundamentalism (Wenzel 2009; Copson 2017). In the wake of the collapse of the Cold War geopolitical system, neoconservative attacks against post-war social justice, welfare state, state intervention in economics, and Enlightenment values gained momentum, replicating the right-wing attacks against modernity and modernization during the Interwar period (Antonio 2000: 47). Thus, postmodernism, delegitimizing facts and relying on interpretations alone, has created the basis for a new right-wing and populist 'reactionary tribalism' in which ethnic origins are retrieved and reinvented as a bulwark against the hegemony and homogeneity of globalization. With the dissolution of modernist epistemological and normative criteria, a 'total critique of modernity' is envisaged and implemented by appealing once again to twentieth-century far-right or conservative, anti-democratic thinkers (Antonio 2000). Coming full circle, 'postmodernism turns out to be a magical antirealism' in which faith and beliefs share the same epistemological truth-values of science (Ferraris 2014: 17), thus destroying the most basic processes of progressive decision-making informed by education as the basis of modern democracy (Antonio 2000: 54).

Within this new framework, if academic and scientific 'objectivity' is merely the result of societal negotiation and power dynamics, the assumption of objective truth achievable by the scholar according to precise, empirical protocols of enquiry is refuted (Strenski 2015: 165, 167). Meanwhile, 'the goal of religious studies should be the appreciation of the distinctiveness of each religion [. . .]. The believer is the equal of the scholar' (Segal 2006: 158). The aim of the new postmodernist RS is a 'comparison of [all] narratives', independent from their degree of truth (Wenzel 2009: 175). In the study of myths and religions, for instance, one might identify a 'polyvalenced discourse' between 'self-affirming fantastical and [. . .] self-negating' elements which, in turn, reveal 'interpretive possibilities that challenge and exceed interpretive, hegemonic frames, past and present' (Taylor 2008: 95). *Discourse* here is the keyword. Originally conceived by French philosopher Michel Foucault (1926–1984), a RS discourse, as Kocku von Stuckrad defines it, is the result of 'communicative structures that organize knowledge in a given community' and which 'establish, stabilize and legitimize systems of meaning and provide collectively shared orders of knowledge in an institutionalized social ensemble' (von Stuckrad 2014: 11; cf. Ambasciano 2016b). In a discursive study of religion, religion and science are reduced to 'the societal organization of knowledge' about those themes (von Stuckrad 2014: 14). But, as Ivan Strenski aptly summarized, 'if all discourse is "constructed" with some human purpose in mind, the job of the religious studies scholar is to "deconstruct" theoretical discourse to get behind the agendas lurking there. The idea of a "science of religion" seems incoherent from the start' (Strenski 2015: 167; see King 2013; for an overview, see also Capps 1995: 238–44).

If no 'scientific approach' can ever be capable of gathering universal knowledge which can be considered valid, compelling, reliable and convincing, and if everything depends on the availability of resources and various kinds of institutional, political or financial capitals to invest in order to assert one's own discursive power over others, then, 'the ambitions of the founders of the study of religion to do general cross-cultural comparison should be abandoned' in favour of specific, local, explorations by scholars engaged in the emic sharing of non-scientific knowledge (Strenski 2015: 167). Even though the scientific ambitions of the comparative science of religion had already been crushed by the advent of the HoR, a fully fledged postmodern approach entails the end of *any* study of religion(s) as we know it or, at the very least, the fractionation of the new RS in a multitude of loosely communicating clusters of research.

From Deconstruction to New Realism

The manifesto of postmodern HoR is to be found in Jonathan Z. Smith's 1982 statement that 'religion is solely the creation of the scholar's study' (Smith 1982: xi). I have already remarked that this is one of the most important statements in the history of the recent HoR. And yet this is also one of the most misunderstood. The usual focus is on the first part of Smith's statement, that is, the non-existence of the essential label 'religion'. Instead, the second section should be considered more important, with its stress on the (critical *and* scientifically based) 'self-consciousness' of the judicious scholar of religion (Smith 1982: xi). This misunderstanding is a consequence of the growing disappointment with the ancestors of the classical, phenomenological HoR in general – and Eliade in particular. As Francis Landy summarized in 2008,

> Eliade insists that there is something suprahistorical and unique about religion
> and *homo religiosus*, while the closest Smith gets to a definition of religion is that it
> 'is the relentlessly human activity of thinking through a "situation"', and as such
> inseparable from all human culture. Smith prefers the 'general' to the 'universal', the
> 'individual' to the 'unique' precisely because they are provisional, are not all-
> encompassing, and admit of exceptions.
>
> Landy 2008: 210; Smith's cit. from Smith 2004: 32

Comparison is allowed when performed between different (religious) cultures if the very act of comparing is strictly and rigorously delimited by the aforementioned 'self-consciousness' (Landy 2008). While Derrida did share the same attention towards differential comparison (cf. Ambasciano 2014: 27), he also asked pessimistically 'what if *religio* [i.e. the Latin term] remains untranslatable?' (Derrida 2002: 67). Even though Smith would not adhere to such a view, many current scholars involved in cultural studies, RS and HoR, have aligned themselves, whether or not consciously, with Derrida's view in that they have problematized the usefulness and applicability of the term 'religion' by differentiating local forms of devotion and beliefs from the Eurocentric concept of 'religion' (cf. Bergunder 2014). For instance, they have questioned whether or not atheism, a label reputed to have been elaborated in a specific, Western and modern timeframe, could really be applied to ancient societies (e.g. Nongbri 2013). Likewise, contemporary scholars have questioned the universal application of such labels as 'religion', the best known being perhaps Talal Asad with his *Genealogies of Religion* (Asad 1993; see also Dubuisson 2003; cf. Segal 2006, and Stausberg and Gardiner 2016).

Even though such scholarly attempts have clarified in unprecedented detail many critical points, highlighting that some terms have a precise cultural history and, therefore, a potentially limited applicability, the loss has sometimes outweighed the possible heuristic benefits. For instance, the idea that the rational values of the

Enlightenment and modern scientific approaches are merely Western inventions at the service of imperialism is a historiographical blunder: given the set of panhuman cognitive universals, we can expect them to arise everywhere and whenever free thinking is sufficiently valued and defended (cf. Chapter 2, *The Deep History of Comparison*). Our knowledge of such instances, instead, is limited by the top-down control exerted on 'cultural technologies of external download' such as literacy, art and writing (cf. Ambasciano 2016a: 188–9). Moreover, as I have written elsewhere,

> in the light of population thinking, and taking into account psychological biases, computational deficits, ontogenetic changes and cognitive mechanisms of relative diffusion and stability of ideas (Sperber 1996; Sørensen 2004), human cultural differences and lack of homogeneity of thought are always to be expected (Richerson and Boyd 2005: 76), even in the context of low within-group variation (Foley and Mirazón Lahr 2012). As to religion(s), openly critical dissent, social protest or merely silent (or repressed) acknowledgement of doubts ultimately depend on whether coercive social control is exercised and to what extent,

in which case, 'alternative or contested narratives, scepticism, agnosticism and atheism [might] have coexisted with religious and cultural options since forever, as a concealed possibility for individuals or, given the right socio-historical conditions as frank expression of critical thinking, doubts, rationalization or mockery' (cf. Ricketts' analysis of the trickster; cit. slightly modified from Ambasciano 2016a: 188–9).

Unconvinced by similar explanations, postmodernists have implemented an *us vs. them* distinction in the field which, in turn, favoured the outright criticism, and ultimately rejection, of scientific tools (e.g. Day 2010; Arnal and McCutcheon 2013: 91–101). Sometimes, postmodern religious scholars have forgotten the necessary reconstruction after the deconstructive enterprise, leaving a wasteland of conceptual ruins behind them (cf. Strenski 2004; Ferraris 2008: 55). The consequence of denouncing and criticizing the past historiography of the discipline as a whole, even those theoretical assumptions and empirical studies which were mostly untouched by ideology, 'is that issues already analysed decades earlier are re-discussed as if they were new, in complete ignorance of the existing literature on the topic' (Spineto 2009: 43; cf. Spineto 2010: 1201 on Brelich's differential comparison).

In particular, when judged on its own terms, there is a serious epistemological flaw that stains the postmodern HoR. Derrida's deconstruction was primarily aimed at enlightening the ideological tenets that underpin half-concealed, philosophical and political thought. When employed in textual analysis, 'a deconstructionist reading of a text subverts its apparent significance by uncovering contradictions and conflict within it' (Blackburn 2016: 121). As such, this approach was meant to identify the regressive ideological value judgements embedded in those discourses and to perform a *reverse-engineering* of someone's ideas, that is, the conceptual 'reproduction of another manufacturer's product following detailed examination of its construction or composition' (Pearsall 1999: 1225). Most of all, notwithstanding its abstruse complexity, Derrida's deconstructionism was also meant to provide a bulwark for *democracy* and *justice* in order to create the preconditions for a durable reconstruction (Derrida 1994:

35; Derrida 1997: 105; Derrida and Ferraris 2001: 56–7; cf. Ferraris 2006: 56). Interestingly, Derrida's works were also free from the 'systematic misuse of [. . .] science' which other postmodernist philosophers were guilty of (Sokal and Bricmont 1998: 7; on Foucault, see Bergunder 2014: 275). In the wake of these most important, yet often neglected, points, the deconstructive approach has been further updated and expanded by a recent philosophical school, i.e. *New Realism*, which saves poststructural critique, includes specifically the ontological, 'undeconstructible' (i.e. not amenable to deconstruction) basis of the natural sciences, and renounces the postmodern interpretive excesses and the 'politicized conception of science' (Spiro 1996: 775; see Ferraris 2014; Ferraris 2015; cf. D'Ancona 2017: 105–7).[8] After all, in Michael Albert's incisive sentence, 'there is nothing truthful, wise, humane or strategic about confusing hostility to injustice and oppression, which is leftist, with hostility to science and rationality, which is nonsense' (Albert 1996: 69; from Sokal 2010: xv).

Unfortunately, this reality turn has failed to appeal to the HoR, and the disciplinary adoption of postmodernist deconstruction has turned out to boost a renewed fideism, strengthening emic approaches, and providing the much needed lifesaver that theological and phenomenological approaches were looking for. Postmodernism has thus become *postmodernity*, in which the sacred has returned and multiple forms of religiosity – finally unshackled from Western imperialism and theological control – have made a successful comeback (Spineto 2010: 1298). Because of the power dynamics involved, those that are now reputed 'delegitimized strands of knowledge', like magic or mysticism, are considered as worthy of sympathetic, emic attention by academics (cf. Ambasciano 2016b). Scientific academic approaches are to be understood at best as equivalent to magic, esotericism, Intelligent Design, paranormal beliefs, conspiracy theories, etc., and, at worst, as the illegitimate, invalid, unjustified imposition of dogmas not much different from any other kind of dogmas (e.g. King 2013; Kripal 2014; von Stuckrad 2014). In other words, the wild, liberating force of poststructuralism has been domesticated into a docile pet *at the service of the same ideological agendas* against which postructuralism was adopted in the HoR in the first place, creating the current post-truth environment characterized by the abuse of *methodological agnosticism or atheism*, according to which critical judgement on religious truths is suspended and formal respect for the topics studied is exploited as an avoidance strategy to cloak one's own beliefs and advance an apologetic agenda (Wiebe 1999: 57; cf. Koertge 2013; Cox 2014b; Stausberg 2014; Cantrell 2016). Myth can be read once again as true story, and Mircea Eliade can be reproposed as an exemplary postmodern scholar who liberates the Humanities from the shackles of Western science (e.g. Rennie 2006: 356–62; Taylor 2008: 107; cf. Ginzburg 2010). There is no crypto-theological agenda in the field any more, for now fideistic commitment is proudly reclaimed as the engine of the enquiry (cf. the field's 'failure of nerve' in Wiebe 1999: 141–62). More worryingly, the whole discipline of historiography becomes virtually pointless. In the final chapter of *Denying History*, a volume on the works of those negationists who deny the historical existence of the Shoah, Michael Shermer and Alex Grobman reflect on the impact of postmodernism in the humanities and note that 'if there is no method of discriminating between true and false interpretations of the past, between history and pseudohistory, between revisionism and denial, then there is no point in even having a discipline of history. With this pseudohistory, historiography becomes hagiography, science becomes ideology, history becomes myth, and revision becomes denial' (Shermer and Grobman 2009: 244; cf. Momigliano 1979: 373).

The impact of postmodernism on the Humanities and the social sciences has baffled many scholars and, most of all, many readers. Is there really no universal basis at all underlying human behaviours and beliefs? Is science, as a collective activity, inherently flawed? Is there always something left unexplained by science, which may open the door for transcendence and, consequently, for the old phenomenology of religion? Is knowledge exclusively dominated by power relationships? Should we accept that there is no discernible, ontological truth? The long answers to these ill-posed questions range from the unexceptionally trivial to the disappointingly bothersome and, to cut a long story short, they are all best resolved by a resounding 'no': despite its many flaws, science remains the best set of practical tools, accumulated knowledge, problem-solving techniques, social networking, rigorous reviewing practices, and critical thinking available to human beings to understand ourselves, the planet, the universe and everything in between. In the words of Maarten Boudry, 'no revelation, *sensus divinitatis*, personal intuition, or inner sense of certainty will succeed where science has failed' (Boudry 2017: 43; cf. Medawar 1986: 86). Unfortunately, this is something that scholars involved in the emic, fideistic, post-truth HoR or postmodern RS have failed to understand.

7

The Cognitive (R)evolution: The End?

False facts are highly injurious to the progress of science, for they often long endure; but false views, if supported by some evidence, do little harm, as every one takes a salutary pleasure in proving their falseness; and when this is done, one path towards error is closed and the road to truth is often at the same time opened.

Charles R. Darwin

Post-truth rules

In November 2016, The Oxford Dictionaries chose as Word of the Year 'post-truth', an adjective which 'relat[es] to or denot[es] circumstances in which objective facts are less influential in shaping public opinion than appeals to emotion and personal belief' (Oxford Dictionaries 2016a). According to Oxford Dictionaries, there has been an astounding increase by 2,000 per cent in the frequency of the word in the past months due to both the Brexit referendum, concerning the 'withdrawal of the United Kingdom from the European Union', and the Presidential campaign in the United States – both of which showed a blatant disregard for the truth and a penchant for deliberate misinformation (Flood 2016; Oxford Dictionaries 2016b; D'Ancona 2017).

What does post-truth mean? As I have anticipated in my *Preface*, it means that what is reputed important is the perception of the 'truthiness' of facts, something that appeals to one's emotions just because it 'feels' right to believe (Oxford Dictionaries 2016b; Zimmer 2010; D'Ancona 2017: 31). Post-truth defines the very act of holding as true whatever it is that someone might believe in notwithstanding the lack of evidence – even in the face of contrary evidence. There is a clear echo of the most extreme trends of postmodern social constructionism, now helped by the overabundance of information typical of the digital age: it is 'easy to caricature [any] torrent of indigestible data as no more than a series of arbitrary claims' (D'Ancona 2017: 17). It is not difficult to imagine what the implications of this post-truth political and cultural environment might be for the comparative and historical study of religion(s).[1] I have already recalled that a similar cultural environment in Interwar Romania fuelled the growth and diffusion of the anti-scientific, anti-democratic, emic, fideistic phenomenological HoR, epitomized by Eliade's works (see Chapter 5, §*Eliade, 1920s–1980s*).

A recent commentary provides us with an interesting perspective on the contemporary relationship between the post-truth cultural environment and the HoR.

The article in question is a book review written by historian of religions Bryan S. Rennie and published in the *Los Angeles Review of Books* in October 2016. The subject of the commentary is the recent translation and publication of the partly autobiographical Romanian novel written by Mircea Eliade when he was just 17 years old, entitled *Diary of a Short-Sighted Adolescent*. However, the novel is not the central theme of Rennie's piece, as the publication provides a pretext for Rennie to offer his musings on the current state of the HoR. His starting point is that three factors have coincided to diminish the alleged importance of the HoR in contemporary academia:

1. a specific set of self-critical disciplinary trends;
2. a strong judgemental approach as to the political aspects of Eliade's works and his political engagement;
3. a general academic culture oriented towards 'hard' science.

In his review, in particular, Rennie focuses on the third point. He calls attention to the fact that

> the real villain of the piece here may be the ongoing and exaggerated valorization of 'science' in the Anglophone academy. Psychology and the social sciences have long coveted the glamour and standing (and grant-attracting prowess) of the 'hard' sciences, but lately even the study of religion hankers for 'scientific' status, with more and more resources being devoted to the cognitive science of religion.
>
> Rennie 2016

Then, Rennie cites the birth of a subfield in medical sciences called 'medical humanities', i.e. a complementary field devoted to researching meaning and experience in cultural media related to medical fields, and he traces the following parallel: 'if the arts and humanities are now deemed important in as practical and technical a field as the medical profession, how much longer can they be minimized, marginalized and neglected in such self-evidently humanistic realms as the study of religion?' (Rennie 2016). The end of the book review is a loud and proud claim of autonomy for the Eliadean HoR, accompanied by quite a disdainful account of scientific research and, perhaps, a bit of envy towards the resources of the cognitive science of religion (CSR): 'science will never formalize a methodology that could generate the sort of knowledge of the self that Eliade sought. The methodical and systematic procedures of science cannot replace the unsystematic and creative flashes of brilliance and insight that constitute revolutionary advances in understanding. These are produced by humanistic interpretation' (Rennie 2016).

It is not that science, as Rennie observes, will never be able to 'formalize' the methodology and the kind of research heralded or pursued by Eliade (whose 'revolutionary advances in understanding' are not specified), it is just that many researchers across the Humanities and social sciences have independently adopted it, tested it, and found it wanting. In epistemological terms, the Eliadean research programme has been falsified and discarded as a degenerating pseudoscience (Ambasciano 2018b). As Lakatos remarked, insisting on the academic endorsement and pursuit of such endeavour drains financial resources in vain (Lakatos 1989: 117). As to Rennie's rigid demarcation between science and the Humanities, with science depicted as a competitor to defeat in order to survive, the image itself is misguided at best, or a straw man at worst: science is neither a precise group of elitist, excluding disciplines, nor a unique set of methods (we will get to this point in a moment). Such antireductionism is a direct heritage of the Eliadean research programme, where reductionism appears to have been confused with *eliminativism*, as if an implicit goal of science would be the eradication of humanistic research (McCauley 2017: 1–24). To achieve the goal of delegitimizing the recourse to science within the Humanities, Rennie resorts to a distorted idea of what science is, exaggerating the most technical and fearful aspects of dehumanization in order to make its destruction acceptable to his readers. The example of the medical humanities should further prove that the Humanities are necessary even in the most arid of scientific fields. However, Rennie forgets to add that the recent implementation of the medical Humanities does not represent a revolutionary change of paradigm, as they are simply an interdisciplinary endeavour dedicated to the impact of medicine on artistic, social, historical and cultural media – no superpowered phenomenologist is expected to assist in the operating theatre (Jack 2015).

Such attacks against science are not rare in these post-truth days. It is true that a general reorientation of academia as a whole towards science and technology after the economic crisis of 2008 has been translated by narrow-minded politicians and administrators into the reckless and grievous cancellation and disappearance of several courses, or even entire chairs, from the Humanities in universities worldwide (markedly in the US), mainly because it is claimed that such knowledge does not contribute to the overall technological and financial growth of post-industrial societies (Pigliucci 2015a; see Flexner 1939 for the brainlessness of such an approach). And yet, this contingent and recent development does not explain the whole story. There has always been more than that behind the whole 'Humanities vs. science' confrontation. As cognitive literary scholar Jonathan Gottschall aptly remarked about the situation in contemporary literary studies,

> the very idea of bringing science – with its sleek machines, its cold statistics, its unlovely jargon – into Neverland makes many people nervous. Fictions, fantasies, dreams – these are, to the humanistic imagination, a kind of sacred preserve. They are the last bastion of magic. They are the one place where science cannot – should not – penetrate, reducing ancient mysteries to electro-chemical storms in the brain or the timeless warfare among selfish genes.
>
> Gottschall 2012: xv–xvi

Literary studies are not alone, and a case study could easily show that this anti-science feeling is widespread in humanistic academia. As I have already shown *ad abundantiam*, something similar characterizes the HoR and RS, where, as Luther H. Martin and Donald Wiebe have remarked, 'researchers systematically avoid critical studies and theoretically based explanations of their subject of study' (Martin and Wiebe 2016: 227). To rephrase this statement, there are indeed countless investigations on theories and methods in modern HoR and RS, but until very recently they have merely proposed sequences of variations on the same theme, that is, redescriptions of religious writings, actions, objects, etc., into an equally religion-friendly academic jargon (Smith 2004: 362–74). For instance, an 'Announcements for Sessions of 1960–1961' provided for and by the same Divinity School of Chicago University where Eliade had been working since the mid-1950s, stated quite apodictically that 'it is the contention of the discipline of History of Religions that a valid case can be made for *the interpretation of transcendence as transcendence*' (Smith 2004: 372; my emphasis). From such a perspective, religious transcendence, that is, everything that is supposed to go beyond ('transcend') everyday, normal, profane life, is considered untranslatable in other terms. As Smith has remarked, HoR turns into the ongoing paraphrase of religious documents not for the sake of understanding them in a scientific way, but, whether or not explicitly, to promote and appreciate religion and the sacred as something separate from reality (Smith 2004: 372). This 'comparative theology' (Gilhus 2014: 200) is at the same time strikingly different from and similar to Derrida's concern about the conceptual embeddedness and problematic application of *religio* (as a term with a history-laden and geographically precise meaning; see previous chapter), in that translatability is allowed, but only within the emic, fideistic, enclosures of theologically friendly HoR. As Rennie has summarized, any other kind of scientific approach (be it social or natural) which does not share the same tenets of the HoR in the academy is excluded, relegating the HoR to its untranslatable, self-referential universe of post-truth transcendence.

What is science, anyway?

The existence of an unbridgeable gap between the Humanities and the social sciences on the one hand, and scientific research on the other hand, constitutes one of the most widespread misconceptions in the study of religions of the past century. This misconception is divisible into three major sub-statements:

1. the *rigidly mathematical nature* of 'hard' or 'precise' sciences;
2. the *exclusion of history* from both 'hard' and 'soft' sciences;
3. the *nature of human culture* as not amenable to scientific analysis.

Let us analyse these ideas in detail, focusing in particular on historiography.

Point (1) is probably one of the most pernicious half-truths of all. Science is not just numbers and equations. And it is not a predetermined set of arid tools which sucks the life away from the Humanities. It is much more than that: it is a mindset that allows human beings to conduct reliable enquiries, to keep our innate biases in check and 'evaluat[e] interpretations of experiments and observations' (Blackburn 2017: 97), and to accumulate coherent and supported knowledge while discarding false ideas and theories (Pigliucci 2013). As such, before, behind and after any *quantitative* hypothesis-testing or computational analysis, there is a slow *qualitative* collection and evaluation of the available data and documents leading to an *explanatory narrative*, feeding into a loop of qualitative, critical evaluation (Leroi 2015: 365–6). Even if the overwhelming majority of historians forget about this detail, historiographical research, when conducted properly, is science. As American anthropologist Leslie White (1900–1975) recalled in the late 1930s, the assumptions that stand behind any idealized systematization of scientific versus nonscientific disciplines (e.g. physics vs. sociology or biology)

> are not only confusing; they are unwarranted. The basic assumptions and techniques which comprise the scientific way of interpreting reality are applicable equally to all of its phases, to the human-social, or cultural, as well as to the biological and the physical. This means that we must cease viewing science as an entity which is divisible into a number of qualitatively different parts.
>
> White 1938: 372

Science, White contended, is best viewed 'as a way of behaving, as a way of interpreting reality, rather than as an entity in itself, as a segment of that reality' (White 1938: 372).[2] Therefore, science is the process of doing science, i.e. 'sciencing'. I cannot delve much deeper into the philosophy of science, but I want to highlight at least three main tenets, starting from *consilience of inductions*, a label advanced by English philosopher of science William Whewell (1794–1866) in his volume *The Philosophy of the Inductive Sciences, Founded Upon Their History* (1840). Whewell's consilience points to the convergence of different, independent classes of proof which might be explained by a common pattern and, therefore, an epistemically warranted explanation (Blackburn 2016: 101). Another important methodological tenet – which is actually more a guideline than a proper law – is what has been called Occam's razor. Originally proposed by Franciscan friar William of Ockham (?1287–1347), this epistemological tool posits that, all else being equal, the most parsimonious explanation of a fact should be preferred among competing hypotheses, or, in other words, there is no need to multiply the factors which explain the fact without a sufficient warrant (Blackburn 2016: 338). Finally, historiography, like any other science, is founded on the progressive accumulation of data and explanations, a process which is empirically grounded on the falsification of previous hypotheses as well as on the continuous *trial-and-error approach* of discarding what does not comply with the advancement of knowledge. These basic tenets were held in great consideration by the most important representatives of the Victorian comparative science of religion, but were subsequently abandoned by their successors.

Point (2) is another classical misrepresentation of what science really is. Quite simply, as White concluded, '"history" is that way of sciencing in which events are dealt with in terms of their temporal relationships alone' (White 1938: 374). This apparently modest statement reveals the grandiose epistemological *fil rouge* that links together the study of the Big Bang, dinosaurs and ancient Mediterranean religions: *time*. The historical study of human cultures, ideas and behaviours is a subset of a larger group of historiographical disciplines, i.e. historiography, evolutionary biology, palaeontology, palaeoanthropology, historical geology, cosmology, cultural geography, epidemiology, linguistics, etc. All these disciplines share one key feature that differentiates them from other natural sciences, i.e. the impossibility to repeat any experiment. We cannot repeat the formation of the universe, we cannot rerun the chain of events that led once to the diffusion of ancient Roman cults in the Italic peninsula, and we are equally unable to replicate in the laboratory the contingent evolution of avian dinosaurs (more prosaically, birds), from their non-avian theropod dinosaurian ancestors. But, as astrophysicist Carl Sagan (1934–1996) remarked, 'in those historical sciences where you cannot arrange a rerun, you can examine related cases and begin to recognize their common components' (Sagan 1996: 242–3) thanks to the comparison of different classes of evidence and identification of common patterns and smoking guns. Historians have social and political documents, astrophysicists have astronomical observatories, and palaeontologists deal with fossils. All these historians might also engage in comparative analyses and computational simulations of various kinds (Diamond and Robinson 2010; Cleland and Brindell 2013).

Point (3), finally, ignores the advances in recent cross-disciplinary explorations. Human beings are primates and, as such, their behaviours can be approached as any other nonhuman animal behaviour, that is, by adopting an *ethological perspective*. In the end, *everything social is biological*, and vice versa (up to a certain point). This does not mean the reduction of cultures to sterile matrices of data (even if the collection of data remains a most important part of any scientific study). It means identifying and explaining the *ultimate evolutionary processes* that underlie such behaviours and the *proximate cognitive outputs* that lead to the enactment of sequences of actions. Even if there is nothing new in this approach, which in its current form dates back at least to the early days of modern comparative zoology, this idea is constantly met with outraged response and open dissent. One example might suffice. Evolutionary biologist Jared Diamond (1937–) published his volume *The World Until Yesterday: What Can We Learn from Traditional Societies?* in 2012. The book was a 'neutral catalogue' of cross-cultural explorations in ecological and social adaptations, mostly dedicated to the observation of the various cultural and religious systems of non-Western, non-Abrahamic societies as social and normative experiments of *Homo sapiens* (Pievani 2013b). *The World Until Yesterday* offered also a specific focus on what can be still useful in terms of survival, health and adaptability for contemporary WEIRD societies (i.e. Western, Educated, Industrialized, Rich and Democratic) and, conversely, what is bad in both kinds of societies (war, violence, bad traditional habits, etc.). Notwithstanding the specific focus of the book, aimed at scientifically investigating all human cultures (WEIRD ones included) as imperfect and perfectible systems of ecological and social relationships, Diamond's efforts were met with mixed to negative criticism from anthropological and humanistic quarters. In the wake of the postmodern criticism of science, the volume has been expectedly accused of Western imperialistic determinism and neo-colonialist racism. The editorial of an academic journal in political ecology was entitled, quite eloquently, *F**k Jared Diamond* (Correia 2013).

Anti-scientific labelling and name-calling apart, the naturalization of human cultures by contemporary cutting-edge inter-disciplinary academic research is going steady. This means that religion is increasingly seen as a part of human ethology (e.g. Bulbulia and Slingerland 2012). However, replicating the postmodern issue of already-addressed topics neglected due to disciplinary ignorance (cf. Spineto 2009: 43), most contemporary researchers tend to forget that many brilliant forerunners in the Humanities or the social sciences have always tried to advance outstanding scientific hypotheses with regard to culture and religion, even when science was ignored or despised by their colleagues (cf. Dennett 2006: 264). In what follows, I would like to present an incomplete and purely indicative selection of some of the most important researchers who dealt with comparative religion, HoR and RS during the past century and anticipated or contributed to the birth of the contemporary cognitive and evolutionary study of religion. As you might recall from the third chapter, the link between evolution and cognition is of paramount importance, and much of what you will find in the paragraphs below reflects this link.

Forgotten forerunners: Baldwin's evolutionary epistemology

Culture changes and develops through time. This much had already been gauged and proposed in myriad ways well before Darwin. However, even after Darwin the nature of culture's development through time and space, as well as the relationship between biological and cultural evolution, remained highly controversial to say the least (e.g. Fracchia and Lewontin 1999 in Lewontin and Levins 2007: 296; Kundt 2015). Innumerable proposals have tried to make sense of the differences that have long prevented an evolutionary synthesis within the social sciences. The main problem was constituted by the discrepancy between the mechanics of cultural development and natural evolution. And without any scientific understanding of how culture works, religion was destined to remain a *sui generis* dominion.

As we have already seen, Darwinian evolution is actually built on a five-fold understanding of the evolutionary process (see Chapter 3, §*The Origin of Species, 1859*). In addition to these conceptual pillars, Ernst Mayr listed several other equally important corollaries of Darwin's theory, including the role of 'effects of use and disuse' on the organism and on the inheritance of acquired characters (Mayr 1991: 35). Such specific corollary was proved wrong after the rediscovery of Mendelian genetics: the transmission of characters is based upon blind, often slow, genetic inheritance of units of information unaffected by life changes (i.e. non-Lamarckian) and later filtered by external factors (see Mesoudi 2011: 40). The original Darwinian theory of evolution has been continuously corrected and updated, maintaining almost untouched its original five-fold core, becoming one of the most resilient and fruitful research programmes ever conceived by humankind (Pievani 2011). Thanks to such ongoing consilient work, the chasm between micro-evolution (genetics) and macro-evolution (palaeontology) was slowly bridged in the mid-twentieth century with the so-called Neo-Darwinian synthesis (Raup and Jablonski 1986; Sepkoski and Ruse 2010). However, this new understanding of evolution made useless any transfer of conceptual and practical evolutionary tools to the study of culture(s). Genetics does not allow any 'differential adoption' and transmission of blended traits (Mesoudi 2011: 79), while cultural selection depends precisely on intentional, often fast, guided inheritance and blending of information arising by choice and ongoing *bricolage* (cf. Mesoudi 2011: 40–6). One way to resolve this conundrum is to acknowledge the difference between cultural and natural evolution within a naturalistic approach. Alex Mesoudi has recently proposed the adoption of the *original* Darwinian framework – effects of use and disuse included – to support theoretically cultural evolution while avoiding the cross-disciplinary hindrances of the Neo-Darwinian approach (Mesoudi 2011; S. J. Gould did something similar almost 10 years earlier; see Gould 2002: 277–9).

However, in 1896 US psychologist James M. Baldwin (1861–1934) had already proposed an evolutionary mechanism which bypassed these constraints, and went even further. The mechanism of 'organic evolution', presented by Baldwin as a 'new factor in evolution', and later called the *Baldwin Effect*, describes how the 'environmentally elicited individual phenotypic adaptations might come under genotypic control and hence [be] transmitted via inheritance to offspring' (Plotkin 2004: 77; Baldwin 1896). In other words, behavioural changes produced by learning during the growth stages of the life history of an organism (i.e. ontogeny) have a selective impact on individual habits (e.g. diet) which, in the long run, produce genetic adaptations. The Baldwin effect dispenses with simple Lamarckian explanations of use and disuse and provides an evolutionary and psychological explanation of how learning mechanisms help to develop cultures which, in turn, help to cope with the environment. Adaptive learning is the keyword here, as it contributes to actively imitating, selecting and adopting behaviours that may have a positive influence on one organism's fitness (that is, the chance to pass its genes to the next generation), and therefore leaving a substantial modified cultural niche to its progeny via 'social heredity', and so on.

What is a cultural niche? As far as human beings are involved, the cultural niche is the result of the peculiar human cognitive prowess. To paraphrase White's definition, it is folk-sciencing writ large: the ability to accumulate knowledge and modify natural and social environments, thereby changing the selective pressures thanks to problem-solving, cooperation, imaginative thinking and abstract reasoning (Pinker 2010; Boyd, Richerson and Henrich 2011; for niche construction theory, see Kendal *et al.* 2011). Religious ideas and behaviours might be pivotal elements for change insofar as the implementation of several technologies, cognitive skills and normative rules (e.g. writing, philological tools, numeracy, marriage rules, etc.) contribute to overcoming social and ecological problems or even modifying the societal selective pressures on individuals (e.g. Purzycki and Sosis 2013).

In 1909, Baldwin published *Darwin and the Humanities*, designed as a consilient interdisciplinary manifesto (Baldwin 1909). In it, Baldwin anticipated the most basic assumption behind the cultural niche and intuited its potential to bridge the gap between nature and culture. He developed what would be later called a fully fledged evolutionary epistemology, that is, a Darwinian selectionist account of how ideas are transmitted and adopted and how they impact on biology.[3] Evolutionary epistemology was thought by Baldwin to be the result of the interaction between two factors:

1. the ontogenetic phase of 'ejective consciousness', in which a child develops the cognitive skills which allow her to 'understand that others have subjective mental states too, and that such understanding is the basis for entering into a community of shared knowledge';
2. the inter-personal mental '"environment of thought" in which ideas are subjected to variation, are selected, and then transmitted and hence conserved' (Plotkin 2004: 76–7).

Within this framework, Baldwin re-contextualized religion as a most important environment of thought: religion provides 'tribal or national self-consciousness' and personal meaning wrapped in a social institution 'handed down by "social heredity"', and supported by symbolic and anthropomorphic thinking, problem-solving social imagination, kin recognition and inter-personal psychological projection (Baldwin 1909: 101, 95, 107, 98). Differences among and within religions were ascribed to 'the power of "variations" in moral and mental characters and products' (Baldwin 1909: 107). Being the result of the psychological 'process of self-consciousness' of each individual, even with the demise of religion human beings might fall *de novo* for other 'sublimated equivalents' in some form of renewed mysticism: 'the man who scoffs at a creed stands in awe before the mysteries of table-turning and spirit-rapping; and the sceptic in the matter of miracles, accepts faith cures, telepathic messages from the unseen world, second sight and other equally miraculous violations of the natural order' (Baldwin 1909: 106).

Forgotten forerunners: Harrison's evolutionary psychology

An evolutionary perspective of this kind represented the most epistemically warranted way to demolish fideistic and theological approaches while relocating the research focus from emically religious understanding to natural processes. In June 1909, the same year in which Baldwin's *Darwin and the Humanities* was published, Cambridge classicist Jane E. Harrison participated in the commemoration of the centenary of the birth of Charles Darwin and the 50th anniversary of the publication of the *Origin of Species*. Although held in a period of anti-Darwinian reaction, it was a spectacle to behold, for this 3-day interdisciplinary celebration was unprecedented in the annals of science in both scope and grandiosity; '235 scientists from 167 different countries and 68 British institutions' (Richmond 2006: 447).[4] In her contribution, Harrison recognized Darwin as the true founder of the 'scientific study of Religions', thanks to which the predominance of revealed theology, as 'a *doctrine*, a body of supposed truths', and the existence of a 'teleological scheme complete and unadulterated, which had been revealed to man once and for all by a highly anthropomorphic God', were finally challenged (Harrison 1909: 494–5). According to Harrison,

> psychology was henceforth to be based on 'the necessary acquirement of each mental capacity by gradation'. With these memorable words [by Darwin] the door closes on the old and opens on the new horizon. The mental focus henceforth is not on the maintaining or refuting of an orthodoxy but on the genesis and evolution of a capacity, not on perfection but on process. Continuous evolution leaves no gap for revelation sudden and complete. We have henceforth to ask, not when was religion revealed or what was the revelation, but how did religious phenomena arise and develop.
>
> Harrison 1909: 497–8; Harrison refers here to the 6th edition
> of the *Origin of Species*; Darwin 1872b: 428

Harrison's evolutionary psychology encapsulated nothing less than a powerful attack against degenerationism, i.e. the idea of the degenerescence and decadence from a primordial and divinely pure monotheism into polytheism. As we have already seen, this idea, originally advanced in its Victorian form by Lang, came to dominate the HoR until the 1950s (Harrison 1909: 498; Wheeler-Barclay 2010: 104–39). Harrison's goal was to dethrone anthropomorphism, with gods in the image of men, as the 'final achievement in religious thought', and make religious scholars recognize that, instead, 'anthropomorphism lies at the very beginning of our consciousness' (Harrison 1909: 508). As she writes, 'the "decadence" theory is dead and should be buried' (Harrison 1909: 498).

After a biographical sketch of Darwin's gradual abandonment of religious faith, Harrison focused on the 'religion of the primitive peoples', in which she included Graeco-Roman religions as well, to highlight that 'vague beliefs necessarily abound' while 'ritual is dominant and imperative' (Harrison 1909: 498). The 'primitive mind', as well as that of a child,[5] is compelled to understand the world not just by empirical, present observation, but also by imaginatively resorting to anthropomorphic thinking, to 'dreams, visions, hallucinations, nightmares', to emotional memories of the past, to problem-solving about the future. The understanding of such a 'supersensuous world' is further elaborated into a properly spiritual mind–body dualism, which brings about 'ghosts and sprites, ancestor worship, the soul, oracles, prophecy' (Harrison 1909: 499–502; a hint of obsessive–compulsive disorder in religious thinking and acting is also available on p. 502). Then, Harrison recognized that such a folk-theory of religious development would have soon fallen into oblivion through scientific trial and error, had it not been for ritual: 'but man has ritual as well as mythology; that is, he feels and acts as well as thinks; nay, he probably feels and acts long before he definitely thinks. This contradicts all our preconceived notions of theology', for worship precedes belief in gods (Harrison 1909: 503). Dance, ecstatic rapture, ritual imitation ('imitation begets custom, custom begets sanctity'; Harrison 1909: 509), emotion, magical 'pan-vitalism' (Harrison 1909: 506), the emotional experience of agency, power and will; all concur to substantiate religion more than any belief. Summarizing the psychosocial features of ritual, Harrison stressed that 'imitation, repetition, uniformity and social collectivity have been found by the experience of all time to have a twofold influence – they inhibit the intellect, they stimulate and suggest emotion, ecstasy, trance' (Harrison 1909: 510, based on Beck 1904).

Forgotten forerunners: Macalister's invention of tradition

What is the ultimate evolutionary explanation, if any, for such cognitive depletion, euphoric arousal and prosociality occurring during religious ritual? Unfortunately, Harrison did not provide any answer. Baldwin, instead, concluded his chapter on religion included in *Darwin and the Humanities* by stating that 'religion is both a moral satisfaction and a social weapon', whose ultimate cultural role was to provide the cognitive scaffolding for implementing normative decisions to strengthen in-group cohesion (Baldwin 1909: 108). Indeed, according to what Darwin himself suggested, and later discussed in evolutionary biology and genetics, selection might be exerted on different, interacting levels, i.e. genes, cells, individuals, family, groups, providing a multilevel environment for selective forces to act upon. Group selection, in particular, is a much discussed theory that sees evolution as acting at the levels of groups instead of that of individuals, with extended cooperation and altruistic behaviour among the members of the group as the driving force to outcompete other groups (Pievani and Parravicini 2016). This most basic Darwinian approach was one of the earliest to inform the scientific study of religion.[6]

An interesting study of in-group cohesion was already advanced in 1882 by Cambridge professor of Anatomy Alexander Macalister (1844–1919), during the Inaugural Meeting of the Dublin Presbyterian Association. Published as *Evolution in Church History*, Macalister's contribution was a historico-philological analysis of the modification through time of the social composition of the 'office-bearers' of the Presbyterian Church, i.e. the presbyters or elders. In an anti-fundamentalist tone, Macalister rejected the Presbyterian claim of historical continuity and preservation of the Presbyterian polity since the primitive church, disassembled Biblical inerrancy, and acknowledged human interventions in religious matters (Macalister 1882: 7–9). Then, he defined evolution as 'a capacity of variation in the train of sequences and external modifying influences, and the latter may be either the direct action of the environments on the phenomena, or may be due to a power from without, overruling and directly ordering the modifications'. In cultural matters, this 'power from without' was human action (Macalister 1882: 10), which, insofar as socio-political environments were changing through time, devised new ways to adapt successfully to them. Therefore, Macalister identified the cultural forces within and without the community of believers, from ancient times to the Reformation, which shaped the creation, divergence, selection and modification of the Presbyterian polity (e.g. the need to institutionalize a 'moral police force' to assist a limited number of pastors during the Reformation, the rapid diffusion of Protestant churches, etc.). Most importantly, and quite remarkably from a cognitive perspective, Macalister highlighted the historical appeal to authority and tradition to justify the invention and the continuous modification of such an institutional office (Macalister 1882: 28, 30–1).

Disappearing without a trace

The three scholars that I have chosen (following a suggestion in L. H. Martin 2014: 163–5) represent only a tiny fraction of possible cases, but no matter how big the sample, whatever the outcome and the brilliance, their scientific hypotheses were soon forgotten. Baldwin's, Harrison's and Macalister's ideas were to leave almost no trace in the HoR. Baldwin fell in disgrace after being found in a brothel in 1908, and 'what was thought to be even more scandalous [...] was that the prostitutes were black' (Plotkin 2004: 88). Notwithstanding the fact that charges against him were eventually dropped, he was forced to resign from Johns Hopkins University, and relocated in France (Richards 1987: 451–502; Wozniak and Santiago-Blay 2013). Baldwin's demise was a setback for the scientific study of culture, and his ideas are just being currently rediscovered (Morgan and Harris 2015; independent theories that described the same mechanism behind the Baldwin Effect had almost no impact on the study of culture). He anticipated many of the current pillars of both CSR and ESR, mainly the elaboration of both the theory of mind (or ToM), which is the innate ability to recognize what other individuals think and what their desires and beliefs behind their behaviours might be, and the epidemiology of cultural representations, which explains the cognitive mechanisms in charge of the selection, retention and transmission of ideas (Plotkin 2004: 77), while providing a unified Darwinian frame. As far as the scientific study of religion is concerned, Baldwin argued for a 'comparison and correlation of the results reached by' the 'anthropo-genetic study of religion', i.e. the comparative and anthropological history of religion(s), and the 'psycho-genetic study', that is, psychological aspects of religion (Baldwin 1909: 90–1). This consilient approach has been reinvigorated only recently (e.g. Burman 2014).

Harrison's evolutionary psychology and her focus on cognition were destined to oblivion too – although her situation is more complex. As seen above, Harrison's evolutionary foray exhibited some of the spiritualist biases that were beginning to characterize the field at that time (Wheeler-Barclay 2010: 235–42). Harrison ended her contribution by citing a complacent passage on magic from esotericist Eliphas Lévi (Harrison 1909: 510) and saving 'mystical apprehension' from criticism, which she thought might reveal the existence of something true, thus crucially undermining her plea for scientific clarity and her lucid evolutionary rebuttal of degenerationism (Harrison 1909: 510–11). Harrison was aware of her self-sabotage: 'I am deeply conscious that what I say here is a merely personal opinion or sentiment, *unsupported and perhaps unsupportable by reason, and very possibly quite worthless*, but for fear of misunderstanding I prefer to state it' (Harrison 1909: 511 n. 2; my emphasis). Harrison also 'accept[ed] uncritically the assumption that bedevilled ritual studies, that what the anthropologists learned about savages was indicative of the early development of human beings everywhere at all times' (Robinson 2002: 209). Finally, a misguided focus on the differences between 'primitive' and modern cultures drew scholarly attention away from the outstanding similarities among religious rites and ethological rituals exhibited by nonhuman animals – a Darwinian point which would only be resurrected more than 80 years later. Admittedly, this mix makes any overall judgement of Harrison's scholarly legacy highly problematic, although, in the end, her penchant for pseudoscience did more harm than good to the comparative science of religion (cf. Wheeler-Barclay 2010). In any case, by highlighting the 'psychosocial role' of ritual in an evolutionary framework, Harrison's work was in line with coeval cutting-edge cross-disciplinary research (Wheeler-Barclay 2010: 230–2).

As to Macalister's ideas, his adaptationist cultural in-group cooperation had no impact on the HoR whatsoever. Again, his ideas have been independently replicated only in recent times (e.g. Wilson 2002).

The Dark Ages: Psychoanalysis, behaviourism and cultural anthropology

During the first decades of the twentieth century, the newly born HoR and the social sciences in general turned away from the Darwinian relation between evolution and cognition, and it became clear that there was no immediate future for the scientific study of religion. In the same decades that followed Baldwin's demise, notwithstanding clear evidence proving the contrary, the ignorant confusion of Darwinian evolution with selective breeding, inhumane eugenics or racial genocide led to the thorough 'de-biologization' of anthropology and psychology (Plotkin 2004: 68–9). As we have seen, most comparative science of religion and phenomenological HoR kept on pursuing the equally fallacious orthogenetic development of human beings towards contemporary Western religions or cultures and other non-Darwinian evolutionary processes.

Sigmund Freud's and Carl G. Jung's schools of psychoanalysis had a remarkable influence on the historical study of religions and on historiography as well, where it generated the psychohistorical offshoot (resp., Merkur 1996; Weinstein 1995). Freud, in particular, was regarded as a towering giant and, notwithstanding his unscrupulous use of others' theses and manipulation of his results (Orbecchi 2015), 'no psychological theory has had a greater cultural impact than that of Freud', upon which Jung's theories rested (Plotkin 2004: 33). As to the scientific underpinnings of the theory, Freud and Jung were both educated and well-read in German evolutionary biology (Sulloway 1992; Noll 1994; Orbecchi 2015), and they both assumed a specific view of the mechanisms of the human unconscious 'in which instincts and developmental experience figured large' (Plotkin 2004: 33). Both Freud and Jung adopted a Lamarckian approach, in that behaviours were thought to become fixated in memory during one's lifetime and passed on to future generations, and embraced recapitulation theory, a theory first proposed by German evolutionary biologist Ernst Haeckel (1834–1919), according to which individual life history (i.e. ontogeny) recapitulates the evolution of the species (i.e. phylogeny). The result was a set of theories that developed into outright pseudoscience and that differed in their aversion or sympathy towards religion (resp., Freud's and Jung's; see Ambasciano 2014: 98–104, 456–7).

Although Freud is credited with having started a revolution in human understanding, the scientific underpinnings of all his theses were lacking. In an article entitled *A Phylogenetic Fantasy*, written in 1915 and published posthumously in 1987, Freud tried to develop a fully fledged evolutionary chain of psychopathological disorders in which psychological disorders (or 'neuroses') 'bear witness to the history of the mental development of mankind' (Freud 1987: 11). Interestingly, even if unpublished, this evolutionary speculation dominates the Freudian psychoanalytic production (cf. Gould 2001: 147–58). Inspired by a close correspondence with Hungarian psychoanalyst Sándor Ferenczi (1873–1933), who 'viewed the full sequence of a human life [...] as a recapitulation of the gigantic tableau of our entire evolutionary past' (Gould 2001: 151), Freud established a developmental sequence of psychosexual neuroses in 'six successive stages', with each neurosis linked to a precise evolutionary stage in human evolution (see Gould 2001: 152–3). In brief, the direst environmental conditions and social problems of the prehistoric Ice Age, as imagined by Freud, impressed the archaic human mind and became engrained in the human psyche as psychopathologies, with repressed sexual urges and desires leading to guilt, vengeance and shameful repair in the forms of societal organization and religious rituals – but also resurfacing occasionally as individual psychopathologies. Freud's reconstructed narrative included a prehistoric family organization dominated by a tyrannical father who controlled resources, limited sexual activity and exerted coercive control over his sons (Freud 1987). Freud's 'fatally and falsely Eurocentric' speculations were unsupported, and the most relevant shortcoming was that 'human evolution was not shaped near the ice sheets of Northern Europe' (Gould 2001: 154).

Jung started his career focusing on biological approaches; later, he embraced a markedly spiritual and occultist stance. He posited the existence of a 'collective unconscious', that is, a racial repository of symbolic, instinctual knowledge accumulated from time immemorial during the proto-history of ethnic groups and which expressed itself in the unconscious of each person. It was a sort of quasi-eternal, and almost divine, wisdom experienced by modern human beings via dreams, archetypes and visions, and re-accessed thanks to a 'transcendent function' aimed at interpreting personal events in the light of the history of mankind, thus overcoming the complementary oppositions of phylogeny (unconscious) and ontogeny (consciousness) (Noll 1994: 218–46). In order to disentangle the messages from the collective unconscious, a comparative historico-religious analysis of myth was to complement the psychoanalytical session: despite the superficial historical differences among cultures, deep similarity across worldwide mythologies was explained by 'inherited mental preferences and images [. . .] deeply and innately embedded in the evolutionary construction of the human brain' (Gould 2001: 277). However, Jungian psychoanalysis slowly morphed into a phenomenological history of religions *sub specie psychologica* – which might appear as complementary to Eliadean psychoanalysis *sub specie theologica* (Filoramo and Prandi 1997: 199–200; see Ambasciano 2014: 457). Indeed, Jung and Eliade initially developed their ideas concurrently and independently, but they later adopted and shared more or less the same framework, both with mystically and spiritually explicit undertones (Spineto 2006: 68).

A peculiar version of psychoanalysis was mixed with other literary sources (e.g. James Joyce) by US scholar of comparative mythology Joseph Campbell (1904–1987) to create the hyper-reductionistic 'mono-myth', a narrative magnification of a basic three-part structure: separation, initiation and return'. According to Campbell, the mono-myth explained every major religious and mythological figure's adventures (the so-called 'hero's journey'; MacWilliams 2005: 1379). Campbell's thesis resonated in the Humanities and the HoR, and rose to great fame notwithstanding its racial, ideological underpinnings, epistemic untenability and methodological uselessness (MacWilliams 2005).

While the HoR and the Humanities were toying with psychoanalysis, the proverbial nail in the coffin of scientific approaches to human cultures and religions came from the diffusion and success of *behaviourism*. Broadly, behaviourism was a psychological school that posited the blank slate of the human mind at birth (in Latin, *tabula rasa*), and that laws of conditioning are the sole engine of human behaviours. US psychologist John B. Watson (1878–1958) published what is generally regarded as the behaviourist manifesto in 1913, in which he excluded cognition and evolution from the study of psychology, focusing instead on the study of responses from peripheral organs:

> psychology as the behaviorist views it is a purely objective experimental branch of natural science. Its theoretical goal is the prediction and control of behavior.

Introspection forms no essential part of its methods, nor is the scientific value of its data dependent upon the readiness with which they lend themselves to interpretation in terms of consciousness. The behaviorist, in his efforts to get a unitary scheme of animal response, recognizes no dividing line between man and brute. The behavior of man, with all of its refinement and complexity, forms only a part of the behaviorist's total scheme of investigation. Similarly, in the study of language, the main paradigm held that external stimuli coming from the surrounding environment were sufficient to prompt an infant to learn to speak.

Watson 1913: 158; from Plotkin 2004: 58

As paradoxical as it may sound, Watson's was a psychology without the brain, without 'ideas, beliefs, desires and feelings', a science based on reflexes and muscular responses due to conditioning and reinforcement (Pinker 2002: 19; Plotkin 2004: 62). Meanwhile, US cultural anthropology converged with this anti-scientific stance by assuming that social stimuli, and culture writ large, are more powerful than biological inheritance (Plotkin 2004: 62–9). This was the main legacy that German-born anthropologist Franz Boas (1858–1942) brought from Germany into the US, having studied under anti-Darwinian or Lamarckian mentors (Plotkin 2004: 65). Boas came from a liberal family which, in the wake of the Revolution of 1848, 'had broken through the shackles of dogma'; he studied mathematics and physics, and received a PhD in physics, while cultivating his interest in geography (Boas 1938). With such a scientific career, Boas initially saw the need for a close study of biology and social anthropology as interacting systems; however, with the alarming rise of eugenics and racial policy, he reacted by adopting a more pronounced form of culturalism. As recapped by Henry Plotkin, in Boas' view 'culture is extragenetic, works identically in all social groups, results in differences in different social groups because of differences in history, and as a force in humans is far, far greater than that of biology' (Plotkin 2004: 68). However, Boas maintained a scientific point of view, albeit sceptical ('I claim that, *unless the contrary can be proved*, we must assume that all complex activities are socially determined, not hereditary'; Boas 1916: 473; my emphasis). His students and successors, instead, moved away from science altogether, bringing to a close any interdisciplinary collaboration (Pinker 2002: 23; see Sharpe 1986: 187–90). Although with laudable and honourable anti-racist intentions, cultural anthropology was cut off from science and set to propagate the historiographically unwarranted and most pernicious idea that Darwinian evolution was a racist tool for domination (Pinker 2002). Eventually, genetics and evolutionary constraints were thought not applicable to human beings, detached from natural constraints (cf. Bloch and Sperber 2002: 725).

These were the most important contributions of psychology and anthropology of the past century, which helped pave the way for postmodern and anti-scientific interpretations. It is no surprise, therefore, that, during the mid-twentieth century, evolution and cognition in the academic study of religion(s) were either misunderstood in a non-Darwinian way and imbued with spiritualism (Jungian psychoanalysis) or rejected in favour of an extreme form of empiricism (behaviourism) or culturalism (cultural anthropology). This schizophrenic attitude led to the HoR mainly holding to the first option, while entire fields of more historically oriented study of ancient religions embraced the other options, even rejecting the notion of 'belief' (e.g. Graeco-Roman history; an exhaustive bibliographical discussion is available in Mackey 2009).

Modern trail-blazers, 1950s–1990s: The slow Renaissance of science

The 1950s showed some signs of slow, scientific resurgence for the Humanities and social sciences, but the shared consensus of researchers highlighted in the previous paragraph remained largely unscathed. Isolated scholars swam against the anti-scientific tide. For instance, in the early 1950s, anthropologist Leslie White, expanding on the feedback loop process between nature and culture, defined culture as 'an extrasomatic mechanism [that is, a process that takes place out of the body] employed by a particular animal species in order to make its life secure and continuous' (White 1952: 8; from Purzycki and Sosis 2013: 99). Following this inter-disciplinary pattern, in 1962 archaeologist Lewis Binford (1930–2011) promoted a cross-disciplinary integration between anthropology, archaeology and biology based on White's earlier definition of culture, remarking once more that *all* technology (ideological justifications included) could be studied as a cultural subset of extrasomatic culture 'which function[s] to adapt the human organism, conceived generally, to its total environment both physical and social' (Binford 1962: 218; cf. Purzycki and Sosis 2013: 99).[7] Still, behaviourism remained a major hindrance. Three years earlier, linguist Noam Chomsky (1928–) focused on the acquisition of language, attacked frontally behaviourism and similar theories

> which favoured environmental causes and their ontogenetic effects on newborns and infants [, and advanced] a nativist viewpoint: he posited that external stimuli were not sufficient to account for the acquisition of language, whose mastery and richness by infants rapidly exceed any possible received environmental information (Chomsky 1959). Thus, Chomsky hypothesized the existence of a Universal Grammar (UG), i.e. the innate knowledge of language as an independent mental module: cultural differences may shape the differential traits, but language domain and its complex rules in their entirety (grammar, syntax, etc.) are something that does not need to be taught to an infant.
>
> Ambasciano 2016a: 168

Although initially based on rigid anti-evolutionary positions, Chomsky's thesis was later reformulated to become a seminal theory in the newborn cognitive sciences (Ambasciano 2016a).

In the following years, steady advances in cybernetics fuelled a cross-disciplinary integration and an epistemological consilience between Artificial Intelligence, psycholinguistics, neuroscience and philosophy of mind. This fruitful collaboration resulted in a new scientific paradigm for cognitive sciences (Bechtel, Abrahamsen and Graham 2001). Mental mechanisms that process information in the brain slowly came back as the focus of scientific investigation. Two examples relevant to the study of religion and culture might suffice here (cf. Stausberg 2009 for a more comprehensive panorama). In 1975 French social and cognitive scholar Dan Sperber (1942–) proposed a critical re-evaluation of *semiotics*, that is, the study of symbols and other signs as vehicles of meaningful communication, anchoring mental mechanisms to cognitive psychology and thus moving away from the blank-slate paradigm and from the non-explanatory cultural reinterpretation of symbolism (Sperber 1975). Twenty years later, Sperber's book *Explaining Culture* collected articles published between the 1980s and the 1990s and offered a comprehensive scientific framework to study both human universals and the diffusion of ideas which proved highly influential: in brief, the transmission of ideas entails (a) the imperfect copy of the content, due to specific cognitive constraints, and (b) the shift of such content towards certain intuitive and stabilizing attractors (Sperber 1996). Sperber's thesis also built on massive modularity, i.e. the idea that cognitive processing of information is due to a system of many specialized, encapsulated, domain-specific computational devices, which in turn built on further elaborations of Chomsky's UG (Barrett and Kurzban 2006: 631; cf. Ambasciano 2016a: 168–9; for a wider methodological description, see Sperber 2011).

In 1980, US anthropologist Stewart E. Guthrie (1941–) published a thought-provoking article entitled 'A Cognitive Theory of Religion' in which he proposed anthropomorphism, that is, the intuitive, inborn identification of human-like agentive intentions and actions in nonhuman environments, as the major trigger behind human reasoning in general, and religious thinking in particular, thus reviving Tylor's animism as an evolutionary strategy evolved to deal with potential threats and updating Frazer's repository of cross-cultural examples (Guthrie 1980). According to Guthrie, '"Religion", then, means applying models to the nonhuman world in whole or in part that credit it with a capacity for language (as do prayer and other linguistic, including some "ritual", action) and for associated symbolic action (as do, e.g. sacrifice for rain and other "rituals")' (Guthrie 1980: 189). With Guthrie's work the distinction between religious and nonreligious behaviour (such as the one posited by the very hypothesis of the *homo religiosus*) becomes cognitively useless: a better definition for our taxon would be *homo semioticus*, because we are bound by 'nature and nurture to interpret and influence the world through language, [and thus] we search for signs, symbols and meanings everywhere' (Guthrie 1993: 198). Guthrie's theory, later expanded in a dedicated monograph (Guthrie 1993), provided the foundational stone for the cognitive science of religion (CSR), with religion intended as an aggregated system of many different components (i.e. social, cognitive, institutional, etc.) put together in distinctive ways in particular cultural and historical settings (cf. Whitehouse 2013a).

In the meantime, specific historico-religious attempts at scientific theorizing remained extremely rare while other approaches flourished (cf. Stausberg 2009: 8–9). Among the few forerunners, and apart from the names that I have already recalled in the previous chapter (e.g. J. Z. Smith's plea for numerical taxonomy dating from the late 1970s), at least the following scholars might be recalled here:

1. Dutch indologist Frits Staal (1930–2012) noticed the theoretical development of a proto-Chomskian UG in ancient India (thus highlighting that science and critical

thinking can flourish whenever and wherever favourable conditions arise), proposed the ideas that mantras used in religious rituals preceded the evolution of proper human languages and that ritual has no rational meaning *prima facie* but should be understood as the biological result of behavioural patterns (Staal 1979; Staal 1988; Staal 1989).

2. US historian of religions William E. Paden (1939–) started in the late 1980s to update some of the classic notions of the Chicago School of the HoR as cultural and cognitive niche construction and imaginary world-making of *H. sapiens* in a way that closely resembled the proposals by White and Binford recalled above (i.e. religion as an evolutionary means of cultural adaptation and an ethological way of organizing human life; see Paden 2016; Ambasciano 2018a).

3. Swiss historian Walter Burkert (1931–2015) began to investigate the similarities between religious rituals and ethological behaviours from a sociobiological perspective in the mid–1990s. In particular, he focused his research on the cultural elaboration of an inborn primate heritage directed at managing social hierarchical relationships between dominant individuals and subordinates (Burkert 1996) – an approach independently adopted earlier by Culianu and recently by Hector A. Garcia (2015).

4. Around the same period, US historian of religions Luther H. Martin (1937–) started showing that evolutionary theory and the cognitive sciences might be successfully combined with the sociological devices of poststructuralism (in particular, Foucault's) to provide a more comprehensive frame for the identification of power dynamics and fictive kinship within religious groups (Martin 1997; L. H. Martin 2014; Martin and Wiebe 2016).

All the aforementioned scholars outgrew the antireductionistic *diktat* of the HoR and moved away from the phenomenological focus on belief, and thus remained more or less isolated in their field. In any case, a common framework to unify all the Humanistic and social scholars interested in an evolutionary and cognitive approach was still lacking. It was only in the early 1990s that a ground-breaking chapter entitled 'The Psychological Foundations of Culture', and published by evolutionary psychologists John Tooby and Leda Cosmides, gave momentum to the various attempts that tried to read cultures and religions through the lenses of evolution. In that contribution, the two authors criticized the blank slate paradigm of the social sciences, which they labelled the Standard Social Science Model (SSSM), and they wrote that

culture is the manufactured product of evolved psychological mechanisms situated in individuals living in groups. Culture and human social behavior is complexly variable, but not because the human mind is a social product, a blank slate, or an externally programmed general-purpose computer, lacking a richly defined evolved structure. Instead, human culture and social behavior is richly variable because it is generated by an incredibly intricate, contingent set of functional programs that use and process information from the world, including information that is provided both intentionally and unintentionally by other human beings.

Tooby and Cosmides 1992: 24; from Geertz 2015: 389

Psychology slowly re-entered the social sciences via biology, and evolution was again the main engine behind evolved cognitive abilities. Eventually, learning was decoupled from hard behaviourism, and deconstructed into a series of cognitive mechanisms: 'under closer inspection, "learning" is turning out to be a diverse set of processes caused by a series of incredibly intricate, functionally organized cognitive adaptations, implemented in neurobiological machinery' (Tooby and Cosmides 1992: 123). By challenging directly the main assumptions of the behaviourist research programme, some of the liveliest actors on the humanistic stage were ready to welcome again new and updated scientific approaches.

Cognition, 2000s: Back to a natural history of religion

The interdisciplinary and scientific study of religion experienced a remarkable increase in both quantity and quality of research between the late 1990s and the early 2000s, which led to the official birth of the CSR as an academic field in 2000, the foundation of the International Association for the Cognitive Science of Religion (IACSR) in 2006, and the birth of dedicated academic journals (e.g. *Journal of Cognition and Culture, Religion, Brain, & Behavior*, and *Journal for the Cognitive Science of Religion*, which is the official journal of the IACSR; see Barrett 2000; Lawson 2000; Lawson 2004; a comprehensive selection of seminal articles is available in Slone 2006; cf. also Tremlin 2006). Today, the CSR is 'actively pursued' in the following research centres: Aarhus University (Denmark); Masaryk University (Brno, Czech Republic); Oxford University; Queen's University (Belfast); the University of British Columbia (Canada); plus, a dedicated chair has been established at California State University, Northridge (Martin and Wiebe 2017: 3).

Scholars involved in the development of the first CSR paradigm, irritated and discouraged by the methodological and epistemological malaise caused by the dominant HoR paradigm (cf. L. H. Martin 2014; McCauley 2017), took an empirical and inter-disciplinary U-turn and headed back to a reductionistic natural history of religion. It is impossible to provide in such a limited space a comprehensive account of the recent history of the CSR; given the high degree of collaboration and inter-disciplinarity, a scholar-based list of advances falls beyond the scope of this brief account (cf. Geertz 2004; Jensen 2014; Geertz 2015). The following is a tentative historical list of the most important tenets from the first wave of CSR investigations (Ambasciano 2017b), obviously without claiming to be exhaustive (overviews are available in Pyysiäinen and Anttonen 2002; Bulbulia *et al.* 2008; cf. Figure 15):

0. *Ground zero*: there is no cognitive justification for a *sui generis* religion. The same mechanisms operate in both religious and non-religious reasoning (Boyer 2001; Barrett 2004); likewise, there is no neuroscientific evidence for a 'God spot' in the brain, and no epistemic warrant for spiritual, fideistic claims (Geertz 2009).
1. *Human ancestry*: *H. sapiens* is a social primate, and its cognitive machinery reflects the constraints of such deep history (Guthrie 1993; Boyer 2001; Atran 2002). The most basic features of human cognition represent an evolutionary, 'good-enough'

trade-off that bartered quick, almost automatic information-processing heuristics to deal with socioecological challenges for firing imprecision. The evolutionary impact of social interactions on cognition is discernible in the following building blocks of cognition:

a. *agency detection* and *mind-reading* (i.e. theory of mind, or ToM), selected for by predator avoidance and social interaction;

b. cognitive biases such as *group conformity* and *prestige bias* represent another legacy of evolutionary social constraints;

c. *causal cognition, imagination* and *conceptual blending,* exploited to cope with problem-solving, might also be affected by intuitive anthropocentric biases such as *teleological reasoning* to infer functionality and *anthropomorphism* to detect purposes and intentions (e.g. Fauconnier 2001);

d. *logic* and *argumentative reasoning* might have evolved to justify one's choices and convince others, thus pointing to its social function (Bloch 2008; Trivers 2011; Mercier and Sperber 2011);

e. intuitive *mind–body dualism,* a folk cognitive universal variously reinterpreted in human cultures which might be an exaptation (i.e. a non-adaptive trait co-opted for a new use) from computational misfiring in social cognition (Bering 2006; cf. Gould 2002, Chapter 11).

2. *Storytelling*: panhuman addiction to storytelling might be the result of evolutionary pressures that have selected for a 'universal grammar in world fiction' (Gottschall 2012: 55), whose themes are ultimately rooted in social cooperation (cf. Ambasciano 2015b). As to contents and diffusion:

a. cross-temporal and cross-cultural similarities in human narrative, whether or not religious, are the result of both 'logical limits to the structure of the stories' and mental preferences for a specific set of contents (cf. Gould 2002: 277);

b. inborn mechanisms by which human beings access and compute information reduce religious variety in both rituals and beliefs to an evolved, psychological grid of five ontological domains (i.e. person, animal, plant, natural object, artefact) and three categories of folk knowledge (i.e. physics, biology, psychology). Minimal breach or violation in the resulting fifteen slots enhances the appeal and success of the final representations (e.g. telepathic mind-reading, talking snake, unconsumed burning bush, crying statue, etc.), making them optimally attention-grabbing, memorable, and prone to diffusion (Boyer 2000; Boyer and Ramble 2001);

c. religious and mythological content is a particularly successful subset of cultural representations. Minimally counterintuitive superhuman agents, ancestors, spirits, and mythical characters happen to tick the right boxes that elicit and support both social cognition and personal meaning (e.g. Pyysiäinen 2009). In the wake of point (0), the same methodology has been fruitfully applied to the successful spread and appeal of urban legends (Eriksson and Coultas 2014), superhero comics (Carney and Mac Carron 2017), and horror stories (Clasen 2017).

3. *Rituals*: human ritual behaviours constitute the culturally elaborated endpoints of evolved, ancient, panhuman universals primarily aimed at boosting prosociality and in-group cooperation. Three main and slightly overlapping theories tackle the development of ritual structures:

   a. *theory of ritual competence*: cross-temporal and cross-cultural similarities in human rituals are explained by the presence of an intuitive, fixed UG behind ritual that presides over action enactment and accounts for a basic systemic interaction between agent, patient and tools (Lawson and McCauley 1990);

   b. *ritual form hypothesis*: the culturally postulated presence of superhuman agents during the ritual is realized via the cognitive occupation of one of the three UG slots (i.e. agent, patient, instrument). Which slot is occupied impacts on the implementation and structure of the ritual itself. If the superhuman entity is an agent, whose immediate presence is maybe perceived through altered states of consciousness, then the ritual exhibits high levels of sensory pageantry and its effects are usually considered irreversible (e.g. baptism, initiation). Conversely, if the superhuman entity is present through a special patient (a representative) or a special tool, then the ritual is emotionally less captivating, routinely repeated and its effects are considered temporary (e.g. Catholic Mass, sacrifices; McCauley and Lawson 2002);

   c. *theory of the modes of religiosity*: socio-political ritual organization differs on the basis of the evoked mnemonic system: (c.1) *episodic* and *flashbulb memory* is stimulated by emotionally arousing rituals which are celebrated seldom, in turn eliciting a spontaneous meaning-making process (i.e. exegetical reflection) and strengthening group cohesion, typically in the absence of fixed orthodoxy and a hierarchical leadership; (c.2) *semantic* and *procedural memory* supports repetitive rituals which take place habitually on the basis of a fixed set of dogmas and thanks to the continuous supervision of a hierarchical priesthood. Mode (c.1), typical of small communities, is called *imagistic*; mode (c.2), which characterizes ultrasocial groups of potential strangers, has been labelled *doctrinal*, although mixed variants and combinations are frequent (Whitehouse 2000; Whitehouse 2002).

One of the major problems faced by the first CSR scholars was to reconcile the presence of superhuman and superpowered agents deprived of religious worship in every human culture (such as comic-book characters like Mickey Mouse, or folk figures like Santa Claus) with the existence of religious demons, spirits, ancestors, gods, goddesses, deities, etc., as subjects of devotion. The solution to this issue, the so-called 'Mickey Mouse problem' (Atran 2002: 13), came only with the combination of different explanations:

1. 'culturally postulated superhuman agents' (Spiro 1971: 96), whether or not inspired by historical figures and resulting from the local minimally counterintuitive elaboration of anthropomorphic intuitions, are constructed so as to possess full access to strategic information relevant for the individuals and the community (Boyer 2001);

2. such agents are reputedly able to generate further commitment through a historical chain of traditional displays (Henrich 2009);

3. the cultural transmission of these imagined agents across generations is guaranteed by vertical indoctrination (e.g. Sunday schools, catechism) or, where formal religious teaching is missing, horizontal-oblique imitation (Cavalli-Sforza and Feldman 1981; Dennett 2006: 326). Such mechanisms piggyback affective attachment and reinforce personal investment in the belief;

4. socio-political institutionalization promotes the alleged antiquity or prestige of the cult itself, a process which in turn contributes to endowing with respect and admiration the political hierarchy, resulting in an intergenerational strengthening of belief-statements (cf. Paden 2016; Ambasciano 2016d);

5. 'belief in belief' in those imagined superhuman agents does pay off socially and politically (Dennett 2006; Roubekas 2015), closing a feedback loop that feeds into point (1).

Indeed, Mickey Mouse was not conceived by a virgin, he does not know your sins, there is no cult devoted to him, certainly no orthodoxy to learn in a catechism-like environment and, as far as I know, no religious institution has ever been founded on his cult (Jediism would be another cup of tea).

Cognition, 2010s: More than meets the eye

Although with some concessions to later research for clarity, the previous paragraph has recapped the main points of the first wave of CSR. CSR 1.0 advanced some detailed meta-theoretical and epistemological reflections that laid the foundations for subsequent in-depth explorations. Four main contemporary branches might be further identified (it goes without saying that all the branches interact in such a way that in some cases a clear-cut demarcation is very difficult to trace; Figure 15):

1. *CSR 2.0*: the most recent trends in the field have privileged a decidedly experimental outlook, more focused on practical collaboration between neurosciences and anthropology, in order to test quantitatively, *in vivo* (in the field), *in silico* (via computer modelling and simulation), or in the lab (via neurocognitive imaging), theories advanced by the first-wave scholars. One of the goals of the CSR 2.0 is to 'code the data and perform statistical analyses of the resulting database to determine which hypotheses most parsimoniously explain the empirical patterns' (Geertz 2015: 393; cf. Xygalatas 2013; see Ambasciano 2017b).

2. *Evolutionary science of religion* (ESR): an area of enquiry closely related to the most recent and updated version of cultural evolution (an international organization, Cultural Evolution Society, was founded in 2015) grounded on the quantitative analysis of the evolutionary benefits provided by the adoption of religious beliefs and behaviours. In other words, religions are considered adaptive insofar as they elicit, boost and support cooperation through belief-statements and

ritual behaviours (e.g. Wilson 2002). A recent focus of the ESR has been the development and maintenance of ultrasociality, i.e. the enormous amount of cooperation between genetically unrelated strangers in groups made of hundreds, thousands or even millions of individuals (e.g. Norenzayan 2013; Turchin 2015). Converging and overlapping with CSR 2.0, ESR teams are building huge historiographical databases with the aim of testing quantitative theories about the evolution of human cultures and religions (e.g. *Seshat: Global History Databank*, http://seshatdatabank.info/; *DRH – The Database of Religious History*, https://religiondatabase.org/);

3. *cognitive historiography*: in a sense the scientific successor of the old HoR, cognitive historiography studies from a qualitative perspective the historical interaction between human universals, cultural variants, socio-political systems and cognitive dynamics, with a special focus on data from 'dead minds' (Eidinow and Martin 2014: 5) and on the religions of the past. The main focus is on the psychological mindsets of historical actors and social groups. The journal of reference, *Journal of Cognitive Historiography*, was launched in 2014;

4. *neurohistory*: a historiographical sub-discipline that focuses on the cultural niche and builds on neuroendocrinology (i.e. the study of the interaction between hormonal, nervous, and cerebral activities) to investigate the historical patterns of conscious and unconscious manipulation of 'moods, emotions and predispositions inherited from the ancestral past', experienced via behaviours (such as rituals and beliefs, gender norms, food consumption, etc.) and modulated or mediated in the brain by neurochemicals (Smail 2008: 117; cf. Shryock and Smail 2011b, and L. H. Martin 2014: 254–71; cf. Geertz 2010).

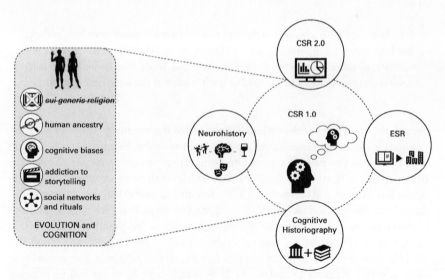

Figure 15 Cognition and evolution reshape the scientific study of religion(s): fundamentals and academic developments

Learning from your mistakes: The usefulness of scientific 'false views'

Apparently, the academic study of the history of religion(s) has finally returned to Darwin's original intuitions in terms of both methodology and epistemology. Some crucial points may still be a matter of contention and discussion, such as the hypothetical derivation of the main contents of worldwide mythologies from a single, prehistoric, African source and spread in subsequent waves via human migrations (Witzel 2012; cf. Geertz 2014b; see also Gould 2002: 277); the effective adaptive value of religion intended either as a cultural taxon or as a loose web of features linked by family resemblances (resp., Richerson and Newson 2009; Sterelny 2017);[8] the lack of a more robust integration between sociology, cultural anthropology, evolution, and cognition (e.g. DiMaggio 1997; Turner *et al.* 2018) or with evolutionary psychology (e.g. Slone and Van Slyke 2015); or the neglect of historiography in the CSR 2.0 (cf. Hughes 2010; Ambasciano 2016e; Ambasciano 2017b). As normal in any progressive research programme, these issues will be assessed and discussed within a reliable scientific framework, and hopefully resolved in due time. According to the same *sciencing* procedure, what has been proved to be useful is revised and reintegrated in the web of knowledge (e.g. Tylor's animism), and what has been falsified is discarded and substituted by a better research programme (e.g. the Eliadean research programme or Jungian archetypes). 'One of the beautiful things about science', writes Martin Schwartz, 'is that it allows us to bumble along, getting it wrong time after time, and feel perfectly fine as long as we learn something each time' (Schwartz 2008: 1771; see Ambasciano 2015a). Darwin captured this process, as well as the risk of post-truth and *bullshit* (*sensu* Frankfurt 2005), with crystal-clear lucidity at the end of the *Descent of Man*:

> false facts are highly injurious to the progress of science, for they often long endure; but false views, if supported by some evidence, do little harm, as every one takes a salutary pleasure in proving their falseness; and when this is done, one path towards error is closed and the road to truth is often at the same time opened.
>
> Darwin 1871, 2: 385

Indeed, the knowledge of the various 'false views' of the 'founding fathers' of the HoR (e.g. the obsessive focus on religious similarities identified as historically homological) might also open new paths or suggest interesting reflections and reanalysis. As historian of science Paolo Rossi (1923–2012) has remarked, the (hi)story of scientific blunders 'is not less relevant, and certainly not less interesting to study, than the statements and discoveries about [scientific] truth' (Rossi 2003: 79). Hypotheses are not accumulated haphazardly in the footnotes of the discipline like a 'neighborhood garage sale' (McCauley 2011: 163), but are assessed, evaluated and eventually welcomed or rejected. And the same applies to both old and new hypothesis and proposals. Two examples of early scientific attempts to update the HoR, which have been falsified by ongoing research, might be useful here: Culianu's cognitive study of religion and J. Z. Smith's proposal of a cultural phenetics.

By the very end of the 1980s, Culianu was well aware that the times were changing. As he acutely commented in a book review published in 1990, 'a unified discipline which would study religion in its context and historical development, but also in its systemic dimension', would necessarily entail a shift from the Chicago phenomenological school to the 'cognitive aspects of the Humanities. If such synthesis fails to be realized [...] *Religionswissenschaft* will perish' (Culianu 1990b: 136). In other words, the entire fate of the comparative HoR depended on a radical scientific update. However, Culianu's own attempts were marred by a confused juxtaposition of far too many topics. As we have seen in the previous chapter, Culianu identified two promising main areas of inter-disciplinary research, i.e. (1) ethology and primatology, and (2) morphology and cognition. Now, area (1) represented the most auspicious interdisciplinary breakthrough at that time, but the heuristic value of Culianu's later speculations as exemplified by point (2) was diminished by a misplaced overconfidence in mathematical Platonism, unnecessary cross-disciplinary complications, epistemological naïvety and a penchant for what today would amount to quantum quackery (on cognitive and mathematical shortcomings, see Ambasciano 2014: 167–3; David 2015; Pigliucci 2015b; cf. Mezei 2014; for a sympathetic reading of Culianu's scientific blunders, see Kripal 2014: 374–6). Despite Culianu's remarkable interest in cognitive studies, like Claude Lévi-Strauss's attempt to derive basic and panhuman cognitive functions from mythology (Boyer 2013; Culianu 2005a: 166), there was little actual cognitive science behind his forays into cognition. Paraphrasing what I have previously written concerning William Robertson Smith's untimely death, we can only hypothesize that the advancements in cognitive sciences, and Culianu's scientific open-mindedness, would have eventually concurred to make him recede from such pseudoscientific theses and develop further his reverse-engineering plan for the HoR.[9]

With regard to J. Z. Smith's interest in numerical taxonomy, his model of preference soon showed its limitations. While pheneticists claimed that their model was objective, repeatable and testable, their approach relied on three *a priori* assumptions: *first*, phenetic analysis was almost entirely methodologically dependent on 'which mathematical tools one employs', especially when there is 'no objective, theory-free way to choose which algorithm to use' (Hamilton and Wheeler 2008: 335); *second*, despite the presence of statistical tools, phenograms are synchronic accounts of affinity between taxa, without historical depth; *third*, according to its opponents, phenetics suffered from a confirmationist approach incapable of discriminating between homologies and analogies, i.e. respectively, traits inherited from common descent or traits independently developed due to similar historical constraints (cf. Rieppel 2008; see Hull 1988). As S. J. Gould summarized, the main problem is that, without historical data, 'morphology is not the best source of data for unraveling history' (Gould 1986: 68). Morphology deceives: crocodiles are more closely related to birds than snakes; sharks, tunas, ichthyosaurs and dolphins share a superficially similar, hydrodynamic body plan because of convergent evolution in a similar environment but they are all distantly related (e.g. dolphins are more closely related to humans than to tunas); the panda is more closely related to bears than to the lesser panda; fungi are not 'plants', and so on. A competing system, called cladistics, proved to be more resilient and epistemically warranted.

Cladistics relied on: (a) differently weighted characters according to their historical states (i.e. *derived*, or more recent; *primitive*, or more ancient); (b) the acknowledgement that taxa branch and modify through time and space, being derived from a common ancestor; (c) the separation between homologies and analogies, with homologies further subdivided into *synapomorphies*, i.e. recent and informative, and *symplesiomorphies*, i.e. primitive and uninformative (Hennig 1966; cf. Saler 2000: 180–96). Finally, in a time when taxonomy was accused of bordering on pseudoscientific status, cladistics implemented a hypothetico-deductive falsificationism as its *modus operandi*, conferring renewed dignity to the sub-field (but cf. Rieppel 2008; on the history of taxonomy, see Hull 1988). Today, cladistics has been successfully adopted to study cultural evolution (cf. Mesoudi 2011: 86–94). It is very likely that Smith's passion for agrostology might have misled him in his attempt to transfer this particular taxonomic model to historiography and cultural studies, for phenetics has been able to resist for a long time within the niche of grass evolution and domestication while it failed to recover homological patterns in angiosperm variation and was substituted by cladistics elsewhere in natural sciences (Chapman 1992; Stuessy 2009: 72; for conflicting topologies in plant taxonomy, cf. Mishler 2000). In hindsight, grass was not a good tool to think with insofar as culture and religion were concerned, and this might even be the main cause of the unanswered theoretical issues left by Smith (Strenski 2016).

This brief digression should be enough to show that it is necessary to know the (hi)story of scientific blunders, the Darwinian 'false views', because if you know the crooked paths pursued with good intentions but which essentially led to nowhere, you are able to avoid them and take the right shortcuts. However, if we put post-Eliadean HoR aside for a moment and focus instead on the whole history of the twentieth-century phenomenological and morphological HoR, we may notice a different sort of pattern altogether.

A short-lived success? The inevitable rise of 'false facts'

As a whole, the institutional HoR of the past provides a clear example of the Darwinian 'false facts [that] are highly injurious to the progress of science, for they often long endure'. And long endure they did. For the most part, the history of the HoR is the account of such false facts, or Frankfurt's *bullshit*, assumed *a priori* and against all evidence as correct. It is the story of an internecine competition with a profoundly admired theology for academic prestige and acknowledgement in the absence of any scientific support (Gilhus 2014). It is also the story of many epistemological blunders and even more disciplinary and methodological dead ends pursued in the service of ideology, with the purpose of exalting its subject of study (cf. Pinker 1997: 555). Today, because of the present post-truth cultural environment, propelled by online social media and institutional disengagement from reality, bullshit is again on the rise in the HoR. In past decades, historians of religions from all over the world have begun to react against the presence of the CSR and other scientific approaches in the academic study of religions (cf. Ambasciano 2014; Ambasciano 2015a; Ambasciano 2016a; Ambasciano 2016b). The ontological nature of the paranormal is slowly becoming the new normal in the HoR. In Eliade's footsteps, Jeffrey J. Kripal advocates the complete disciplinary acceptance of *psi* – that is to say the alleged 'paranormal powers of mind' such as psychokinesis (PK) and extrasensory perception (ESP) (cf. Humphrey 1999: 116) – against the materialism of science and the dogmatism of religious theologies (Kripal 2011a; Kripal 2011b; Kripal 2014). Kocku von Stuckrad has advanced a postmodern scaffolding of science and religion as mere discourses on reality, each one equally valid on its own, yet ambiguously proposing science as an oppressive force that unjustly rejects mysticism, magic and the paranormal (von Stuckrad 2014). Postmodern critiques of 'mainstream "secularist" historiographies' and pleas for 'alternative models' (i.e. alternative to science) abound (e.g. King 2013). While the phenomenon is widespread, some national schools seem to be more prone to assume *in toto* such an anti-scientific stance. Rennie notes correctly that in Germany and Italy 'scholars, [. . .] continue to appreciate both the distinction between *Naturwissenschaft* and *Geisteswissenschaft* and the value of the latter. In Italy, *Storia delle religioni* remains a creative and interpretive discipline, one that may *use* science but is still self-consciously humanist – philosophical, creative and interpretative – and openly appreciative of Eliade's contributions' (Rennie 2016; see, for instance, Sfameni Gasparro 2016: 83 for a powerful reaffirmation of the 'validity and scientific autonomy on historical grounds' of Bianchi's post-historicist HoR against CSR). Obviously, such trends do reflect the local history of the HoR that we have reviewed so far. Many more examples might be added, but these should suffice. To cut a very long story short, Eliadological approaches are being constantly revived, revered and celebrated in the HoR as a bulwark against science.

Evolutionary perspectives still struggle to spread evenly in the social sciences and the Humanities (Rosengren *et al.* 2012). Folk cognition and intuitive thinking reign supreme. It is not just that postmodernism has left an indelible, permanent mark on those academic domains. It is that science is fragile, and its mastery requires a long and slow process of ongoing update and constant study because science goes against the grain of intuitive thinking. During each and every generation, science can progress and build upon previous conquests if, and only if, the socio-political system allows the newest generation to enjoy the intellectual support, the freedom to pursue a scientific path, and the institutional means to achieve the required level of knowledge. The same intuitive, cognitive devices which make religious ideas so easy to grasp and so immediate to understand are a constant stumbling block for the academic and scientific study of religion. As philosopher of science and cognitive scholar Robert N. McCauley has nicely summed up,

> religions share the same cognitive origins and vary within the same limited framework of natural cognitive constraints. Science overturns those constraints and regularly produces new, original ideas. Religion mainly obeys those constraints and replays minor variations on the same ideas time and time again. The sciences inevitably generate radically counterintuitive representations. Religions inevitably traffic in representations whose counterintuitiveness is quite modest. This is the sense in which [psychologist Steven] Pinker states that 'compared to the mind-bending ideas of modern science, religious beliefs are notable for their lack of imagination'.
>
> McCauley 2011: 152; cit. from Pinker 1997: 555

And, to be sure, the idea that the more science there is in any given humanistic research, the more arid and unimaginative said research becomes, is utterly false. Like Gottschall wrote, 'science adds to wonder, it doesn't dissolve it. Scientists always report that the more they discover, the more lovely and mysterious things become. As the great novelist and distinguished lepidopterist Vladimir Nabokov once put it, "The greater one's science, the deeper the sense of mystery"' (Gottschall 2012: xvi).

However, political actors, on the basis of biases and logical fallacies, influenced by theological or political ideologies, might mistakenly decide that science is not necessary, that science is superfluous if not dangerous to one's own worldview. Indeed, science is fragile and might disappear. It has already happened in past human history, and it can happen now (see McCauley 2011: 270–2, 279–86). In his last interview delivered in 1996, one of the most eminent scientists of the past century, Carl Sagan, made exactly this point: 'We've arranged a society on science and technology in which nobody understands anything about science and technology, and this combustible mixture of ignorance and power sooner or later is going to blow up in our faces. I mean, who is running the science and technology in a democracy if the people don't know anything about it?' (Sagan 1996b; cf. Sagan 1996a: 28). Today, the presence of theological, spiritual or religious agendas in contemporary international, democratic and public education is threatening to sever the already fragile link between institutions and scientific literacy. In this regard, the case of faith schools in the United Kingdom is a tragic reminder of the degrading effect of religious agendas and religious financial sponsorship meddling in public education, with the resulting adoption of school subjects (e.g. creationism, literalism) and practices (e.g. mysoginistic sex segregation or homophobia) contrary to science and democracy (Gillard 2002; Gillard 2007; Gillard 2016; Adams 2017; Copson 2017; Marsh 2018).

Science and democracy are intertwined. They both empower people – provided that an education in critical thinking has been implemented. As Sagan remarked,

> science thrives on, indeed requires, the free exchange of ideas; its values are antithetical to secrecy. Science holds to no special vantage points or privileged positions. Both science and democracy encourage unconventional opinions and vigorous debate. Both demand adequate reason, coherent argument, rigorous standards of evidence and honesty. Science is a way to call the bluff of those who only pretend to knowledge. It is a bulwark against mysticism, against superstition, against religion misapplied to where it has no business.
>
> Sagan 1996a: 40–1

No wonder that so many past right-wing reactionary or conservative historians of religions and disciplinary schools have embraced essentialist, anti-scientific, theological postulates and devoted so much effort to delegitimize science and its role in human knowledge. In the past few decades, however, the ever-critical and self-aware stance of left-wing postmodernism, once a healthy immunological system of defence, has gone awry, and now acts more like an autoimmune disease whose life-threatening attacks are triggered by innocuous stimuli. For instance, the fact that cognitive science has tried to carve for itself an academic niche within HoR and RS has been decried as another camouflaged attempt at reinstating a *sui generis* approach. Elaborating on this point, historian of religions Matthew Day 'believe[s that] the ambition to erect a *science of religion* threatens to undo much of what has been accomplished in the way of establishing a non-confessional academic field of broadly naturalistic inquiry' (Day 2010: 5). *First,* I sincerely doubt that modern HoR and a significant part of RS can be labelled as 'non-confessional' and 'naturalistic'. *Second,* approaching religion as a semi-autonomous *mythological machine* theorized and implemented as such in ancient, modern and contemporary societies does not imply an abdication of criticism and a return to a *sui generis* study, as this provides a convenient starting point to undertake scientific reverse-engineering. *Third,* the fact that there might be no justifiable cognitive divide within and between any cultural or religious domain envisaged by *H. sapiens* should lead to the complete dissolution of every humanistic academic approach, to be substituted by a diffused, eminently interdisciplinary web of knowledge. Indeed, other contemporary disciplines like Big History and Deep History are investigating all historiographical disciplines as one, treating them as a unique subject matter, so that each historiographical human event is the result of the combination, interplay and interconnection of various contingent historiographical levels, from cosmology to evolution (Smail 2008; Christian 2011). A few years ago, Finland was already on the verge of a radical restructuring of its school system with the inclusion of broader, interdisciplinary subjects, 'such as the European Union, community and climate change, or 100 years of Finland's independence, which would bring in multidisciplinary modules on languages, geography, sciences and economics' (Pasi Sahlberg, quoted in Strauss 2015). Pending a complete overhaul of global academia, this bold experiment should at least give pause to anti-scientific postmodern criticism. And yet, given the aforementioned presence of theological and intuitive biases towards an appreciative attitude of religions, Day's criticism might not be too far off the mark.

Epilogue:

The Night of Pseudoscience

Too much historical research is being done by people who do not know why they are doing it and without regard to the limits imposed by the evidence. An improvement in this respect is both possible and desirable.

Arnaldo Momigliano

A critical milestone in the study of comparative religion was the publication in 2012 of *Religious Studies as a Scientific Discipline: The Persistence of a Delusion* by Luther H. Martin and Donald Wiebe, in which they confessed their historically justified belief that, notwithstanding the diffusion of the CSR and due to the overwhelming presence of cognitive biases, logical fallacies and theological or ideological beliefs, a completely independent, fully implemented scientific study of religion will never take place in contemporary academia (article republished in Martin and Wiebe 2016: 221–30). Three years later, Armin W. Geertz aptly summarized the many ways in which theological or ideological biases and *a priori* assumptions have infiltrated the scientific research on religions, e.g. favouring pro-social and cooperative tendencies over assortative trends in religious behaviours or highlighting the usefulness of religion in promoting better health (Geertz 2015: 392–3). Even though the ongoing process of revision, falsification and peer review should guarantee the highest standard for scientific research, the continuous struggle with theological, spiritual or religion-friendly perspectives threatens the existence of any scientific study of religion. Donald Wiebe has highlighted the alarming correlation of fideistic biases with the recent rise of private financial support from religious institutions in the field (Wiebe 2009).

When the watchmen who should preside over the reliability of the scientific process are themselves influenced by those biases, who is going to watch over them? For instance, the recently launched European Academy of Religion (EuARE; https://www.europeanacademyofreligion.org/), an academic association strongly inclined to provide a religion-friendly, inter-faith, ecumenical dialogue, whose first conference, held in Bologna, Italy, on 18–22 June 2017, prompted almost immediately an extremely critical 'Joint statement' by the presidents of the IAHR and the European Association for the Study of Religions (EASR). In their declaration, these associations denounced the confessional nature of the EuARE as an attempt to divert the perception of the study of religion in the public sphere, as well as its sources of funding, in a direction that is detrimental to the study of religions as an academically rigorous field of research, as well as to the pursuit of unbiased knowledge about religions which is needed if the challenges of contemporary societies as well as their historical roots are to be correctly understood.

Thomassen and Jensen 2017

The paradoxical fact that such associations are to a large extent dominated by a fideistic, spiritual, emic perspective is another matter of confusion (see Wiebe 1999; Martin and Wiebe 2016: 9–23, 36–41). In the postmodern digital reincarnation of post-truth, filtering out reliable data from the background noise is almost hopeless, for there is no straightforward way to demarcate between science and pseudoscience: everything coexists with anything. In such a state of epistemic crisis, EuARE, EASR, IAHR and innumerable other organizations do coexist and they all share a chair at the postmodern High Table, no matter how epistemically warranted their programmes. As we have seen in this book, the study of religion(s) has been mostly influenced by affective attachment to, if not emic sharing of, theological and ideological ideas, and fideistic factoids are the most emotionally resilient form of post-truth there is (cf. D'Ancona 2017: 126; Blackburn 2017: 126). Such factoids are the fuel for post-truth, and once they have penetrated academia, fooling the immune system of scientific control, there is very little hope of eradicating them, as we have seen in the effect Eliade had on the HoR.

More precisely, the 'Eliade effect' describes the self-feeding, ongoing and intuitive appeal of the charismatic Eliadean HoR: the constant, inter-generational susceptibility demonstrated by intelligent people (e.g. students, early-career researchers, etc.) to fall for pseudo-profound bullshit that has been created by scholars unconcerned with truth and/or fallen themselves for 'deep spiritual meaning' (McCutcheon 2004: 323; Sperber 2010; Pennycook *et al.* 2015). Intelligent people caught in this vicious network, because of the academic prestige of their mentors or HoR readings assigned for their syllabi, will defend more aptly their convictions, spreading bullshit to other individuals thanks to their intelligent charisma, and so on and so forth. In a self-sustaining academic loop, disciplinary momentum is never lost thanks to group conformity and a penchant for both authoritative argument and bandwagon effect, resulting in an 'unwillingness and a (learned) incapacity to engage in reflexivity, a partial closing of the mind, freezing of the intellectual effort, a narrowed focus and an absence of requests for justification' (Alvesson and Spicer 2012: 1213). Thanks to such 'functional stupidity', to the in-built system of defence from criticism, and to its persuasive claim of total coherence, the Eliadean HoR is one of the most resilient belief systems ever created within academia and probably the most successful and conscious institutional attacks on the rational values of the Enlightenment (see Ambasciano 2015a). The success of the Eliadean HoR, in turn, legitimizes the constellation of pseudoscience in which the sub-field is inscribed. But the fact that, for instance, Eliadean HoR, Guénonian Perennialism, van der Leeuw's phenomenology and Schmidtian *Kulturkreislehre* converged on some crucial points is not an example of consilience of induction. It is, rather, an exceptional case study of an ideological, extra-epistemic accumulation of knowledge in which *content biases* (intuitive psychological shortcuts such as essentialism, teleology and intuitive design stance) had been bridled and led through a set of *context biases* (e.g. deference to authority, appeal to tradition and authority) to reinforce *a priori* intuitions through a sharpshooter fallacy (cf. Eco 1989; Boudry and Braeckman 2012). When Stephen J. Gould wrote about the (in)famous fossil fraud of the Piltdown man, he concluded that 'we cannot simply laugh and forget. Piltdown absorbed the professional attention of many fine scientists. It led millions of people astray for 40 years. It cast a false light upon the basic processes of human evolution. Careers are too short and time too precious to view so much waste with equanimity' (Gould 1990: 225–6). The same, I contend, could be stated almost word for word for most of the modern anti-scientific, fideistic, emic HoR. Religious, spiritual or fideistic truths cannot coexist with science; the entire history of the HoR reveals that when they do, they phagocytize and neutralize science (*contra* Gould 1999, and Baggini 2017; see Coyne 2015; cf. Alles 2008; McGrath 2015).

And yet, thanks to the recent rise of cognitive and evolutionary approaches, there is hope that, once implemented in public and compulsory education, science literacy could do much to fix this disciplinary spread of false facts. On the contrary, postmodernists claim that RS can do away with science entirely and still manage to thrive (for both points of view, see, for instance, the responses by D. Zbíral, H. G. Hödl, H. Seiwert, R. Kundt, T. Bubík, K. von Stuckrad, N. Frankenberry, R. N. McCauley and E. Slingerland collected in Martin and Wiebe 2016: 236–78, 291–301). Which group has the best chance to accomplish its agenda? As Frazer had already envisaged at the beginning of the twentieth century, cognitive biases, fideistic partiality and religious interests would always threaten, and eventually trump, any rational education. In this sense, the history of the comparative HoR reveals a series of disheartening episodes in *agnotology*, i.e. the epistemologic study of ignorance, with scientific evidence continually downplayed, manipulated or silenced (Proctor 2008). Even if the many approaches advocated within the CSR could eventually bring to reason the comparative study of religion(s), the main problem of the current digital post-truth era is the epistemic abdication of the powers that be at institutional level, which is reflected in the social and political endorsement of post-truth beliefs and the para-institutional establishment of a new 'grey power' made up by social networks, gigantic online retailers and digital streaming services, overseeing or allowing the creation and diffusion of bullshit information with a *laissez-faire* attitude (Floridi 2015; D'Ancona 2017). The interaction between politics, the new social information system and the financial disruption of academia has already had a most worrisome impact on Western scientific and democratic literacy as a whole (Copson 2017; D'Ancona 2017; see Lincoln 2012: 135). What is happening in the current neo-Eliadean HoR and the more general reaction against science is just the epiphenomenon of something described with outstanding perspicuity by Sagan in the mid–1990s:

I have a foreboding of an America in my children's or grandchildren's time – when the United States is a service and information economy; when nearly all the key manufacturing industries have slipped away to other countries; when awesome technological powers are in the hands of a very few, and no one representing the public interest can even grasp the issues; when the people have lost the ability to set their own agendas or knowledgeably question those in authority; when, clutching our crystals and nervously consulting our horoscopes, our critical faculties in decline, unable to distinguish between what feels good and what's true, we slide, almost without noticing, back into superstition and darkness.

Sagan 1996a: 28

That time is now: Sagan's America has morphed into the current globalized, digitally connected, post-truth world. The *unnatural history of religion*, i.e. HoR and the postmodern RS, is successfully adapted to survive in such a cultural environment, while I doubt that an epistemically warranted scientific study of religion, with its dependence on financial resources, extended scholarly networks and scientific infrastructures, which ultimately depend on political activity and institutional support, might outlive this period (cf. Talmont-Kaminski 2013). Obviously, this does not mean that we should go gently into the night of pseudoscience. Sooner or later, everything eventually changes, and we can still hope that the flickering light of the modern Enlightenment will lead soon to a new dawn. But even if Frazer's and Sagan's gloomy predictions are right, even if scientific knowledge will eventually be lost against the rising tide of post-truth ignorance and superstition, even in the darkest days of anti-democratic recrudescence, we will still have a moral and personal choice. Therefore, when the chips are down, the epistemological and ethical question is pretty simple. *Science & democracy* or *post-truth & pseudoscience*: whose side are you on?

Notes

Preface: Ghosts, Post-truth Despair, and Brandolini's Law

1   The software–hardware metaphor might seem a bit passé. Indeed, studies in neuroplastic adaptability have shown the limits of this analogy: the brain changes and adapts on several different scales (e.g. neural networks, cortical re-mapping) during the entire human lifetime, even in response to brain trauma and injury. Consequently, a cross- and inter-disciplinary revision has updated this metaphoric parallel (e.g. Costandi 2016). If the brain is some sort of physical hardware, it is nonetheless a *peculiar*, self-repairing, ever-adapting kind of hardware. However, neuroplasticity has obvious limits: the brain itself is still a physical hardware that, for all its might, can fail and affect the functioning of the software, the mind (think about strokes or neurodegenerative disorders). Thus, the metaphor can still be used to describe in general terms the scientific study of human and nonhuman cognitive abilities and physiology, with cognitive sciences and philosophy of mind studying the *software* 'mind' and the neurosciences tackling the *hardware* 'brain and body'.

2   *Family Guy*, 'Airport '07', Season 5, Episode 12, first aired 4 March 2007. Directed by J. Holmquist, written by T. Devanney.

3   For a cognitive and historical rebuttal of the idea that agnosticism and atheism are merely modern ideas inapplicable to ancient cultures, see Geertz and Markússon 2010; Whitmarsh 2015; and Ambasciano 2016a.

4   This is why I have sometimes resorted to an inclusive *religion(s)* in the course of the book.

5   One might argue that, according to Sturgeon's Law, if '90 per cent of everything is crap' anyway, then my critical point of view on the current status of the HoR is unfounded (Dennett 2013: 36–7). Ten per cent, according to such hypothetical rebuttal, is the normal, physiological percentage that keeps research going. There is, quite undoubtedly, excellent research in the field, as we will see shortly. But let us not forget the demarcation issue at stake here. I propose a postulate to Sturgeon's Law: when the remaining 90 per cent of an academic, purportedly science-based discipline is filled with bullshit (Frankfurt 2005), that field has become *pseudoscience*. We will discuss this issue at length in the following chapters.

1. An Incoherent Contradiction

1   'Religious Studies' is also used generally to group different disciplines interested in the study of religion(s) (cf. the opening epigraph).

2   Even though there is no neurophysiological distinction between non-religious and religious ways of thinking (e.g. Boyer 2001; Barrett 2004; cf. Bloch 2008), a fact that might account for a common cross-disciplinary ground, there are other significant differences between RS and CSR, as we will see in the following chapters.

3 The most remarkable among these cognitive mechanisms are teleological reasoning, over-detection of agency, mind/body dualism, tool cognition, precautionary routines, imitation, etc. See Whitehouse 2013a: 36.

4 A flourishing industry of bio-bibliographical and epistolary studies devoted to past HoR scholars is increasingly seen as a convenient retreat immunized from criticism (cf. Spineto 2010: 1302; Spineto 2013).

2. The Deep History of Comparison

1 Of course, this does not mean that everything was tolerated or tolerable. For instance, the Romans were very strict about the exclusion of what they considered 'superstition', that is, an incorrect way of being religious according to their own culture.

2 A most dramatic clash between these theological perspectives materialized in the harsh dispute between Ambrosius, bishop of *Mediolanum* (modern-day Milan, Italy), and Roman senator and prefect Quintus Aurelius Symmachus, concerning the removal of the Altar of Victory (i.e. a goddess) from the Senate. In a vibrant epistle dated from 384 CE and addressed to the emperor Valentinian II, Symmachus wrote that 'it is reasonable to regard as identical that which all worship. We look at the same stars; we share the same sky; the same world enfolds us. What difference does it make by what system of knowledge each man sees the truth? Man cannot come to so profound a mystery by one road alone' (Symmachus, *Relatio* 3.10; modified from Salzman 2011: 122). However, Symmachus' plea was ultimately unsuccessful and the request bluntly denied as suggested by the powerful bishop (see Sogno 2006: 45–57).

3 Although still used indiscriminately, 'myth' is an all-encompassing label, or a wastebasket taxon, which includes cultural representations whose origins, functions, aims and morphology can differ dramatically. Pending a reorganization of the concept, cf. Wayland Barber and Barber 2004, and Masse *et al.* 2007.

4 A recent example of this Us/Them division is the rise of the so-called toxic fandoms in blockbuster films, exemplified by the Marvel Cinematic Universe vs. the DC Extended Universe feud or the backlash against the new Star Wars sequel trilogy.

5 *Homo religiosus* mimicks Linnean binomial nomenclature, but it is not a scientific concept. Therefore, being not regulated as such by the *International Code of Zoological Nomenclature*, the initial letter of the 'genus' *homo* should not be capitalized (except after a full stop, of course).

6 Today, thanks to interdisciplinary advances in genetics, archaeology and linguistics, there are two main hypotheses that try to pinpoint the spatio-temporal point of departure for the divergence and spread of Indo-European languages: the Anatolian hypothesis, strictly tied to the diffusion of agriculture (*ca.* 9,000 years ago), and the Pontic steppe hypothesis, whose supposed vehicles of diffusion are thought to be horse-riding nomads (*ca.* 6,000 years ago). The first hypothesis is supported by scientific and phylogenetic analyses, the second one is the traditional one. Cf. resp., Bouckaert *et al.* 2012, and Pereltsvaig and Lewis 2015.

7 As recalled by Cristiano Grottanelli and Bruce Lincoln (1998: 313), 'in contrast to Marx, who dismissed religion as false consciousness *tout court*, Engels perceived that within any society and any historical moment, there may be multiple competing religious attitudes and movements, which express, maintain and even (at times) exacerbate the other tensions and conflicts within that society'.

8   It is fair to say that Durkheim's sociological take on religion has aged remarkably better than Freud's psychoanalysis (cf. resp., Whitehouse 2013b, and Bloch 2015; Paden 2016, and Cioffi 2013). And yet, Freud's ideas, falsified as a set of non-scientific and dogmatic heuristics, were soon to re-enter HoR as a re-theologized tool with Carl G. Jung's own version of psychoanalysis. Some other eclectic psychoanalytical movements value religious constructions openly (e.g. Roberto Assagioli's 'higher unconscious' and the positive role played by ideas 'such as the "inner Christ" in certain forms of Christian piety'); see Brooke 1991: 325.

9   However, it is important to note that Lévy-Bruhl stated in a posthumous work that what he labelled as logical and prelogical mentalities do coexist in every society. See Lévy-Bruhl 1949, and Smith 1993: 265–88; cf. Sørensen 2007: 25–6, 187 for a re-evaluation of Lévy-Bruhl's concept of magic. Interestingly, the idea of two independent modes of thinking is still a central topic in cognitive science (e.g. Kahneman 2011).

10   However, considering James' penchant 'to accept the genuineness of the transcendental reference in religious experience' (Sharpe 1986: 104, 108–12), his role in the development of a truly scientific study of religion remains highly debatable. See also Martin and Pyysiäinen 2013: 220–1 on Luhrmann 2013.

3. The Darwinian Road Not Taken

1   This historical attention to apparently useless traits, while moving glaringly adaptive, successful traits in the background, has led palaeontologist Elliot Sober to baptize this gold standard of the new evolutionary natural history as 'Darwin's principle', to indicate 'that selectively advantageous traits are "almost valueless" as evidence of common ancestry' (Sober 2008: 297).

2   On theodicy:

> one word more on 'designed laws' & 'undesigned results'. I see a bird which I want for food, take my gun & kill it, I do this *designedly*. An innocent & good man stands under a tree & is killed by flash of lightning. Do you believe (& I really shd like to hear) that God *designedly* killed this man? Many or most persons do believe this; I can't & don't. If you believe so, do you believe that when a swallow snaps up a gnat that God designed that that particular swallow shd. snap up that particular gnat at that particular instant? I believe that the man & the gnat are in same predicament. If the death of neither man or gnat are designed, I see no good reason to believe that their first birth or production shd. be necessarily designed. Yet, as I said before, I cannot persuade myself that electricity acts, that the tree grows, that man aspires to loftiest conceptions all from blind, brute force.
>
> Letter of Darwin to Asa Gray, 3 July 1860, *Darwin Correspondence Project*,
> DCP-LETT–2855

3   Frazer's evolutionism was a second-hand concept: it came through the filters provided by Tylor's and Robertson Smith's approaches (Ackerman 1975). Consequently, Frazer's cultural evolutionary model, while apparently more akin to pre-Darwinian progressionism, might be considered a parallel enterprise with regard to other, more informed, evolutionary anthropologists' frameworks (cf. Ackerman 1990: 77). Privately, as testified to in a letter of his, Frazer even seemed to have poorly understood Darwin's theory (Frazer to W. J. Lewis, 28 December 1919, in Ackerman 2005b: 362).

4   A third strategy to assuage such dissonance was provided by eclectic religious and esoteric reinterpretation, and manipulation, of scientific contents (Bowler 1987: 9).

5   The well-known motto *survival of the fittest* was originally alien to Darwin's vocabulary: it was coined by English philosopher Herbert Spencer (1820–1903) to recap the concept of natural selection (1864, 1: 444), and only later adopted by Darwin himself. It is misleading because it fails to capture the variety of mechanisms implied.

6   Mayr listed also a series of corollaries to Darwin's basic core: 'sexual selection, pangenesis, effect of use and disuse, and character divergence' (Mayr 1991: 35). This whole group of theories represents altogether what Mayr labelled as the 'first Darwinian revolution' (Mayr 1991: 12ff.). Sexual selection and character divergence still stand as two of the most important drives in evolutionary biology. Pangenesis was later disproved in favour of Mendelian genetics, and the study of the effects of use and disuse has been reprised in a completely new genetic frame (i.e. epigenetics and evolutionary developmental biology, or *evo-devo*). Also, gradualism has been flanked by evolutionarily rapid (geologically speaking) outbursts (Eldredge and Gould 1972; Gould and Eldredge 1977; Gould and Eldredge 1993). Of these revisions, none had any significant impact on the five-fold Darwinian hard core (Pievani 2011).

7   Unfortunately, when Darwin published the *Descent of Man* in 1871, 12 years after the *Origin of Species* (a volume upon which he had worked for 20 years), he was old, sick and quite exhausted from all the religious discussions and personal attacks led against him. Therefore, he adopted a slightly modified version of the dominant anthropological view in vogue, that is, a 'linear human scale' of beliefs with contemporary Europeans at the top, who had left behind 'the remnants of former false religious beliefs'. It was a Darwinian development which the Darwin of 1859, fed up with unscientific identification of 'high' or 'low' forms of life, would have not accepted (Desmond and Moore 2009: 365; cf. Plotkin 2004: 64).

8   In the words of Darwin himself: 'the difference in mind between man and the higher animals, great as it is, is certainly one of degree and not of kind' (Darwin 1871, 1: 105). This statement has been recently accused of fostering a fallacious understanding of animal cognition as anthropomorphically tailored (Penn 2011: 257; see also Penn, Holyoak and Povinelli 2008). However, new primatological and comparative cognitive research has vindicated Darwin's original approach; for an overview, see de Waal 2013; Ambasciano 2016a: 167–72, *passim*; de Waal 2016.

9   The striking analogy behind the mechanisms of development and differentiation through time in both philology and natural sciences caught also the attention of evolutionary biologists. As early as 1854, Thomas H. Huxley, having recalled the common ancestry of '*unus, uno, un, one, ein*', and '*Hemp, Hennep, Hanf, and Cannabis, Canapa and Chanvre*', wrote that 'Philology demonstrates that the words are the same by a reference to the independently ascertained laws of change and substitution for the letters of corresponding words, in the Indo-Germanic tongues: by showing in fact, that though these words are not the same, yet they are modifications by known developmental laws of the same root' (Huxley 1854: 283).

10 Max Müller's anti-evolutionary criticism was just one voice in a field dominated by human exceptionalism; cf. Ambasciano 2016a: 167–72.

11 On Frazer's criticism of Max Müller's divinely inspired 'natural' reason, see his letter to Henry Jackson, 22 August 1888, in Ackerman 2005b: 47.

12 As Darwin wrote in 1837 in one of his notebooks, 'the tree of life should perhaps be called the coral of life, base of branches dead, so that passages cannot be seen' (Barrett *et al.* 1987: 177; cf. Eldredge 2005).

4. Goodbye Science

1   However, *mana* implies the positive outcome of a certain action. Commenting on Ann
    Taves' Marettian definition of *mana* as something which is '*beyond the ordinary* power
    of men, *outside the common* processes of nature' (Taves 2013: 145), Luther H. Martin
    has recently remarked on an important aspect concerning spiritual and ethnocentric
    distortions:

> during my travels, however, I once asked the chief of a traditional Melanesian
> village how he would define *mana*. He thought for a moment and then answered:
> 'If a fisherman goes out to fish and he wants to catch a big fish and he does – that's
> mana.' A good day, perhaps, but hardly a feat '*beyond the ordinary* power of men,
> *outside the* common processes of nature'. In other words, do we (modern Western
> scholars) really understand what people in other times and places consider to be
> 'nonordinary', or do such inferences reflect our cultural biases? It might be noted
> that Marett's description of mana was based upon an Anglican missionary's
> account of Melanesians (Taves 2013: 145, citing Marett 1914b: 104) and, although I
> am not a trained ethnologist, it is at least questionable whether this missionary's
> anecdotal report is of more scientific significance than my own.
>
> Martin 2015: 126

    See also Sharpe 1986: 69–71.

2   Although the double surname has stuck in the disciplinary historiography, the
    surname was Smith, Robertson being the maiden name of his mother (Turner 2014:
    448: note 13).

3   'Natural Philosophy' had been defined as follows by physicist Lord Kelvin (William
    Thomson; 1824–1907) and Tait himself:

> the term Natural Philosophy was used by Newton, and is still used in British
> Universities, to denote the investigation of laws in the material world, and the
> deduction of results not directly observed. Observation, classification, and
> description of phenomena necessarily precede Natural Philosophy in every
> department of natural science. The earlier stage is, in some branches, commonly
> called Natural History.
>
> Tait and Thomson 1912 [1879]: v

4   One of Smith's colleagues in the Physical Laboratory was future Scottish novelist
    Robert Louis Stevenson (1850–1894), who took pleasure in debating amicably with
    Smith and future oceanographer John Murray (1841–1914) 'on the age of the earth
    and the foundations of Christianity' – this when not 'plaguing Robertson Smith with
    irrelevant metaphysical questions' (resp., Knott 1911: 72; Booth 2010).

5   As to how could Frazer the atheist be such a close and intimate friend of Free Church
    minister Smith, it should be remarked that the two never discussed *personal* religious
    beliefs. Frazer wrote in a letter to John F. White, dated 15 December 1897, what
    follows: 'I confess I never understood his [i.e. Smith's] inmost views on religion. On
    this subject he maintained a certain reserve which neither I nor (so far as I know) any
    of his intimates cared to break through. I never even approached, far less discussed,
    the subject with him' (Ackerman 2005b: 109). On his part, Smith was 'reluctant to
    disclose, let alone discuss his religious convictions' (Maier 2009: 227).

6   It is interesting to note that Darwin knew the works of Smith, as evident by a citation
    included in the posthumous second edition of *The Expression of the Emotions in Man*

and Animals (1890). In that passage, Darwin was exploring the ethnographic data pertaining to the universal emotion of shame accompanied by a more or less visible blushing of the face. The philological authority of Smith is summoned to correct a passage of the first edition with regard to the apparent absence of blushing as inferred from a passage in the Bible (i.e. Jeremiah 6:15): 'according to Professor Robertson Smith, these words do not imply blushing. It seems possible that pallor is meant. There is, however, a word [*haphar*] occurring in *Psalm* xxxiv. 5, which probably means to blush' (1890: 335 n. 11). Archival research might provide more information about direct or indirect epistolary contacts or links between Darwin and Smith (cf. the letter sent from J. V. Carus to Darwin, 24 October 1872, *Darwin Correspondence Project*, DCP-LETT–8574, where the above issue had been mentioned with an explicit reference to Smith's predecessor at the Cambridge chair of Arabic, i.e. William Wright).

7   The unprecedented attention towards social aspects made James Turner claim that Smith is 'the founder of the comparative study of religion' (Turner 2014: 294; see also Warburg 1989 and Segal's insistence on Smith's 'revolutionary' role in Segal 2002b: ix–xi, xiv).

8   For the early influence of Tylor's works on Tiele, see Platvoet 1998a: 143 n. 41.

9   As recapped by Arie L. Molendijk (2004: 333–5), these laws described the universal and panhuman progressive development of religion: (1) 'law of the unity of the human mind', according to which cultural evolution precedes (and accompanies) religious development; (2) 'law of balance' between authority and freedom, by which, through stable socio-political conditions, religious advancement is achieved. Three corollaries completed Tiele's scheme (i.e. 'law of reformation'; 'law of [Tylorian] survival'; 'law of advancement by reaction'). In any case, it is fair to add that Tiele's general scheme and laws changed through the years; please refer to Molendijk 2004 and Molendijk 2005 for further information on the topic.

10 Later on, his disciplinary proposal would become further subdivided into a nomothetic history of *religion* (singular) and a descriptive history of *religions* (plural; relabelled as 'hierography'; cf. Tiele 1877: 1–2; see also Wiebe 1999: 35–6).

11 Tiele wrote the entry 'Religions' for the *Encyclopaedia Britannica*, in which, according to Tiele's plan to conflate 'natural' and 'revealed' religions, the concept 'world religions' was criticized (Tiele 1886; the aforementioned letter is archived in the Tiele Collection of the Leiden University Library, BPL 2710. See Molendijk 2016: 174 for further critical and biographical references; cf. also Molendijk 2004: 340). However, by 1893, Smith acknowledged that Tiele's conception had become more favourable to belief than to ritual: 'for my own part I am inclined to think that you give too great prominence to *gods*, while you on the other hand will think that I give too much prominence to *institutions*' (Smith's own emphasis; Smith to Tiele, 23 October, BPL 2710; from Molendijk 2005: 125 n. 7).

12 On Chantepie's fervent anti-Darwinism, see Platvoet 1998a: 124.

13 Also, as Eric Sharpe acutely noted, an 'objective eidetic vision is quite literally a contradiction in terms' (Sharpe 1986: 224).

14 Interestingly, some *prima facie* epistemologically sound understanding of 'phenomenology' within an empirical framework is also well attested (Tuckett 2016a: 78).

15 Wagner's law stated that geographical isolation of a group made up by a limited number of individuals was necessary to prevent inbreeding with the ancestral community and trigger the evolution of a new species. Wagner failed to conceive of

this process as complementary to selection (in fact, he highlighted it as exclusive), and his thesis has since been revised and expanded to include genetic mechanisms within populational thinking (i.e. founder effect and genetic drift). On the discussions between Wagner and Darwin about the role of isolation in evolution, see Mayr 1982: 562–6, Bowler 1992: 32, and Milner 2009: 433.

16 As Woodruff Smith has observed, 'today, we might say that Ratzel was looking for central statistical tendencies in sets of phenomena, but he did not see it that way. He had some notion that ethnological "truth" depended on the perspective of the observer, but he did not develop the idea' (Smith 1991: 145).

17 Ratzel was not interested in human creativity, and his model scarcely accounted for innovation *per se*. According to him, 'humans are naturally uncreative. They have an inherent tendency toward inertia, from which they must be forced in order for any significant change to occur' (Smith 1991: 144).

18 See also Andriolo 1979: 137 and 143 n. 11 for Schmidt's use of *Geist* as another impersonal, creative force as historical agent.

19 A term coined by philosopher Friedrich Schelling (1775–1854) and adopted by Max Müller to indicate those polytheistic religious systems in which many deities are subordinated to a main divinity (see Yusa 2005).

20 Anyway, Lang did not discard socio-cultural progressive evolutionism; see Wheeler-Barclay 2010: 131. Tylor replied by pointing out that distortions and misinterpretations were the results of missionary zeal and local proselytism (cf. Smith 1982: 67–8).

21 Schmidt, in turn, was ready to acknowledge the similarities between his method and Pettazzoni's – at least before 1922. In 1913, for instance, Schmidt praised the 'exemplary improvement with regards to the previous evolutionary theories' in Pettazzoni's monograph on 'primitive' religion in Sardinia (Schmidt 1913: 575; from Gandini 1996: 116).

22 Pettazzoni, who recognized the 'practical-political value' of the ethnological component of the new HoR to instruct colonial administrators, took part in colonial policy, organizing and proposing the topic (i.e. colonial Africa) of the 8th Convegno Volta (4–11 October 1938), where Schmidt, as delegate for the Vatican, vehemently advocated a *Kulturkreislehre* justification for imperial control in Oriental Africa (see resp., Gandini 2003: 195–202 and Conte 1988: 126–7; cf. Spöttel 1998, Ciurtin 2008: 338–9 and Stausberg 2008: 388–92).

23 Croce's ongoing opposition to Pettazzoni's appointment might also be understood as a way to get back at Gentile's institutional support of the HoR (Gandini 1999b: 118). However, Gentile also established compulsory Catholic education, which ran counter to Pettazzoni's expectation.

24 A few years earlier, Pettazzoni also depicted Aristotle as the founding father of the science of religion, the 'Master' precursor of the union of historicism and phenomenology as well as the designer of a '"history" of theology, conceived as an exposition of theological and theogonic systems of various peoples' (Pettazzoni 1954b).

5. Eliadology

1 Pettazzoni's criticism of Eliade might be compared to Croce's refusal to acknowledge the academic status of the HoR, seen as a sort of 'bibliographical collecting'; cf. Spineto 2006: 110; for HoR's 'parallelomania', see L. H. Martin 2014: 95.

2   Apparently, de Martino was to acknowledge the pitfalls of such endeavour after an unsuccessful field trip in the early 1960s; cf. Angelini 2012.

3   For the political viewpoints held by these thinkers, see Ambasciano 2014. More generally, the biological and evolutionary Romanian panorama of the period was mainly influenced by German and French anti-Darwinist approaches (Tatole 2008). Germany and France were also the main sources of inspiration as far as literature and culture were concerned, and this is also reflected in Eliade's own education and intellectual formation (Werblowsky 2006: 300; Țurcanu 2007).

4   Eliade began writing about the paranormal in 1925, promoting its allegedly positive role in questioning and extending scientific knowledge (Eliade 1996: 243–54).

5   During his Romanian years, influenced by Nae Ionescu, Eliade wrote political articles characterized by an abundance of religious keywords as if they were spiritual sermons, in which he supported a raucous elitism; an antidemocratic pledge 'to redeem the [Romanian] race'; the exaltation of dictatorship; the unconditioned submission to the political leader; the glorification of the fundamentalist and xenophobic Orthodox legionarism as a superior political option with regards to Nazism, fascism and communism; the celebration of local folklore and peasant traditions (reputed to be extremely ancient and thus particularly prestigious); the fanciful claim for a Romanian Orthodox and fundamentalist revolution which would have ultimately conquered Europe; a sense of revanche born of the narrow-minded provincialism which he nourished; and the usual attacks against the bourgeoisie and most of Western science and democracy (reputed fifth columns of the Western powers which had weakened the Romanian population), as well as anti-Semitism, bulgarophoby, and magiarophoby (references and comments in Ambasciano 2014: 277–80).

6   Eliade's signed document dating from his detention has been recently rediscovered and published (Handoca 2008: 382–3). In that document, Eliade renounced the political activity tied to the extremist *Legiune*. However, he publicly denied having signed such a document in his published memoirs, for the very act of signing would have confirmed his political activity (see bibliographical discussion in Ambasciano 2014: 356 n. 614).

7   As a result of his participation in foreign propaganda, among other activities, in 1942 Eliade devoted a celebratory pamphlet to the dictatorship of Portuguese statesman António de Oliveira Salazar (1889–1970). The following year, Eliade rewrote the history of Romania reimagining ancient, pre-Roman Dacia as the 'California of its times', whose Shangri-La opulence attracted many foreign invaders, and depicted early modern Romanian states as the last European defence against the Ottomans. Thus, Eliade's reconstruction drew an explicit parallel with the Eastern Front during the war, with Romania as the last dam against Communist Russia (works collected in Eliade 2007; for a historiographical rebuttal, see Alexandrescu 2006, and Ambasciano 2014: 199–203, 398–401, 410–11).

8   For Lecomte du Noüy, see Ricketts 2000a: 304 (Eliade's note dated 30 January 1948). Eliade and de Chardin met on 23 January 1950 (Eliade 1990a: 102; further personal and philosophical reflections in Eliade 1989a: 99, 170–1, 190, 261). On de Chardin's pseudoscientific and spiritual renovation for modern man, cf. Eliade 1967d; on de Chardin's pseudoscientific eschatology (as well as for Eliade's idea that science and technology show an archetypical religious structure), cf. Eliade 1985.

9   The Dechardinian reflections that underpin these positions are available in Eliade 1989b: 186 (20 January 1975).

10 Eliade revised the English translations of some among his most famous works (e.g. Eliade 1958a and 1958b; Eliade 1961) to expunge the references to the recurrent 'apish imitation' by the unconscious ('imitation simiesque de l'incosciente') with regard to a transcendent reality. This change was occasioned by a letter sent on 19 January 1955 by Carl G. Jung, in which the psychoanalyst scolded Eliade and noted the psychological inconsistency of such interpretation. Jung's letter is in Adler and Jaffè 1976: 220–1; Eliade's two replies are dated 22 January 1955 (from which the aforementioned citation comes), and 11 February 1955 (in which Eliade promises to modify the forthcoming English translation of *Yoga: Immortality and Freedom*), resp. from Handoca 2004b: 84–7, and 89–91.

11 More precisely, according to Natale Spineto, there are three instances of the Eliadean archetypes: (1) archetype as an expression of a metaphysical, Platonic 'archaic ontology'; (2) archetype as an existential intuition about the location of (the male members of the taxon) *H. sapiens* in the cosmos (for Eliadean patriarchal sexism, cf. Kinsley 2002); (3) archetype as a morphological, quintessential structure of religious phenomena *tout court* (Spineto 2006: 179–201).

12 In this case, the exploitation of the historicist mantra of 'myth as true story' is evident: the primacy of the attention-grabbing, hagiographic *historical document* is assumed as truthful because it is reported by second-hand witnesses. On such gullibility, see Hume 2008b: 79–96, and Law 2018.

13 On Eliade's contact with Racoviţă, see Oişteanu 2010: 335.

14 E.g. one might cite the Native Australians' technological mastery and regular control of 'fire-stick farming' as a 'resource management strategy', which was culturally astonishing and evolutionarily unprecedented (Bliege Bird *et al.* 2008). Indeed, this innovation (coupled with other hunting techniques) might have even been *too successful*, contributing to drastic ecological changes as well as to the anthropogenic extinction of the local megafauna (Martin and Klein 1989; Flannery 2002; Tuniz, Gillespie and Jones 2009). However, such hypothesis remains contentious, for the cause–effect relationship is still debated (e.g. fire-stick farming might have been implemented *after* the extinction of the local megafauna).

15 Even though the ethnographic analogy between modern-day peoples and reconstructed past human populations can found some epistemic justifications (e.g. Currie 2016), the fact still remains that we ignore the actual cultural and social environments of the different species of the genus *Homo* as well as their intra-specific and diachronic variability. All we have is reconstructions, which might be epistemically warranted until new palaeoanthropological information is discovered – i.e. they are provisional. Even in the most charitable reading of the case-by-case viability of the ethnographic analogy, one cannot escape the possibility that in-group cultural variation might have been high since the very beginning (cf. Foley and Mirazón Lahr 2012). Expanding on this issue, Peter J. Richerson has commented that

> whatever *Homo erectus* was doing (likely rather different things at different times and places) it was not what the San or Shoshoni were doing, perhaps not even close. [. . .] Some ancient *Homo* might have 'experimented' with forms of social organization that are quite deviant from a straight-line extrapolation from chimps to ethnographic [hunter-gatherers]. Who can say? There is no harm in telling stories about what might have been going on in the Pleistocene. The danger comes if you believe them and try to build a scientific argument on such a foundation of quicksand.

Richerson 2014, reproduced with permission; cf. Richerson and Boyd 2013: 293,
                                    295; Ambasciano 2016a: 187–8.

For a recent scientific enquiry about early human religiosity, see Geertz 2013.

16 Elsewhere in Eliade's academic production, Racoviță's living fossils are strategically
    invoked with regard to the historical existence and resistance of yoga (Eliade 2009:
    361).

17 'If we assume that the human race[s] originated in Asia, and other sciences like
    physical anthropology and prehistory confirm this assumption, then they must have
    migrated from here to all the other parts of the earth' (Schmidt 1926: n.p.; from
    Brandewie 1983: 145–6).

18 Moreover, many recent discoveries have ascertained that Neandertals were capable of
    cognitive abilities on a par with those exhibited by *H. sapiens*, such as the elaboration
    of new lithic techniques (Riel-Salvatore 2010), the exploitation of avian plumage for
    ornamental or symbolic use (Peresani *et al.* 2011), the use of ochre for hygienic and/or
    symbolic purposes during burials (Roebroeks *et al.* 2012), the ability to process and
    cook plant-based food (thus debunking the die-hard myth of Neandertals as meat-
    eater brutes; Hardy *et al.* 2012), and the creation of cave art (Hoffmann *et al.* 2018).
    The list is growing, with new evidence disconfirming the old 'brutal' paradigm added
    on a regular basis (e.g. the alleged Neandertal inability to produce proper speech;
    Krause *et al.* 2007; see also Ambasciano 2014: 126–7).

19 A general discussion of the ideological tenets behind Eliade's perspective is provided
    in Ambasciano 2014.

20 Lessa also remarked how Eliade's 'literary efforts' might mislead 'hasty and
    undiscriminating' readers. Those were the same criticisms Haydon levelled against
    Schmidt; see Chapter 4, §*Schmidt's legacy.*

21 This section has been elaborated upon thanks to the invaluable help of (in alphabetical
    order) Roberto Alciati, Sergio Botta, Francesco Cassata, Enrico Manera and Emiliano
    Rubens Urciuoli.

22 Because of space limitation, I am obviously unable to expand on the history of
    shamanism. For a detailed account of previous, and much debated, historical accounts,
    I refer to DuBois (2009) and Ambasciano (2014).

23 As it happened, Eliade did with ancient Thracian and Romanian folkloric rituals the
    same thing Ohlmarks did for the *seiðr*, differentiating them from shamanic practices,
    often on equally shaky grounds; see Ambasciano 2014: 342–6.

24 The same remarks previously reported in n. 12 apply here.

25 On the Eliadean *scala religionum* made up of different, more or less pristine and
    prestigious 'shamanisms', see Ambasciano 2014: 53, 61–2, 223 (e.g. the allegedly more
    primitive and 'Neandertal' Australian *medicine men* were never identified as 'shamans';
    cf. Znamenski 2007, Znamenski 2009).

6. The Demolition of the *Status Quo*

1   Eliade's tortuous response, dated 3 July 1972 (available in Handoca 2004a: 122–40),
    baffled Scholem, who asked in vain for more clarifications in another letter, dated 28
    March 1973 (in Handoca 2006: 318–19).

2   However, Eliade had already been proposed, and his nomination discarded, as early as
    1957 by Ernesto Koliqi (1903–1975), Professor of Albanian literature at the University
    of Rome (*Nomination Database*, Nobelprize.org. Nobel Media AB 2014).

3   For instance, in the aftermath of World War II, Eliade became involved with Romanian expats' journals culturally tied to the previous national regime (Berger 1994: 64–5). In 1952, Eliade signed an article for the journal *Destin*, entitled *Catastrofă și mesianism. Note pentru o Teologie a Istoriei* ('Catastrophe and Messianism: Notes for a Theology of History'), in which, adopting an anti-historical two-fold fallacy (i.e. assuming the literal reality of sacred texts and judging history teleologically on the basis of such beliefs), he justified the ancient tragedies suffered by the Jews because those events were reputed teleologically necessary to lead their elite towards a purer form of monotheism and, later on, to finally give rise to Christianity – thus reiterating Nae Ionescu's theological anti-Semitism (Eliade 1992b; see Ambasciano 2014: 296–9).

4   To avoid further complications, and for the sake of brevity, I cannot discuss here whether or not shamanism is an adequate tool to analyse Japanese religion.

5   'Power' in classical HoR was understood along the emic reading exemplified by Otto's systematization. Eliade, building on van der Leeuw's meta-theological 'manifestation of power' in religion(s) (Tuckett 2016b: 27–31), wrote about the 'kratophany–hierophany dialectic', that is, the potentially dangerous and profane manifestation of sacred power correctly developed by *homo religiosus* into a symbol which, by its integration in a religious system, extends this sacred, powerful manifestation in time (Eliade 1958a: 20; cf. Spineto 2006: 208).

6   At the same time, Eliade discarded the Schmidtian, theological-value judgement of a generalized 'decadence of the shaman' (Eliade 1964: 125 n. 36, 258; however, this survival does not indicate 'any priority of women in the earliest shamanism'; Eliade 1964: 258).

7   Lincoln was considered by Eliade himself to be his most brilliant student (Doniger 1991: xi). Lincoln shares this record with Culianu who, by the way, was not a direct student of Eliade. Eliade wrote in his journal on 1 August 1984 that 'my admiration for Ioan [P. Culianu] is sincere and without limits' (Eliade 2004: 482).

8   Culianu's admiration for Derrida (2005a: 165–7) might explain his later attention to reality as a linguistic game.

7. The Cognitive (R)evolution: The End?

1   In the wake of the success of the Brexit referendum and the Trump presidential campaign, various mass media, political movements, and social organizations all across Europe have resorted to the same strategy in order to foster what Oxford Dictionaries president Casper Grathwohl has called 'a growing distrust of facts offered up by the establishment' to gain public visibility and reclaim social and political power, which will undoubtedly affect the future of the HoR (Flood 2016).

2   Cf. the following statement in Sagan 1996a: 28: 'science is more than a body of knowledge; it is a way of thinking.'

3   In the 1960s, Donald T. Campbell (1916–1996), a social psychologist, resorted to Baldwin's ideas to devise what is perhaps the catchiest Darwinian formula to describe cultural selection in the arts as a combination of 'blind variation and selective retention', or BVSR (Campbell 1960; Plotkin 2004: 87). In the long run, as evolutionary biologist David Sloan Wilson told historian Daniel Lord Smail, 'even intentions become a form of blind variation when they interact with other intentions and produce unforeseen consequences' (Smail 2008: 91).

4   Frazer contributed with an analysis of ancient Greek anthropogony:

in a volume dedicated to the honour of one who has done more than any other in modern times to shape the ideas of mankind as to their origin it may not be out of place to recall this crude Greek notion of the creation of the human race, and to compare or contrast it with other rudimentary speculations of primitive peoples on the same subject, if only for the sake of marking the interval which divides the childhood from the maturity of science.

Frazer 1909: 153

5 'The child helps us understand our own primitive selfs' (Harrison 1909: 508).
6 It was also one of the most prone to ideological misappropriation. In 1869, German naturalist Gustav Jäger (1832–1917) published *Die Darwin'sche Theorie und ihre Stellung zu Moral und Religion* ('The Darwinian Theory and its Relation to Morals and Religion'). In this work, aimed at reconciling Christianity with an aggressive form of Darwinian evolutionism, Jäger wrote that religion boosts social cohesion in view of out-group hostility and wrote that 'religion is a weapon in the struggle of survival'. However, Jäger's militant viewpoint led him to devise a Christianized Darwinism which justified an aggressive Christian imperialism (Jäger 1869: 114–15, 119; from Achtner 2009: 266). No wonder, thus, that Darwin, who read in part the volume in that same year, ignored Jäger's forays and merely cited an ornithological note about pheasants' plumage and sexual selection from Jäger's work in his *Descent of Man* (Darwin to Gustav Jäger, 9 September 1869, *Darwin Correspondence Project*, Letter no. DCP-LETT-6885).
7 The list, which focuses on consilient approaches in the study of culture and cognition, is illustrative and omits many potential additions, such as US anthropologist A. Irving Hallowell's (1892–1974) take on human culture in a phylogenetic perspective (see Saler 2009: 82–91). A more famous case is perhaps provided by American sociologist Robert N. Bellah (1927–2014). In 1964, Bellah published an article entitled 'Religious Evolution', in which he posited a cultural evolutionary development of social organization through different stages (Bellah 1964), starting a project which will be eventually completed only in the 2010s (Bellah 2011). Bellah, however, was only tangentially concerned with cognition and his reconstruction hinged on obsolete concepts (Ambasciano 2016a: 181–5; see also Boy and Torpey 2013; cf. Turchin 2015 for an updated take on Bellah's theory).
8 Even if religious beliefs and behaviours did not evolve for any particularly adaptive purpose – i.e. they can be a by-product of other cognitive abilities evolved to deal with other purposes (see Ambasciano 2016a: 177 n. 24) – they might have been co-opted and exploited socially and politically, thus re-entering into an evolutionary feedback loop (see Pievani, Girotto and Vallortigara 2014).
9 In 1990, Culianu's interest led him to establish *Incognita: International Journal for Cognitive Studies in the Humanities*, an interdisciplinary academic journal published by Brill, and discontinued after his death.

Bibliography

Achtner, W. (2009). 'The Evolution of Evolutionary Theories of Religions.' In E. Voland and W. Schiefenhovel (eds), *The Biological Evolution of Religious Mind and Behavior*, 257–75. Berlin and Heidelberg: Springer.

Ackerman, R. (1975). 'Frazer on Myth and Ritual.' *Journal of the History of Ideas* 36(1): 115–34.

Ackerman, R. (1990[1987]). *J. G. Frazer: His Life and Work*. Cambridge and New York: Cambridge University Press.

Ackerman, R. (2005a[1987]). *s.v.* 'Frazer, James George.' In L. Jones (ed.), *Encyclopedia of Religion. Second Edition*, vol. 5, 3190–3. Detroit: Thomson Gale–MacMillan.

Ackerman, R. (ed.) (2005b). *Selected Letters of Sir J. G. Frazer*. Cambridge and New York: Cambridge University Press.

Adams, R. (2017). 'Islamic School's Gender Segregation Is Unlawful, Court of Appeal Rules.' *The Guardian*, 13 October. https://www.theguardian.com/education/2017/oct/13/islamic-school-gender-segregation-unlawful-court-of-appeal (accessed 20 December 2017).

Adler, G. and A. Jaffé (eds) (1976). *C. G. Jung Letters*, trans. R. F. C. Hull, vol. 2: 1951–1961. London: Routledge & Kegan Paul.

Åkerlund, A. (2006). 'Åke Ohlmarks and the "Problem" of Shamanism.' *Archæus* 10(1–2): 200–20.

Albert, M. (1996). 'Science, Post-Modernism and the Left.' *Z Magazine* 7–8: 64–9.

Alexandrescu, S. (2006). *Mircea Eliade, dinspre Portugalia*. Bucharest: Humanitas.

Allen, D. and D. Doeing (1980). *Mircea Eliade: An Annotated Bibliography*. New York and London: Garland.

Alles, G. D. (ed.) (2008). *Religious Studies. A Global View*. London and New York: Routledge.

Allmon, W. D. (2002). 'The Structure of Gould: Happenstance, Humanism, History, and the Unity of His View of Life.' In W. D. Allmon, P. H. Kelley and R. M. Ross (eds), *Stephen Jay Gould: Reflections on His View of Life*, 3–68. Oxford and New York: Oxford University Press.

Alter, S. G. (2009). 'Darwin and the Linguists: The Coevolution of Mind and Language, Part 2. The Language–Thought Relationship.' *Studies in History and Philosophy of Science Part C: Studies in History and Philosophy of Biological and Biomedical Sciences* 39(1): 38–50.

Altizer, T. J. J. (2016). *Living the Death of God: A Theological Memoir*. Albany: State University of New York.

Alvesson, M. and A. Spicer (2012). 'A Stupidity-Based Theory of Organizations.' *Journal of Management Studies* 49(7): 1194–220.

Ambasciano, L. (2013). 'Tempi profondi. Geomitologia, storia della natura e studio della religione.' *Studi e Materiali di Storia delle Religioni* 79(1): 152–214.

Ambasciano, L. (2014). *Sciamanesimo senza sciamanesimo. Le radici intellettuali del modello sciamanico di Mircea Eliade: evoluzionismo, psicoanalisi, te(le)ologia*. Rome: Nuova Cultura.

Ambasciano, L. (2015a). 'Mapping Pluto's Republic: Cognitive and Epistemological Reflections on *Philosophy of Pseudoscience: Reconsidering the Demarcation Problem.' Journal for the Cognitive Science of Religion* 3(2): 183–205.

Ambasciano, L. (2015b). '*Homo mendax*, Fictional Heroes and Self-Deception: A Brief Commentary on *The Storytelling Animal: How Stories Make Us Human.' Studi e Materiali di Storia delle Religioni* 81(1): 239–47.

Ambasciano, L. (2016a). 'Mind the (Unbridgeable) Gaps: A Cautionary Tale about Pseudoscientific Distortions and Scientific Misconceptions in the Study of Religion.' *Method & Theory in the Study of Religion* 28(2): 141–225.

Ambasciano, L. (2016b). '(Pseudo)science, Religious Beliefs, and Historiography: Assessing *The Scientification of Religion*'s Method and Theory.' *Zygon* 51(4): 1062–6.

Ambasciano, L. (2016c). 'The Goddess Who Failed? Competitive Networks (or the Lack Thereof), Gender Politics, and the Diffusion of the Roman Cult of Bona Dea.' *Religio: revue pro religionistiku* 24(2): 111–65.

Ambasciano, L. (2016d). 'The Gendered Deep History of the Bona Dea Cult.' *Journal of Cognitive Historiography* 3(1–2): 134–56.

Ambasciano, L. (2016e). 'Achilles' Historiographical Heel, or the Infelicitous Predominance of Experimental Presentism in Ara Norenzayan's *Big Gods.' Studi e Materiali di Storia delle Religioni* 82(2): 1045–68.

Ambasciano, L. (2017a). '"The Persistence of a Delusion": Review of *Conversations and Controversies in the Scientific Study of Religion: Collaborative and Co-authored Essays by Luther H. Martin and Donald Wiebe*, edited by L. H. Martin and D. Wiebe (Brill, 2016).' *Endeavour* 41(3): 71–2.

Ambasciano, L. (2017b). 'Exiting the *Motel of the Mysteries*? How Historiographical Floccinaucinihilipilification Is Affecting CSR 2.0.' In L. H. Martin and D. Wiebe (eds), *Religion Explained? The Cognitive Science of Religion after Twenty-Five Years*, 107–22. London and New York: Bloomsbury.

Ambasciano, L. (2018a). 'Comparative Religion as a Life Science: William E. Paden's Neo-Plinian "New Naturalism".' *Method & Theory in the Study of Religion* 30(2): 141–9.

Ambasciano, L. (2018b). 'Politics of Nostalgia, Logical Fallacies, and Cognitive Biases: The Importance of Epistemology in the Age of Cognitive Historiography.' In A. K. Petersen, I. S. Gilhus, L. H. Martin, J. S. Jensen and J. Sørensen (eds), *Evolution, Cognition, and the History of Religion: A New Synthesis. Festschrift in Honour of Armin W. Geertz*, 280–96. Leiden and Boston: Brill.

Anati, E. (1999[1995]). *La religion des origines. Traduit par P. Michel*. Paris: Bayard.

Anderson, B. (1991[1983]). *Imagined Communities: Reflections on the Origin and Spread of Nationalism. Revised Edition*. London: Verso.

Andrews, K. (2011). 'Beyond Anthropomorphism: Attributing Psychological Properties to Animals.' In T. L. Beauchamp and R. G. Frey (eds), *The Oxford Handbook of Animal Ethics*, 469–94. Oxford: Oxford University Press.

Andriolo, K. R. (1979). '*Kulturkreislehre* and the Austrian Mind.' *Man. New Series* 14(1): 133–44.

Angelini, P. (2005). *L'uomo sul tetto. Mircea Eliade e la 'storia delle religioni'*. Turin: Bollati Boringhieri.

Angelini, P. (2012). 'Eliade, de Martino e il problema dei poteri magici.' In G. Casadio and P. Mander (eds), *Mircea Eliade. Le forme della Tradizione e del Sacro*, 11–38. Rome: Edizioni Mediterranee.

Angermuller, J. (2015). *Why There Is No Poststructuralism in France: The Making of an Intellectual Generation*. London: Bloomsbury.

Anonymous (1894). 'Professor Robertson Smith.' *Nature* 49(1279): 557.

Anton, T. (2005[1996]). *Eros, magie și asasinarea profesorului Culianu. Ediția a II-a românească revăzută*, trans. C. Felea. Iași: Polirom.

Antonio, R. J. (2000). 'After Postmodernism: Reactionary Tribalism.' *American Journal of Sociology* 106(1): 40–87.

Arnal, W. E. and R. T. McCutcheon (2013). *The Sacred Is the Profane: The Political Nature of 'Religion'*. Oxford and New York: Oxford University Press.

Arvidsson, S. (2006[2000]). *Aryan Idols: Indo-European Mythology as Ideology and Science*. Chicago and London: The University of Chicago Press.

Asad, T. (1993). *Genealogies of Religion: Discipline and Reasons of Power in Christianity and Islam*. Baltimore and London: The Johns Hopkins University Press.

Asprem, E. (2016). 'How Schrödinger's Cat Became a Zombie.' *Method & Study in the Study of Religion* 28(2): 113–40.

Atran, S. (2002). *In Gods We Trust: The Evolutionary Landscape of Religion*. Oxford and New York: Oxford University Press.

Baár, M. (2010). *Historians and Nationalism: East-Central Europe in the Nineteenth Century*. Oxford: Oxford University Press.

Bae, C. J., K. Douka and M. D. Petraglia (2017). 'On the Origin of Modern Humans: Asian Perspectives.' *Science* 358(6368): eaai9067.

Baggini, J. (2017). *A Short History of Truth: Consolations for a Post-Truth World*. London: Quercus.

Bahn, P. H. (2010). *Prehistoric Art: Progress and Polemics*. Cambridge and New York: Cambridge University Press.

Baldwin, J. M. (1896). 'A New Factor in Evolution.' *The American Naturalist* 30(354): 441–51.

Baldwin, J. M. (1909). *Darwin and the Humanities*. Baltimore: Williams & Wilkins.

Bamberger, J. (1974). 'The Myth of Matriarchy: Why Men Rule in Primitive Society.' In L. Lamphere and M. Zimbalist Rosaldo (eds), *Women, Culture, and Society*, 263–80. Stanford: Stanford University Press.

Barrett, D. (2010). *Supernormal Stimuli: How Primal Urges Overran Their Evolutionary Purpose*. New York: W.W. Norton & Co.

Barrett, H. C. and R. Kurzban (2006). 'Modularity in Cognition: Framing the Debate.' *Psychological Review* 113(3): 628–47.

Barrett, J. L. (2000). 'Exploring the Natural Foundations of Religion.' *Trends in Cognitive Sciences* 4(1): 29–34.

Barrett, J. L. (2004). *Why Would Anyone Believe in God?* Lanham, MD and Plymouth, UK: AltaMira Press.

Barrett, J. L. (2017). 'Could We Advance the Science of Religion (Better) Without the Concept "Religion"?' *Religion, Brain & Behavior* 7(4): 282–4.

Barrett, P. H., P. J. Gautrey, S. Herbert, D. Kohn and S. Smith (eds) (1987). *Charles Darwin's Notebooks, 1836–1844: Geology, Transmutation of Species, Metaphysical Enquiries*. Cambridge and New York: Cambridge University Press.

Beard, C. (2004). *The Hunt for the Dawn Monkey: Unearthing the Origins of Monkeys, Apes, and Humans*. Berkeley, Los Angeles and London: University of California Press.

Beard, M. (1992). 'Frazer, Leach, and Virgil: The Popularity (and Unpopularity) of the *Golden Bough*.' *Comparative Studies in Society and History* 34(2): 203–24.

Bechtel, W., A. Abrahamsen and G. Graham (2001). 'Cognitive Science, History.' In N. J. Smelser and P. B. Baltes (eds), *International Encyclopedia of the Social and Behavioral Sciences*, 2154–8. Oxford: Elsevier Science.

Beck, P. (1904). *Die Nachahmung und ihre Bedeutung für Psychologie und Völkerkunde.* Leipzig: Kröner.

Bediako, G. M. (1997). *Primal Religion and the Bible: William Robertson and His Heritage.* Sheffield: Sheffield Academic Press.

Beidelman, T. O. (1974). *W. Robertson Smith and the Sociological Study of Religion. With a Foreword by E. E. Evans-Pritchard.* Chicago: The University of Chicago Press.

Bellah, R. N. (1964). 'Religious Evolution.' *American Sociological Review* 29(3): 358–74.

Bellah, R. N. (2011). *Religion in Human Evolution: From the Paleolithic to the Axial Age.* Cambridge, MA: Harvard University Press.

Bentlage, B., M. Eggert, H.-M. Krämer and S. Reichmuth (eds) (2016). *Religious Dynamics under the Impact of Imperialism and Colonialism: A Sourcebook.* Leiden and Boston: Brill.

Berger, A. (1994). 'Mircea Eliade: Romanian Fascism and the History of Religions in the United States.' In N. A. Harrowitz (ed.), *Tainted Greatness: Antisemitism and Cultural Heroes*, 51–74. Philadelphia: Temple University Press.

Bergunder, M. (2014). 'What Is Religion? The Unexplained Subject Matter of Religious Studies.' *Method & Theory in the Study of Religion* 26(3): 246–86.

Bering, J. M. (2006). 'The Folk Psychology of Souls.' *Behavioral and Brain Sciences* 29(5): 453–98.

Bettini, M. (2014). *Elogio del politeismo. Quello che possiamo imparare dalle religioni antiche.* Bologna: il Mulino.

Bianchi, U. (1958). *Il dualismo religioso. Saggio storico ed etnologico.* Rome: L'Erma di Bretschneider.

Bianchi, U. (1961). 'Après Marbourg: (Petit discours sur la méthode).' *Numen* 8(1): 64–78.

Bianchi, U. (1975a). *The History of Religions.* Leiden: Brill.

Bianchi, U. (1975b). 'La religione nella storia delle religioni.' In A. Babolin (ed.), *Il metodo della filosofia della religione*, vol. I, 17–46. Padua: La Garangola.

Bianchi, U. (1989). 'The Study of Religion in the Context of Catholic Culture.' In M. Pye (ed.), *Marburg Revisited: Institutions and Strategies in the Study of Religion*, 49–54. Marburg: Diagonal.

Bianchi, U. and M. J. Vermaseren (eds) (1982). *La soteriologia dei culti orientali nell'Impero romano. Atti del Colloquio Internazionale su La soteriologia dei culti orientali nell'Impero Romano, Roma 24–28 Settembre 1979.* Leiden: Brill.

Binford, L. R. (1962). 'Archaeology as anthropology.' *American Antiquity* 28(2): 217–25.

Black, J. S. and G. Chrystal (1912a). *The Life of William Robertson Smith.* London: Adam & Charles Black.

Black, J. S. and G. Chrystal (eds) (1912b). *Lectures & Essays of William Robertson Smith.* London: Adam & Charles Black.

Blackburn, S. (2016). *The Oxford Dictionary of Philosophy. Third Edition.* Oxford: Oxford University Press.

Blackburn, S. (2017). *Truth.* London: Profile.

Blancke, S. and J. De Smedt (2013). 'Evolved to Be Irrational? Evolutionary and Cognitive Foundations of Pseudosciences.' In M. Pigliucci and M. Boudry (eds), *Philosophy of Pseudoscience: Reconsidering the Demarcation Problem*, 361–80. Chicago and London: The University of Chicago Press.

Blancke, S., H. H. Hjermitslev and P. C. Kjærgaard (eds) (2014). *Creationism in Europe.* Baltimore: Johns Hopkins University Press.

Bliege Bird, R., D. W. Bird, B. F. Codding, C. H. Parker and J. H. Jones (2008). 'The "Fire Stick Farming" Hypothesis: Australian Aboriginal Foraging Strategies, Biodiversity, and Anthropogenic Fire Mosaics.' *Proceedings of the National Academy of Sciences* 105(39): 14796–801.

Bloch, M. (2008). 'Why Religion Is Nothing Special But Is Central.' *Philosophical Transactions of the Royal Society B* 363(1499): 2055–61.

Bloch, M. (2015). 'Durkheimian Anthropology and Religion: Going In and Out of Each Other's Bodies.' *HAU: Journal of Ethnographic Theory* 5(3): 285–99.

Bloch, M. and D. Sperber (2002). 'Kinship and Evolved Psychological Dispositions: The Mother's Brother Controversy Reconsidered.' *Current Anthropology* 43(5): 723–48.

Boas, F. (1916). 'Eugenics.' *The Scientific Monthly* 3(5): 471–8.

Boas, F. (1938). 'An Anthropologist's Credo.' *The Nation* 147: 201–4.

Bod, R. (2015[2010]). *A New History of the Humanities: The Search for Principles and Patterns from Antiquity to the Present*, trans. L. Richards. Oxford: Oxford University Press.

Booth, G. (2010). 'A Trio of London Clubmen: Edmund Gosse, W. Robertson Smith and R. L. Stevenson.' *Literary London* 8(2). http://www.literarylondon.org/london-journal/september2010/booth.html (accessed 15 September 2017).

Bordaş, L. (2012). 'Ioan Petru Culianu, Mircea Eliade and Felix Culpa. With the I. P. Culianu–M. L. Ricketts Correspondence.' In M. Gligor (ed.), *Mircea Eliade between the History of Religions and the Fall into History*, 67–178. Cluj-Napoca: Presa Universitară Clujeană.

Borgeaud, P. (2013). *L'Histoire des religions*. Gollion: inFolio.

Bornemann, F. (1979). 'Aus den Studienjahren P. W. Schmidts.' *Verbum SVD* 20(3): 243–70.

Bouckaert, R., P. Lemey, M. Dunn, S. J. Greenhill, A. V. Alekseyenko, A. J. Drummond, R. D. Gray, M. A. Suchard and Q. Atkinson (2012). 'Mapping the Origins and Expansion of the Indo-European Language Family.' *Science* 337(6097): 957–60.

Boudry, M. (2017). '*Plus Ultra*: Why Science Does Not Have Limits.' In M. Boudry and M. Pigliucci (eds), *Science Unlimited? The Challenges of Scientism*, 31–52. Chicago and London: The University of Chicago Press.

Boudry, M. and J. Braeckman (2012). 'How Convenient! The Epistemic Rationale of Self-validating Belief Systems.' *Philosophical Psychology* 25(3): 341–64.

Boudry, M., S. Blancke and M. Pigliucci (2015). 'What Makes Weird Beliefs Thrive? The Epidemiology of Pseudoscience.' *Philosophical Psychology* 28(8): 1177–98.

Boule, M. (1913). 'L'homme fossile de la Chapelle-aux-Saints.' *Annales de Paléontologie* 8: 209–79.

Bourdieu, P. (1971). 'Genèse et structure du champ religieux.' *Revue française de sociologie* 12(3): 295–334.

Bowler, P. J. (1987[1986]). *Theories of Human Evolution: A Century of Debate 1844–1944*. Oxford: Blackwell.

Bowler, P. J. (1989). 'Holding Your Head Up High: Degeneration and Orthogenesis in Theories of Human Evolution.' In J. R. Moore (ed.), *History, Humanity and Evolution: Essays for John C. Greene*, 329–54. Cambridge and New York: Cambridge University Press.

Bowler, P. J. (1992[1983]). *The Eclipse of Darwinism: Anti-Darwinian Evolution Theories in the Decades around 1900*. Baltimore and London: The Johns Hopkins University Press.

Bowler, P. J. and I. R. Morus (2005). *Making Modern Science: A Historical Survey*. Chicago and London: The University of Chicago Press.

Boy, J. D. and J. Torpey (2013). 'Inventing the Axial Age: The Origins and Uses of a Historical Concept.' *Theory and Society* 42(3): 241–59.

Boyd, R., P. J. Richerson and J. Henrich (2011). 'The Cultural Niche: Why Social Learning Is Essential for Human Adaptation.' *Proceedings of the National Academy of Sciences* 108(26): 10918–25.

Boyer, P. (1983). 'Le status des forgerons et ses justifications symboliques: une hypothèse cognitive.' *Africa: Journal of the International African Institute* 53(1): 44–63.

Boyer, P. (2000). 'Functional Origins of Religious Concepts: Ontological and Strategic Selection in Evolved Minds.' *Journal of the Royal Anthropological Institute* 6(2): 195–214.

Boyer P. (2001). *Religion Explained: The Evolutionary Origins of Religious Thought*. New York: Basic Books.

Boyer, P. (2013). 'Explaining Religious Concepts: Lévi-Strauss, the Brilliant and Problematic Ancestor.' In D. Xygalatas and W. W. McCorkle Jr (eds), *Mental Culture: Classical Social Theory and the Cognitive Science of Religion*, 164–75. Durham and Bristol, CT: Acumen.

Boyer, P. and C. Ramble (2001). 'Cognitive Templates for Religious Concepts: Cross-Cultural Evidence for Recall of Counter-Intuitive Representations.' *Cognitive Science* 25(4): 535–64.

Brandewie, E. (1982). 'Wilhelm Schmidt: A Closer Look.' *Anthropos* 77(1–2): 151–62.

Brandewie, E. (1983). *Wilhelm Schmidt and the Origin of the Idea of God*. Boston and London: University Press of America.

Brelich, A. (1979). *Storia delle religioni, perché?* Naples: Liguori editore.

Bremmer, J. (2011). 'Religious Violence and Its Roots: A View from Antiquity.' *Asdiwal* 6: 71–9.

Brooke, J. H. (1991). *Science and Religion: Some Historical Perspectives*. Cambridge and New York: Cambridge University Press.

Brown, D. M. (1993). 'Introduction.' In J. W. Hall (ed.), *The Cambridge History of Japan. Vol. I: Ancient Japan*, 1–47. Cambridge: Cambridge University Press.

Buekens, F. (2013). 'Agentive Thinking and Illusions of Understanding.' In M. Pigliucci and M. Boudry (eds), *Philosophy of Pseudoscience: Reconsidering the Demarcation Problem*, 439–58. Chicago and London: The University of Chicago Press.

Bulbulia, J. and E. Slingerland (2012). 'Religious Studies as a Life Science.' *Numen* 59(5–6): 564–613.

Bulbulia, J., R. Sosis, E. Harris, R. Genet, C. Genet and K. Wyman (eds) (2008). *The Evolution of Religion: Studies, Theories, & Critiques*. Santa Margarita, CA: Collins Foundation Press.

Burgess, P. (ed.) (1989). *The Annual Obituary 1986*. Chicago and London: St James Press.

Burkert, W. (1996). *Creation of the Sacred: Tracks of Biology in Early Religions*. Cambridge, MA and London: Harvard University Press.

Burman, J. T. (2014). 'Bringing the Brain into History: Behind Hunt's and Smail's Appeals to Neurohistory.' In C. Tileagă and J. Byford (eds), *Psychology and History: Interdisciplinary Explorations*, 64–82. Cambridge and New York: Cambridge University Press.

Campbell, D. T. (1960). 'Blind Variation and Selective Retentions in Creative Thought as in Other Knowledge Processes.' *Psychological Review* 67(6): 380–400.

Cantrell, M. A. (2016). 'Must a Scholar of Religion Be Methodologically Atheistic or Agnostic?' *Journal of the American Academy of Religion* 84(2): 373–400.

Capps, W. H. (1995). *Religious Studies: The Making of a Discipline*. Minneapolis: Fortress Press.

Carney, J. and P. MacCarron (2017). 'Comic-Book Superheroes and Prosocial Agency:

A Large-Scale Quantitative Analysis of the Effects of Cognitive Factors on Popular Representations.' *Journal of Cognition and Culture* 17(3–4): 306–30.

Carozzi, P. A. (1994). 'Due maestri di fenomenologia storica delle religioni: Uberto Pestalozza e Mircea Eliade.' In G. Sfameni Gasparro (ed.), *Agathē Elpis. Studi storico-religiosi dedicati a Ugo Bianchi*, 35–62. Rome: L'Erma di Bretschneider.

Casadio, G. (1993). '"Das Ewig Weibliche zieht uns hinan". Archetipi e storia nell'opera di Uberto Pestalozza: la formazione di uno storico delle religioni.' *Torricelliana* 44: 255–75.

Casadio, G. (2002). 'The Study of Ancient Mediterranean Religions in Eliade's *Encyclopedia of Religion* (1987).' In G. Sfameni Gasparro (ed.), *Themes and Problems of the History of Religions in Contemporary Europe: Proceedings of the International Seminar, Messina, March 30–31, 2001*, 79–110. Cosenza: Edizioni Giordano.

Casadio, G. (2005a). *s.v.* 'Historiography [Further Considerations].' In L. Jones (ed.), *Encyclopedia of Religion*, vol. 6, 4042–52. Detroit: Thomson Gale–MacMillan.

Casadio, G. (2005b). *s.v.* 'Bianchi, Ugo.' In L. Jones (ed.), *Encyclopedia of Religion*, vol. 2, 862–5. Detroit: Thomson Gale–MacMillan.

Casadio, G. (2011). 'Mircea Eliade visto da Mircea Eliade.' In C. Scarlat (ed.), *Mircea Eliade Once Again*, 103–43. Iași: Lumen.

Cassirer, E. A. (1945). 'Structuralism in Modern Linguistics.' *Word* 1(2): 99–120.

Cavalli-Sforza, L. L. (2010). *L'evoluzione della cultura*. Turin: Codice edizioni.

Cavalli-Sforza, L. L. and M. Feldman (1981). *Cultural Transmission and Evolution: A Quantitative Approach*. Princeton: Princeton University Press.

Cavalli-Sforza, L. L., P. Menozzi and A. Piazza (1994). *The History and Geography of Human Genes*. Princeton: Princeton University Press.

Chantepie de la Saussaye, P. D. (1871). *Methodologische bijdrage naar den oorsprong van den godsdienst; academisch proefschrift Utrechtse Hoogeschool*. Utrecht: P. W. van de Weijer.

Chantepie de la Saussaye, P. D. (1887). *Lehrbuch der Religionsgeschichte*. Freiburg im Breisgau: Mohr.

Chapman, G. P. (ed.) (1992). *Grass Evolution and Domestication*. Cambridge: Cambridge University Press.

Chatfield, T. (2018). *Critical Thinking: Your Guide to Effective Argument, Successful Analysis and Independent Study*. Los Angeles and London: SAGE.

Chidester, D. (2009). 'Darwin's Dogs: Animals, Animism, and the Problem of Religion.' *Soundings* 92(1–2): 51–75.

Chidester, D. (2014). *Empire of Religion: Imperialism and Comparative Religion*. Chicago and London: The University of Chicago Press.

Chomsky, N. (1959). 'A Review of B. F. Skinner's Verbal Behavior.' *Language* 35(1): 26–58.

Christian, D. (2010). 'The Return of Universal History.' *History and Theory* 49(4): 6–27.

Christian, D. (2011[2004]). *Maps of Time: An Introduction to Big History. With a New Preface*. Berkeley, Los Angeles and London: University of California Press.

Cioffi, F. (2013). 'Pseudoscience: The Case of Freud's Sexual Etiology of the Neuroses.' In M. Pigliucci and M. Boudry (eds), *Philosophy of Pseudoscience: Reconsidering the Demarcation Problem*, 321–40. Chicago and London: The University of Chicago Press.

Ciurtin, E. (ed.) (2000). *Zalmoxis. Revista de studii religioase. Volumele I–III (1938–1942) publicată sub direcția lui M. Eliade*. Iași: Polirom.

Ciurtin, E. (2008). 'Raffaele Pettazzoni et Mircea Eliade. Historiens des religions généralistes devant les fascismes (1933–1945).' In H. Junginger (ed.), *The Study of Religion Under the Impact of Fascism*, 333–63. Leiden and Boston: Brill.

Clark, E. A. (2005). 'Engaging Bruce Lincoln.' *Method & Theory in the Study of Religion* 17(1): 11–17.

Clark, R. (2015). *Holy Legionary Youth. Fascist Activism in Interwar Romania.* Ithaca and London: Cornell University Press.

Clarkson, C., Z. Jacobs, B. Marwick, R. Fullagar, L. Wallis, M. Smith, R. G. Roberts, E. Hayes, K. Lowe, X. Carah, S. A. Florin, J. McNeil, D. Cox, L. J. Arnold, Q. Hua, J. Huntley, H. E. A. Brandl, T. Manne, A. Fairbairn, J. Shulmeister, L. Lyle, M. Salinas, M. Page, K. Connell, G. Park, K. Norman, T. Murphy and C. Pardoe (2017). 'Human Occupation of Northern Australia by 65,000 Years Ago.' *Nature* 547(7663): 306–10.

Clasen, M. (2017). *Why Horror Seduces.* Oxford: Oxford University Press.

Cleland, C. E. and S. Brindell (2013). 'Science and the Messy, Uncontrollable World of Nature.' In M. Pigliucci and M. Boudry (eds), *Philosophy of Pseudoscience: Reconsidering the Demarcation Problem*, 183–202. Chicago and London: The University of Chicago Press.

Clottes, J. (2011). *Pourquoi l'art préhistorique?* Paris: Gallimard.

Coley, J. D. (2007). 'The Human Animal: Developmental Changes in Judgments of Taxonomic and Psychological Similarity among Humans and Other Animals.' *Cognition, Brain, Behavior* 11(4): 733–56.

Coley, J. D. and K. Tanner (2015). 'Relations between Intuitive Biological Thinking and Biological Misconceptions in Biology Majors and Nonmajors.' *CBE Life Sciences Education* 14(1), ar8. http://doi.org/10.1187/cbe.14–06–0094

Collingwood, R. G. (1946). *The Idea of History.* Oxford: Oxford University Press.

Connelly, J. (2007). 'Catholic Racism and Its Opponents.' *The Journal of Modern History* 79(4): 813–47.

Conte, E. (1988). 'Le confesseur du dernier Habsbourg et les "nouveaux païens" allemands: À propos de Wilhelm Schmidt.' *Ethnologie française*, n.s. 18(2): 120–30.

Copson, A. (2017). *Secularism: Politics, Religion, and Freedom.* Oxford: Oxford University Press.

Correia, D. (2013). 'F**k Jared Diamond.' *Capitalism Nature Socialism* 24(4): 1–6.

Corsi, P. (2005). 'Before Darwin: Transformist Concepts in European Natural History.' *Journal of the History of Biology* 38(1): 67–83.

Costandi, M. (2016). *Neuroplasticity.* Cambridge, MA and London: The MIT Press.

Cox, J. L. (2002). *A Guide to the Phenomenology of Religion: Key Figures, Formative Influences and Subsequent Debates.* London and New York: Bloomsbury.

Cox, J. L. (2014a). *The Invention of God in Indigenous Societies.* London and New York: Routledge.

Cox, J. L. (2014b). 'Phenomenological Perspectives on the Social Responsibility of the Scholar of Religion.' In Á. Kovács and J. L. Cox (eds), *New Trends and Recurring Issues in the Study of Religion*, 133–51. Paris: L'Harmattan.

Coyne, J. A. (2015). *Faith vs. Fact: Why Science and Religion Are Incompatible.* New York: Viking.

Creech, J., P. Kamuf and J. Todd (1985). 'Deconstruction in America: An Interview with Jacques Derrida.' *Critical Exchange* 17: 1–32.

Cresci, E. (2017). 'Why Fans Think Avril Lavigne Died and Was Replaced by a Clone Called Melissa.' *The Guardian*, 15 May. https://www.theguardian.com/lifeandstyle/shortcuts/2017/may/15/avril-lavigne-melissa-cloning-conspiracy-theories (accessed 15 May 2017).

Croce, B. (1947[1921]). *Storia della storiografia italiana nel secolo decimonono*, vol. 2. Bari: Laterza.

Culianu, I. P. (1978). *Mircea Eliade*. Assisi: Cittadella.

Culianu, I. P. (1981). 'Religione e accrescimento del potere.' In G. Romanato, M. G. Lombardo and I. P. Culianu, *Religione e potere*, 173–251. Turin: Marietti.

Culianu, I. P. (1987[1984]). *Eros and Magic in the Renaissance*, trans. M. Cook. Chicago and London: The University of Chicago Press.

Culianu, I. P. (1990a). 'Dr. Faust, Great Sodomite and Necromancer.' *Revue de l'histoire des religions* 207(3): 261–88.

Culianu, I. P. (1990b). Review of Kitagawa, J. M. (1987). 'The History of Religions: Understanding Human Experience.' Atlanta: Scholars Press. *Church History* 59(1): 135–6.

Culianu, I. P. (1991a). 'A Corpus for the Body.' *The Journal of Modern History* 63(1): 61–80.

Culianu, I. P. (1991b). *Out of This World: Otherworldly Journeys from Gilgamesh to Albert Einstein*. Boston: Shambhala Press.

Culianu, I. P. (2002[1992]). 'Religia ca sistem.' In *Jocurile minții. Istoria ideilor, teoria culturii, epistemologie. Ediție îngrijită de M. Antohi și S. Antohi*, 339–48. Iași: Polirom.

Culianu, I. P. (2003[1988]). *s.v.* 'Trickster.' In *Cult, magie, erezii. Articole din enciclopedii ale religiilor. Traduceri de M.-M. Anghelescu și D. Petrescu*, 151–3. Iași: Polirom.

Culianu, I. P. (2005a). *Păcatul împotriva spiritului. Scrieri politice. Ediția a II-a adăugită*. Iași: Polirom.

Culianu, I. P. (2005b[1992]). *Arborele Gnozei. Mitologia Gnostică de la creștinismul timpuriu la nihilismul modern. Ediția a II-a adăugită*, trans. C. Popescu. Iași: Polirom.

Culianu-Petrescu, T. (2003). 'O biografie.' In S. Antohi (ed.), *Ioan Petru Culianu: omul și opera*, trans. C. Popescu, C. Dumitriu, I. Ieronim, C. Ionică and T. Culianu-Petrescu, 54–82. Iași: Polirom.

Culianu-Petrescu, T. and D. Petrescu (eds) (2004). *Dialoguri întrerupte. Corespondență Mircea Eliade -Ioan Petru Culianu*. Iași: Polirom.

Culver, D. C. and T. Pipan (2009). *The Biology of Caves and Other Subterranean Habitats*. Oxford and New York: Oxford University Press.

Currie, A. (2016). 'Ethnographic Analogy, the Comparative Method, and Archaeological Special Pleading.' *Studies in History and Philosophy of Science Part A* (55): 84–94.

Czaplicka, M. A. (1914). *Aboriginal Siberia: A Study in Social Anthropology*. Oxford: Clarendon Press.

Daintith, J. and E. Martin (eds) (2010[1984]). *Oxford Dictionary of Science. Sixth Edition*. Oxford: Oxford University Press.

D'Ancona, M. (2017). *Post-Truth: The New War on Truth and How to Fight Back*. London: Ebury Press.

Daniélou, J., A. H. Couratin and J. Kent (1969). *The Pelican Guide to Modern Theology. Volume 2: Historical Theology*. Harmondsworth: Penguin.

Darwin, C. R. (1859). *On the Origin of Species by Means of Natural Selection, or the Preservation of Favoured Races in the Struggle for Life*. London: John Murray.

Darwin, C. R. (1861). *On the Origin of Species by Means of Natural Selection, or the Preservation of Favoured Races in the Struggle for Life. Third Edition, with Additions and Corrections*. London: John Murray.

Darwin, C. R. (1871). *The Descent of Man, and Selection in Relation to Sex*. 2 vols. London: John Murray.

Darwin, C. R. (1872a). *The Expression of the Emotions in Man and Animals*. London: John Murray.

Darwin, C. R. (1872b). *The Origin of Species by Means of Natural Selection, or the Preservation of Favoured Races in the Struggle for Life*. 6th edn. London: John Murray.

Darwin, C. R. (1874). *The Descent of Man, and Selection in Relation to Sex. Second Edition, Revised and Augmented*. London: John Murray.

Darwin, C. R. (1890). *The Expression of the Emotions in Man and Animals. Second Edition*, edited by F. Darwin. London: John Murray.

David, D. (2015). 'Life Is Not a Fractal in Hilbert Space.' *Studii de Știință și Cultură* 11(2): 95–9.

Davis, J. R. and A. Nicholls (2016). 'Friedrich Max Müller: The Career and Intellectual Trajectory of a German Philologist in Victorian Britain.' *Publications of the English Goethe Society* 85(2–3): 67–97.

Dawkins, R. (1986). *The Blind Watchmaker*. New York: Norton.

Day, M. (2008). 'Godless Savages and Superstitious Dogs: Charles Darwin, Imperial Ethnography, and the Problem of Human Uniqueness.' *Journal of the History of Ideas* 69(1): 49–70.

Day, M. (2010). 'The Educator Must Be Educated: The Study of Religion at the End of the Humanities.' *Method & Theory in the Study of Religion* 22(1): 1–8.

de Camp, L. S. (1970[1954]). *Lost Continents: The Atlantis Theme in History, Science, and Literature. Revised Edition*. New York: Dover.

De Cruz, H. (2017). *s.v.* 'Religion and Science.' In E. N. Zalta (ed.), *The Stanford Encyclopedia of Philosophy* (Spring 2017 Edition). https://plato.stanford.edu/archives/ spr2017/entries/religion-science/ (accessed 20 January 2018).

De Cruz, H. and J. De Smedt (2015). *A Natural History of Natural Theology: The Cognitive Science of Theology and Philosophy of Religion*. Cambridge, MA and London: The MIT Press.

De Matteis, S. (1997). 'La tradizione dell'Occidente e il paradosso del primitivo: dall'etnologia storicista ai poteri magici.' In E. de Martino, *Naturalismo e storicismo nell'etnologia*, edited by S. de Matteis, 9–49. Lecce: Argo.

Dennett, D. C. (2006). *Breaking the Spell: Religion as a Natural Phenomenon*. New York: Viking.

Dennett, D. C. (2013). *Intuition Pumps and Other Tools for Thinking*. London: Penguin.

De Quincey, T. (2013[1821–1856]). *The Confessions of an English Opium-Eater and Other Writings*. Oxford: Oxford University Press.

Derrida, J. (1994). *Force de loi*. Paris: Galilée.

Derrida, J. (1997[1994]). *Politics of Friendship*. London: Verso.

Derrida, J. (2002). *Acts of Religion. Edited and with an Introduction by G. Anidjar*. New York and London: Routledge.

Derrida, J. and M. Ferraris (2001[1997]). *A Taste for the Secret*. Cambridge: Polity Press.

Desmazières, A. (2009). 'La psychanalyse à l'Index: Sigmund Freud aux prises avec le Vatican (1921–1934).' *Vingtième Siècle. Revue d'histoire* 102(2): 79–91.

Desmond, A. J. (2009). *s.v.* 'Huxley, Thomas Henry (1825–1895).' In M. Ruse and J. Travis (eds), *Evolution: The First Four Billion Years*, 649–51. Cambridge, MA and London: The Belknap Press of Harvard University Press.

Desmond, A. J. and J. Moore (1991). *Darwin*. London: Michael Joseph.

Desmond, A. J. and J. Moore (2009). *Darwin's Sacred Cause: Race, Slavery and the Quest for Human Origins*. London: Penguin.

de Waal, F. B. M. (2001). *The Ape and the Sushi Master: Cultural Reflections by a Primatologist*. New York: Basic Books.

de Waal, F. B. M. (2013). *The Bonobo and the Atheist: In Search of Humanism among the Primates*. New York and London: W.W. Norton & Co.

de Waal, F. B. M. (2016). *Are We Smart Enough to Know How Smart Animals Are?* London: Granta.

Diamond, J. (2012). *The World until Yesterday: What Can We Learn from Traditional Societies?* New York: Viking Press.

Diamond, J. (2015). *Da te solo a tutto il mondo. Un ornitologo osserva le società umane.* Turin: Einaudi.

Diamond, J. and J. Robinson (eds) (2010). *Natural Experiments of History.* Cambridge, MA and London: The Belknap Press of Harvard University Press.

Di Donato, R. (2013). *s.v.* 'de Martino, E.' In *Il Contributo italiano alla storia del Pensiero – Storia e Politica.* http://www.treccani.it/enciclopedia/ernesto-de-martino_%28Il-Contributo-italiano-alla-storia-del-Pensiero:-Storia-e-Politica%29/ (accessed 15 December 2017).

Di Donato, R. (2015). *s.v.* 'Pestalozza, Uberto.' In *Dizionario Biografico degli Italiani* 82. http://www.treccani.it/enciclopedia/uberto-pestalozza_(Dizionario_Biografico)/ (accessed 8 December 2017).

Dietrich, S. (1992). 'Mission, Local Culture and the "Catholic Ethnology" of Pater Schmidt.' *Journal of the Anthropological Society of Oxford* 23(2): 111–25.

DiMaggio, P. (1997). 'Culture and Cognition.' *Annual Review of Sociology* 23(1): 263–87.

di Nola, A. M. (1977a). *s.v.* 'Religioni, Storia delle.' In A. M. di Nola (ed.), *Enciclopedia di storia delle religioni*, vol. 5, coll. 260–309. Firenze: Vallecchi.

di Nola, A. M. (1977b). *Mircea Eliade e l'antisemitismo. La Rassegna Mensile di Israel, terza serie* 43(1–2): 12–15.

di Nola, A. M. (1977c). 'Ancora antisemitismo.' *La Rassegna Mensile di Israel, terza serie* 43(5–6): 212–14.

di Nola, A. M. (1989). 'Mircea Eliade tra scienza delle religioni e ideologia "guardista".' *Marxismo oggi* 3(5–6): 66–71.

Doležalová, I., L. H. Martin and D. Papoušek (eds) (2001). *The Study of Religion during the Cold War, East and West.* New York: Peter Lang.

Doniger, W. (1991). 'Foreword.' In B. Lincoln, *Death, War, and Sacrifice: Studies in Ideology & Practice*, ix–xi. Chicago and London: The University of Chicago Press.

DuBois, T. A. (2009). *An Introduction to Shamanism.* Cambridge and New York: Cambridge University Press.

DuBois, T. A. (2011). 'Trends in Contemporary Research on Shamanism.' *Numen* 58(1): 100–28.

Dubuisson, D. (2003[1998]). *The Western Construction of Religion: Myths, Knowledge, and Ideology*, trans. W. Sayers. Baltimore: The Johns Hopkins University Press.

Dubuisson, D. (2005). *Impostures et pseudo-science. L'oeuvre de Mircea Eliade.* Villeneuve d'Ascq: Presses Universitaires du Septentrion.

Dubuisson, D. (2014[1993]). *Twentieth Century Mythologies: Dumézil, Lévi-Strauss, Eliade. Second Edition, Augmented and Revised*, trans. M. Cunningham. London and New York: Routledge.

Dudley, G. III (1976). 'Mircea Eliade as the "Anti-Historian" of Religions.' *Journal of the American Academy of Religion* 44(2): 345–59.

Dumitru, D. (2016). *The State, Antisemitism, and Collaboration in the Holocaust: The Borderlands of Romania and the Soviet Union.* Cambridge: Cambridge University Press.

Eco, U. (1989). 'Introduzione. La semiosi ermetica e il "paradigma del velame".' In M. P. Pozzato (ed.), *L'idea deforme. Interpretazioni esoteriche di Dante.* Milan: Bompiani.

Eco, U. (1995). 'Ur-Fascism.' *The New York Review of Books* 42(11), 22 June: 12–15.

Eco. U. (2008[2006]). *Turning Back the Clock: Hot Wars and Media Populism.* London: Vintage.

Eidinow, E. and L. H. Martin (2014). 'Editors' Introduction.' *Journal of Cognitive Historiography* 1(1): 5–9.

Eldredge, N. (2005). *Darwin: Discovering the Tree of Life*. New York: W.W. Norton & Co.

Eldredge, N. and S. J. Gould (1972). 'Punctuated Equilibria: An Alternative to Phyletic Gradualism.' In T. J. M. Schopf (ed.), *Models in Paleobiology*, 82–115. San Francisco: Freeman, Cooper & Co.

Eliade, M. (1936). *Yoga. Essai sur les origines de la mystique indienne*. Bucharest and Paris: Fundaţia pentru literatură şi artă 'Regele Carol II'/Librairie orientaliste Paul Geuthner.

Eliade, M. (1939). 'Les livres populaires dans la littérature roumaine.' *Zalmoxis* (2): 63–78.

Eliade, M. (1946). 'Le problème du chamanisme.' *Revue de l'histoire des religions* 131(1): 5–52.

Eliade, M. (1948a). 'Science, idéalisme et phénomènes paranormaux.' *Critique* (23): 315–23.

Eliade, M. (1948b). 'La mythologie primitive.' *Critique* (27): 708–17.

Eliade, M. (1949). 'Préface.' In P. Laviosa Zambotti, *Les origines et la diffusion de la civilisation. Introduction à l'histoire universelle*, i–iv. Paris: Payot.

Eliade, M. (1950). Review of G. van der Leeuw (1948). *La religion dans son essence et ses manifestations. Édition française refondue et mise à jour par l'auteur avec la collaboration du traducteur Jacques Marty*. Paris: Payot. *Revue de l'histoire des religions* 138(1): 108–11.

Eliade, M. (1951). *Le chamanisme et les techniques archaïques de l'extase*. Paris: Payot.

Eliade, M. (1954[1949]). *Cosmos and History: The Myth of the Eternal Return*, trans. W. R. Trask. Princeton: Princeton University Press.

Eliade, M. (1958a[1949]). *Patterns in Comparative Religion*, trans. R. Sheed. New York: Sheed & Ward.

Eliade, M. (1958b[1954]). *Yoga: Immortality and Freedom*, trans. W. R. Trask. Princeton: Princeton University Press.

Eliade, M. (1959[1957]). *The Sacred and the Profane: The Nature of Religion*, trans. W. R. Trask. New York: Harcourt, Brace & Co.

Eliade, M. (1960[1954]). 'Sense-experience and Mystical Experience among Primitives.' In *Myths, Dreams, and Mysteries: The Encounter Between Contemporary Faiths and Archaic Realities*, 73–98. New York: Harper & Row.

Eliade, M. (1961[1952]). *Images and Symbols: Studies in Religious Symbolism*, trans. P. Mairet. New York: Sheed & Ward.

Eliade, M. (1964). *Shamanism: Archaic Techniques of Ecstasy*, trans. W. R. Trask. New York: Pantheon-Bollingen Foundation.

Eliade, M. (1966). 'Australian Religions: An Introduction. Part I.' *History of Religions* 6(2): 108–34.

Eliade, M. (1967a). *s.v.* 'Ionescu, Nae.' In P. Edwards (ed.), *Encyclopedia of Philosophy*, vol. 4, 212. London and New York: MacMillan–The Free Press.

Eliade, M. (1967b). *s.v.* 'Rădulescu-Motru, Constantin.' In P. Edwards (ed.), *Encyclopedia of Philosophy*, vol. 7, 63–4. London and New York: MacMillan–The Free Press.

Eliade, M. (1967c). *s.v.* 'Rumanian Philosophy.' In P. Edwards (ed.), *Encyclopedia of Philosophy*, vol. 7, 233–4. London and New York: MacMillan–The Free Press.

Eliade, M. (1967d). 'Cultural Fashions and the History of Religions.' In J. M. Kitagawa, M. Eliade and C. H. Long (eds), *The History of Religions: Essays on the Problem of Understanding*, 20–38. Chicago: The University of Chicago Press.

Eliade, M. (1971). 'Preface.' In W. K. Bolle, *The Persistence of Religion: An Essay on Tantrism and Sri Aurobindo's Philosophy*, ix–xi. Leiden: Brill.

Eliade, M. (1976). *Occultism, Witchcraft and Cultural Fashions: Essays in Comparative Religions*. Chicago and London: The University of Chicago Press.

Eliade, M. (1978a[1976]). *History of Religious Ideas. Vol. I: From the Stone Age to the Eleusinian Mysteries*, trans. W. R. Trask. Chicago: The University of Chicago Press.

Eliade, M. (1978b). 'Foreword.' In D. Allen, *Structure and Creativity in Religion: Hermeneutics in Mircea Eliade's Phenomenology and New Directions*, vii–ix. The Hague: Mouton.

Eliade, M. (1980). 'History of Religions and "Popular" Cultures.' *History of Religions* 20(1–2): 1–26.

Eliade, M. (1981). *Autobiography, Volume 1: 1907–1937, Journey East, Journey West*, trans. M. L. Ricketts. Chicago and London: The University of Chicago Press.

Eliade, M. (1982a[1978]). *Ordeal by Labyrinth: Conversations with Claude-Henri Rocquet. With the Essay 'Brancusi and Mythology'*, trans. D. Coltman. Chicago: The University of Chicago Press.

Eliade, M. (1982b[1978]). *A History of Religious Ideas. Vol. II: From Gautama Buddha to the Triumph of Christianity*, trans. W. R. Trask. Chicago: The University of Chicago Press.

Eliade, M. (1984[1969]). *The Quest: History and Meaning in Religion*. Chicago and London: The University of Chicago Press.

Eliade, M. (1985). '*Homo faber* and *homo religiosus*.' In J. M. Kitagawa (ed.), *The History of Religions: Retrospect and Prospect*, 1–12. New York: MacMillan.

Eliade, M. (1987[1937]). 'Introducere.' In M. Handoca (ed.), *Despre Eminescu și Hasdeu*, 59–104. Iași: Junimea.

Eliade, M. (1989a[1973]). *Journal II: 1957–1969*, trans. F. H. Johnson Jr. Chicago: The University of Chicago Press.

Eliade, M. (1989b[1981]). *Journal III: 1970–1978*, trans. T. Lavender Fagan. Chicago: The University of Chicago Press.

Eliade, M. (1990a[1973]). *Journal I: 1945–1955*, trans. M. L. Ricketts. Chicago: The University of Chicago Press.

Eliade, M. (1990b). *Journal IV: 1979–1985*, trans. M. L. Ricketts. Chicago: The University of Chicago Press.

Eliade, M. (1991). *Drumul spre centru. Antologie alcătuită de Gabriel Liiceanu și Andrei Pleșu*. Bucharest: Univers. Contains partial or unabridged edition of the following works by Mircea Eliade: *Oceanografie, Fragmentarium, Insula lui Euthanasius, Mitul reintegrării, Comentarii la Legenda Meșterului Manole, Alchimia asiatică*.

Eliade, M. (1992a[1952]). 'Europa și cortina de fier.' *In Împotriva deznădejdii. Publicistica exilului. Ediție îngrijită de M. Handoca*, 151–60. Bucharest: Humanitas.

Eliade, M. (1992b[1952]). 'Catastrofă și mesianism. Note pentru o Teologie a Istoriei.' *In Împotriva deznădejdii. Publicistica exilului. Ediție îngrijită de M. Handoca*, 120–6. Bucharest: Humanitas.

Eliade, M. (1993[1937]). 'Și un cuvânt al editorului.' In M. Eliade (ed.), *Nae Ionescu. Roza vânturilor 1926–1933*, 270–84. Chișinău: Hyperion.

Eliade, M. (1996). *Cum am găsit piatra filozofală. Scrieri de tinerețe 1921–1925. Îngrijirea ediției și note de M. Handoca*. Bucharest: Humanitas.

Eliade, M. (1998). *Misterele și inițierea orientală. Scrieri de tinerețe, 1926. Îngrijirea și note de M. Handoca*. Bucharest: Humanitas.

Eliade, M. (2000[1972]). 'Foreword.' In B. Feldman and R. D. Richardson, Jr. (eds), *The Rise of Modern Mythology 1680–1860*, xiii–xvi. Bloomington and Indianapolis: Indiana University Press.

Eliade, M. (2001[1935]). 'L'alchimia asiatica.' In *Il mito dell'alchimia* seguito da *L'alchimia asiatica*, 41–114. Turin: Bollati Boringhieri.

Eliade, M. (2003). *Itinerariu spiritual. Scrieri de tinerețe 1927. Îngrijirea ediției și note de M. Handoca*. Bucharest: Humanitas.

Eliade, M. (2004). *Jurnal 1970–1985. Ediție îngrijită de M. Handoca*. Bucharest: Humanitas.

Eliade, M. (2006[1937]). 'Folklore as an Instrument of Knowledge', trans. M. L. Ricketts. In B. S. Rennie (ed.), *Mircea Eliade: A Critical Reader*, 25–37. London and Oakville, CT: Equinox.

Eliade, M. (2007). *Jurnalul portughez și alte scrieri. Ediția a* II-*a revăzută și adăugită, prefață și îngrijire de ediție de S. Alexandrescu, studii introductive, note și traduceri de S. Alexandrescu, F. Țurcanu, M. Zamfir, traduceri din portugheză si glosare de nume de M. Zamfir*, 2 vols. Bucharest: Humanitas.

Eliade, M. (2008[1939]). *Fragmentarium. Edizione italiana a cura di R. Scagno*, trans. C. Fantechi. Milan: Jaca Book.

Eliade, M. (2009[1954]). *Yoga: Immortality and Freedom*. Princeton: Princeton University Press.

Eliade, M. (2010). *The Portugal Journal*, trans. M. L. Ricketts. Albany: State University of New York Press.

Ellwood, R. (2001). 'Is Mythology Obsolete?' *Journal of the American Academy of Religion* 69(3): 673–86.

Eriksson, K. and J. C. Coultas (2014). 'Corpses, Maggots, Poodles and Rats: Emotional Selection Operating in Three Phases of Cultural Transmission of Urban Legends.' *Journal of Cognition and Culture* 14(1): 1–26.

Evans, E. M. (2000). 'Beyond Scopes: Why Creationism Is Here to Stay.' In K. Rosengren, C. Johnson and P. Harris (eds), *Imagining the Impossible: Magical, Scientific and Religious Thinking in Children*, 305–31. Cambridge: Cambridge University Press.

Evans-Pritchard, E. E. (1965). *Theories of Primitive Religion*. Oxford: Clarendon Press.

Falk, N. A. and R. M. Gross (1980). *Unspoken Worlds: Women's Religious Lives in Non-western Cultures*. San Francisco: Harper & Row.

Fauconnier, G. (2001). 'Conceptual Blending.' In N. J. Smelser and P. B. Baltes (eds), *International Encyclopedia of the Social & Behavioral Sciences*, 2495–8. Oxford: Elsevier.

Fellmeth, A. X. and M. Horwitz (2009). *Guide to Latin in International Law*. Oxford: Oxford University Press.

Ferrari, F. M. (2014[2012]). *Ernesto De Martino on Religion: The Crisis and the Presence*. Abingdon and New York: Routledge.

Ferraris, M. (2006). *Jackie Derrida. Ritratto a memoria*. Turin: Bollati Boringhieri.

Ferraris, M. (2008[2003]). *Introduzione a Derrida*. Rome and Bari: Laterza.

Ferraris, M. (2014[2012]). *Manifesto of New Realism*, trans. S. De Sanctis. Albany: State University of New York Press.

Ferraris, M. (2015). *Introduction to New Realism*, trans. S. De Sanctis. London and New York: Bloomsbury.

Ferretti, F. and I. Adornetti (2014). 'Biology, Culture and Coevolution: Religion and Language as Case Studies.' *Journal of Cognition and Culture* 14(3–4): 305–30.

Ferrone, V. (2015[2010]). *The Enlightenment: History of an Idea. With a New Afterword by the Author*, trans. E. Tarantino. Princeton and Oxford: Princeton University Press.

Feuerbach, L. (1890[1841]). *The Essence of Christianity. Translated from the Second German Edition by Marian Evans*. London: Kegan Paul, Trench, Trübner & Co.

Filoramo, G., and C. Prandi (1997). *Le scienze delle religioni. Terza edizione riveduta e ampliata*. Brescia: Morcelliana.

Fincher, C. and R. Thornhill (2008). 'Assortative Sociality, Limited Dispersal, Infectious Disease and the Genesis of the Global Pattern of Religion Diversity.' *Proceedings of the Royal Society B: Biological Sciences* 275(1651): 2587–94.

Fincher, C. and R. Thornhill (2012). 'Parasite-Stress Promotes In-Group Assortative Sociality: The Cases of Strong Family Ties and Heightened Religiosity.' *Behavioral and Brain Sciences* 35(2): 61–79.

Finkelstein, G. (2000). 'Why Darwin Was English.' *Endeavour* 24(2): 76–8.

Fischer, D. H. (1970). *Historians' Fallacies: Toward a Logic of Historical Thought*. New York and London: Harper Perennial.

Flannery, T. F. (2002[1994]). *The Future Eaters: An Ecological History of the Australasian Lands and People*. New York: Grove Press.

Flexner, A. (1939). 'The Usefulness of Useless Knowledge.' *Harper's Magazine* 179, November: 544–52.

Flood, A. (2016). '"Post-truth" Named Word of the Year by Oxford Dictionaries.' *The Guardian*, 15 November. https://www.theguardian.com/books/2016/nov/15/post-truth-named-word-of-the-year-by-oxford-dictionaries (accessed 20 December 2017).

Floridi, L. (2015). 'The New Grey Power.' *Philosophy & Technology* 28(3): 329–32.

Foley, R. A. and M. Mirazón Lahr (2012). 'The Evolution of the Diversity of Cultures.' In A. Whiten, R. A. Hinde, C. B. Stringer and K. N. Laland (eds), *Culture Evolves*, 251–68. Oxford: Oxford University Press.

Forrest, B. (2013). 'Navigating the Landscape between Science and Religious Pseudoscience: Can Hume Help?' In M. Pigliucci and M. Boudry (eds), *Philosophy of Pseudoscience: Reconsidering the Demarcation Problem*, 263–83. Chicago and London: The University of Chicago Press.

Foucault, M. (1978[1976]). *The History of Sexuality, Volume I: An Introduction*, trans. R. Hurley. New York: Pantheon Books.

Frankfurt, H. G. (2005). *On Bullshit*. Princeton and Oxford: Princeton University Press.

Fraser, R. (2009[1994]). 'Introduction.' In J. G. Frazer, *The Golden Bough: A Study in Magic and Religion. A New Abridgement from the Second and Third Editions. Edited and with an Introduction and Notes by R. Fraser*, ix–xxxix. Oxford: Oxford University Press.

Frazer, J. G. (1890). *The Golden Bough: A Study in Comparative Religion*. 2 vols. London: Macmillan.

Frazer, J. G. (1900). *The Golden Bough: A Study in Magic and Religion. Second Edition, Revised and Enlarged*. 3 vols. London: Macmillan.

Frazer, J. G. (1909). 'Some Primitive Theories of the Origin of Man.' In A. C. Seward (ed.), *Darwin and Modern Science: Essays in Commemoration of the Centenary of the Birth of Charles Darwin and the Fiftieth Anniversary of the Publication of the Origin of Species*, 152–70. Cambridge: Cambridge University Press.

Frazer, J. G. (1910). *Totemism and Exogamy: A Treatise on Certain Early Forms of Superstition and Society*. 4 vols. London: Macmillan.

Freud, S. (1987). *A Phylogenetic Fantasy: Overview of the Transference Neuroses by Sigmund Freud*, edited by I. Grubrich-Simitis, trans. A. Hoffer and P. T. Hoffer. Cambridge, MA: Harvard University Press.

Galef, J. (2011). 'The Straw Vulcan: Hollywood's Illogical Approach to Logical Decisionmaking.' *Measure of Doubt*, 26 November. https://measureofdoubt.com/2011/11/26/the-straw-vulcan-hollywoods-illogical-approach-to-logical-decisionmaking (accessed 20 December 2017).

Galef, J. (2016). 'Debunking Straw Vulcan Rationality.' *Big Think*. http://bigthink.com/in-their-own-words/debunking-straw-vulcan-rationality (accessed 20 December 2017).

Gandini, M. (1993). 'Raffaele Pettazzoni dall'archeologia all'etnologia (1909–1911). Materiali per una biografia IV.' *Strada Maestra* 34: 95–227.

Gandini, M. (1994). 'Raffaele Pettazzoni nell'anno cruciale 1912. Materiali per una biografia V.' *Strada Maestra* 36/37: 177–298.

Gandini, M. (1996). 'Raffaele Pettazzoni dalla libera docenza nell'Università di Roma all'incarico nell'Ateneo bolognese (1913–1914). Materiali per una biografia IX.' *Strada Maestra* 40: 63–205.

Gandini, M. (1998). 'Raffaele Pettazzoni nel primo Dopoguerra (1919–1922). Materiali per una biografia VII.' *Strada Maestra* 44: 97–214.

Gandini, M. (1999a). 'Raffaele Pettazzoni negli anni 1926–1927. Materiali per una biografia XI.' *Strada Maestra* 47: 95–226.

Gandini, M. (1999b). 'Raffaele Pettazzoni negli anni del noviziato universitario romano (1924–1925). Materiali per una biografia X.' *Strada Maestra* 46: 77–223.

Gandini, M. (2000). 'Raffaele Pettazzoni negli anni 1928–1929. Materiali per una biografia XII.' *Strada Maestra* 48: 81–249.

Gandini, M. (2001a). 'Raffaele Pettazzoni nelle spire del fascismo (1931–1933). Materiali per una biografia XIV.' *Strada Maestra* 50: 19–183.

Gandini, M. (2001b). 'Raffaele Pettazzoni dal gennaio 1934 all'estate 1935. Materiali per una biografia XV.' *Strada Maestra* 51: 81–212.

Gandini, M. (2003). 'Raffaele Pettazzoni negli anni 1937–1938. Materiali per una biografia XVII.' *Strada Maestra* 54: 53–232.

Gandini, M. (2005). *s.v.* 'Pettazzoni, Raffaele.' In L. Jones (ed.), *Encyclopedia of Religion. Second Edition*, vol. 10, 7072–7. Detroit: Thomson Gale–MacMillan.

Gandini, M. (2006). 'Raffaele Pettazzoni negli anni 1949–1950. Materiali per una biografia XXIII.' *Strada Maestra* 60: 19–237.

Gandini, M. (2008). 'Raffaele Pettazzoni negli anni 1956–1957. Materiali per una biografia XXVII.' *Strada Maestra* 64: 1–247.

Gandini, M. (2009). 'Raffaele Pettazzoni negli anni 1958–1959. Materiali per una biografia XXVIII.' *Strada Maestra* 65: 1–230.

Garcia, H. A. (2015). *Alpha God: The Psychology of Religious Violence and Oppression*. New York: Prometheus.

Gardaz, M. (2012). 'Mircea Eliade et le «nouvel homme» à la chemise verte.' *Numen* 59(1): 68–92.

Geertz, A. W. (2004). 'Cognitive Approaches to the Study of Religion.' In P. Antes, A. M. Geertz and R. R. Warne (eds), *New Approaches to the Study of Religion. Vol. 2: Textual, comparative, sociological, and cognitive approaches*, 347–99. Berlin and New York: de Gruyter.

Geertz, A. W. (2009). 'When Cognitive Scientists Become Religious, Science Is in Trouble: On Neurotheology from a Philosophy of Science Perspective.' *Religion* 39(4): 319–24.

Geertz, A. W. (2010). 'Brain, Body and Culture: A Biocultural Theory of Religion.' *Method & Theory in the Study of Religion* 22(4): 304–21.

Geertz, A. W. (2013). 'Whence Religion? How the Brain Constructs the World and What This Might Tell Us about the Origins of Religion, Cognition and Culture.' In A. W. Geertz (ed.), *Origins of Religion, Cognition and Culture*, 17–70. Durham and Bristol, CT: Acumen.

Geertz, A. W. (2014a). 'Long-lost Brothers: On the Co-histories and Interactions between the Comparative Science of Religion and the Anthropology of Religion.' *Numen* 61(2–3): 255–80.

Geertz, A. W. (2014b). 'Cladistics in the Study of World Mythologies: A Review of E. J. Michael Witzel, *The Origins of the World's Mythologies*.' *Cosmos* 30: 1–6.

Geertz, A. W. (2015). 'Religious Belief, Evolution of.' In J. D. Wright (editor-in-chief), *International Encyclopedia of the Social & Behavioral Sciences*, 2nd edn, vol. 20, 384–95. Oxford: Elsevier.

Geertz, A. W. and G. I. Markússon (2010). 'Religion Is Natural, Atheism Is Not: On Why Everybody Is Both Right and Wrong.' *Religion* 40(3): 152–65.

Gil-White, F. J. (2001). 'Are Ethnic Groups Biological "Species" to the Human Brain?: Essentialism in Our Cognition of Some Social Categories.' *Current Anthropology* 42(4): 515–53.

Gilhus, I. S. (2014). 'Founding Fathers, Turtles and the Elephant in the Room: The Quest for Origins in the Scientific Study of Religion.' *Temenos* 50: 193–214.

Gillard, D. (2002). 'Creationism: Bad Science, Bad Religion, Bad Education.' *Education in England: The Story of Our Schools.* http://www.educationengland.org.uk/articles/20creationism.html (accessed 20 December 2017).

Gillard, D. (2007). 'Never Mind the Evidence: Blair's Obsession with Faith Schools.' *Education in England: The Story of Our Schools.* http://www.educationengland.org.uk/articles/26blairfaith.html (accessed 20 December 2017).

Gillard, D. (2016). 'Chapter 13: 2010–2015. Gove v The Blob.' *Education in England: The Story of Our Schools.* http://www.educationengland.org.uk/history/chapter13.html (accessed 20 December 2017).

Ginzburg, C. (2010). 'Mircea Eliade's Ambivalent Legacy.' In C. K. Wedemeyer and W. Doniger (eds), *Hermeneutics, Politics, and the History of Religions: The Contested Legacies of Joachim Wach and Mircea Eliade*, 307–23. Oxford and New York: Oxford University Press.

Girotto, V., T. Pievani and G. Vallortigara (2014). 'Supernatural Beliefs: Adaptations for Social Life or By-products of Cognitive Adaptations?' In F. B. M. de Waal, P. Smith Churchland, T. Pievani and S. Parmigiani (eds), *Evolved Morality: The Biology and Philosophy of Human Conscience*, 249–66. Leiden and Boston: Brill.

Glick, T. F. (2008). 'The Anthropology of Race across the Darwinian Revolution.' In H. Kuklick (ed.), *A New History of Anthropology*, 225–41. Malden, MA and Oxford: Blackwell.

Goethe, J. W. von (1893[1870]). *The Maxims and Reflections of Goethe. Translated by Baily Saunders. With a Preface.* London and New York: Macmillan.

Goldenberg, J. L., T. Pyszczynski, J. Greenberg, S. Solomon, B. Kluck and R. Cornwell (2001). 'I Am *Not* an Animal: Mortality Salience, Disgust, and the Denial of Human Creatureliness.' *Journal of Experimental Psychology: General* 130(3): 427–35.

Gottschall, J. (2012). *The Storytelling Animal: How Stories Make Us Human.* Boston and New York: Houghton Mifflin Harcourt.

Gould, R. A., D. A. Koster and A. H. L. Sontz (1971). 'The Lithic Assemblage of the Western Desert Aborigines of Australia.' *American Antiquity* 36(2): 149–69.

Gould, S. J. (1977). *Ever Since Darwin: Reflections in Natural History.* New York: W.W. Norton & Co.

Gould, S. J. (1980). *The Panda's Thumb: More Reflections in Natural Histories.* New York: W.W. Norton & Co.

Gould, S. J. (1986). 'Evolution and the Triumph of Homology, or Why History Matters.' *American Scientist* 74(1): 60–9.

Gould, S. J. (1989). *Wonderful Life: The Burgess Shale and the Nature of History.* New York and London: W.W. Norton & Co.

Gould, S. J. (1990[1983]). *Hen's Teeth and Horse's Toes*. London: Penguin.

Gould, S. J. (1991[1985]). *The Flamingo's Smile: Reflections in Natural History*. London: Penguin.

Gould, S. J. (1996). *Full House: The Spread of Excellence from Plato to Darwin*. New York: Harmony Books.

Gould, S. J. (1999). *Rocks of Ages: Science and Religion in the Fullness of Life*. New York: Ballantine.

Gould, S. J. (2001). *I Have Landed: The End of a Beginning in Natural History*. New York: Harmony Books.

Gould, S. J. (2002). *The Structure of Evolutionary Theory*. Cambridge, MA and London: The Belknap Press of Harvard University Press.

Gould, S. J. and N. Eldredge (1977). 'Punctuated Equilibria: The Tempo and Mode of Evolution Reconsidered.' *Paleobiology* 3(2): 115–51.

Gould, S. J. and N. Eldredge (1993). 'Punctuated Equilibrium Comes of Age.' *Nature* 6452(366): 223–7.

Gracián, B. (2015[1647]). *How to Use your Enemies*, trans. J. Robbins, selection taken from *The Pocket Oracle and the Art of Prudence*. London: Penguin.

Grapard, A. (1991). 'Visions of Excess and Excesses of Vision: Women and Transgression in Japanese Myth.' *Japanese Journal of Religious Studies* 18(1): 3–22.

Gregory, W. K. (1927). 'Two Views on the Origin of Man.' *Science* 65(1695): 601–5.

Griffin, R. (1993). *The Nature of Fascism*. London and New York: Routledge.

Gross, R. M. (1977). *Beyond Androcentrism: New Essays on Women and Religion*. Chico, CA: Scholars Press.

Gross, R. M. (1980). 'Women's Studies in Religion: The State of the Art, 1980.' In P. Slater and D. Wiebe (eds), *Traditions in Contact and Change: Selected Proceedings of the XIVth Congress of the International Association for the History of Religions*, 579–92. Waterloo, ON: Wilfrid Laurier University Press.

Gross, R. M. (1994). 'Studying Women and Religion: Conclusions Twenty-Five Years Later.' In A. Sharma (ed.), *Today's Woman in World Religions*, 327–63. Albany: State University of New York.

Gross, R. M. (2005). 'Methodology – Tool or Trap? Comments from a Feminist Perspective.' In R. Gothoni (ed.), *How to Do Comparative Religion? Three Ways, Many Goals*, 149–66. Berlin and New York: de Gruyter.

Gross, R. M. (2009). *A Garland of Feminist Reflections: Forty Years of Religious Exploration*. Berkeley, Los Angeles and London: University of California Press.

Gross, R. M. and R. R. Ruether (2016). *Religious Feminism and the Future of the Planet: A Christian–Buddhist Conversation*. London and New York: Bloomsbury.

Grottanelli, C. (1983). 'Tricksters, Scape-Goats, Champions, Saviors.' *History of Religions* 23(2): 117–39.

Grottanelli, C. (2000). 'Alfonso M. di Nola e Mircea Eliade.' In A. Spirito and I. Bellotta (eds), *Antropologia e storia delle religioni. Studi in onore di Alfonso M. di Nola*, 35–40. Rome: Newton Compton.

Grottanelli, C. and B. Lincoln (1998[1984–1985]). 'A Brief Note on (Future) Research on the History of Religions.' *Method & Theory in the Study of Religion* 10(3): 311–25.

Guthrie, S. E. (1980). 'A Cognitive Theory of Religion.' *Current Anthropology* 21(2): 181–94.

Guthrie, S. E. (1993). *Faces in the Clouds: A New Theory of Religion*. Oxford and New York: Oxford University Press.

Guthrie, S. E. (2000). 'On Animism.' *Current Anthropology* 41(1): 106–7.

Guthrie, S. E. (2002). 'Animal Animism: Evolutionary Roots of Religious Cognition.' In I. Pyysiäinen and V. Anttonen (eds), *Current Approaches in the Cognitive Science of Religion*, 38–67. London and New York: Continuum.

Guthrie, S. E. (2013). 'Early Cognitive Theorists of Religion: Robin Horton and His Predecessors.' In D. Xygalatas and W. W. McCorkle Jr (eds), *Mental Culture: Classical Social Theory and the Cognitive Science of Religion*, 33–51. Durham and Bristol, CT: Acumen.

Hamayon, R. N. (1993). 'Are "Trance", "Ecstasy" and Similar Concepts Appropriate in the Study of Shamanism?' *Shaman* 1(2): 3–26.

Hamilton, A. and Q. D. Wheeler (2008). 'Taxonomy and Why History of Science Matters for Science: A Case Study.' *Isis* 99(2): 331–40.

Hampson, R. (1991). 'Frazer, Conrad and the "Truth of Primitive Passion".' In R. G. Fraser (ed.), *Sir James Frazer and the Literary Imagination: Essays in Affinity and Influence*, 172–91. New York: St Martin's.

Handoca, M. (1997). *Mircea Eliade. 1907–1986. Biobibliografie*, vol. I. Bucharest: Editura Jurnalul literar.

Handoca, M. (ed.) (1998). *Convorbiri cu și despre Mircea Eliade*. Bucharest: Humanitas.

Handoca, M. (ed.) (2001). *Textele 'legionare' și despre 'românism'*. Cluj-Napoca: Dacia.

Handoca, M. (ed.) (2004a). *Mircea Eliade. Europa, Asia, America . . . Corespondență R-Z*. Bucharest: Humanitas.

Handoca, M. (2004b). *Mircea Eliade. Europa, Asia, America . . . Corespondență I-P*. Bucharest: Humanitas.

Handoca, M. (ed.) (2006). *Mircea Eliade și corespondenții săi. Vol. 4 (R–S)*. Bucharest: Criterion.

Handoca, M. (2008). *Mircea Eliade. Pagini regăsite*. Bucharest: Lider.

Hardy K., S. Buckley, M. J. Collins, A. Estalrrich, D. Brothwell, L. Copeland, A. García-Tabernero, S. García-Vargas, M. de la Rasilla, C. Lalueza-Fox, R. Huguet, M. Bastir, D. Santamaría, M. Madella, J. Wilson, A. F. Cortés and A. Rosas (2012). 'Neanderthal Medics? Evidence for Food, Cooking, and Medicinal Plants Entrapped in Dental Calculus.' *Naturwissenschaften* 99(8): 617–26.

Harnack, A. von (1906). *Reden und Aufsätze*, vol. 2. Giessen: Töpelmann.

Harrison, J. E. (1909). 'The Influence of Darwinism on the Study of Religions.' In A. C. Seward (ed.), *Darwin and Modern Science: Essays in Commemoration of the Centenary of the Birth of Charles Darwin and the Fiftieth Anniversary of the Publication of the Origin of Species*, 494–511. Cambridge: Cambridge University Press.

Harva, U. (1922). *Der Baum des Lebens*. Helsinki: Annales Academiae Scientiarum Fennicae B 16(3).

Hawes, G. (2014). *Rationalizing Myth in Antiquity*. Oxford: Oxford University Press.

Haydon, A. E. (1931). 'The Origin of Religion.' *The Journal of Religion* 11(4): 610–11.

Heinen, A. (2006[1986]). *Legiunea «Arhangelul Mihail»: mișcarea socială și organizație politică. O contribuție la problema fascismului internațional*. Bucharest: Humanitas.

Helmore, E. (2017). 'CDC Banned Words Include "Diversity", "Transgender" and "Fetus" – Report.' *The Guardian*, 16 December. https://www.theguardian.com/us-news/2017/dec/16/cdc-banned-words-fetus-transgender-diversity (accessed 16 December 2017).

Hennig, W. (1966[1950]). *Philogenetic Systematics*, trans. D. D. Davis and R. Zangerl. Urbana, Chicago and London: University of Illinois Press.

Henninger, J. and A. Ciattini (2005[1987]). *s.v.* 'Schmidt, Wilhelm.' In L. Jones (ed.), *Encyclopedia of Religion. Second Edition*, vol. 12, 8167–71. Detroit: Thomson Gale–MacMillan.

Henrich, J. (2009). 'The Evolution of Costly Displays, Cooperation, and Religion: Credibility Enhancing Displays and their Implications for Cultural Evolution.' *Evolution and Human Behavior* 30(4): 244–60.

Hobsbawm, E. and T. Ranger (eds) (1992[1983]). *The Invention of Tradition.* Cambridge: Cambridge University Press.

Hoffmann D. L., C. D. Standish, M. García-Diez, P. B. Pettitt, J. A. Milton, J. Zilhão, J. J. Alcolea-González, P. Cantalejo-Duarte, H. Collado, R. de Balbín, M. Lorblanchet, J. Ramos-Muñoz, G.-Ch. Weniger and A. W. G. Pike (2018). 'U-Th Dating of Carbonate Crusts Reveals Neandertal Origin of Iberian Cave Art.' *Science* 359(6378): 912–15.

Horder, T. J. (2009). *s.v.* 'On Growth and Form (D'Arcy Wentworth Thompson).' In M. Ruse and J. Travis (eds), *Evolution: The First Four Billion Years,* 768–70. Cambridge, MA and London: The Belknap Press of Harvard University Press.

Hughes, A. W. (2010). 'Science Envy in Theories of Religion.' *Method & Theory in the Study of Religion* 22(4): 293–303.

Hughes, A. W. (2012). *Abrahamic Religions: On the Uses and Abuses of History.* Oxford and New York: Oxford University Press.

Hull, D. L. (1988). *Science as a Process. An Evolutionary Account of the Social and Conceptual Development of Science.* Chicago and London: The University of Chicago Press.

Hume, D. (2008a[1757]). *Dialogues and Natural History of Religion. Edited and with an Introduction and Notes by J. C. A. Gaskin.* Oxford: Oxford University Press.

Hume, D. (2008b[1748]). *An Enquiry Concerning Human Understanding. Edited and with an Introduction and Notes by P. Millican.* Oxford: Oxford University Press.

Humphrey, N. (1999[1996]). *Leaps of Faith: Science, Miracles, and the Search for Supernatural Consolation. With a New Foreword by D. C. Dennett.* New York: Copernicus.

Huxley, T. H. (1854). 'Abstract on the Common Plan of Animal Forms.' *Notices of the Proceedings at the Meetings of the Royal Institution with Abstracts of the Discourses Delivered at the Evening Meetings* 1(1851–1854): 444–6.

Idel, M. (2014). *Mircea Eliade de la magie la mit.* Iaşi: Polirom.

Idinopulos, T. A. and E. A. Yonan (eds) (1994). *Religion and Reductionism: Essays on Eliade, Segal, and the Challenge of the Social Sciences for the Study of Religion.* Leiden, New York and Cologne: Brill.

Ioanid, R. (2005). 'The Sacralised Politics of the Romanian Iron Guard.' In R. Griffin (ed.), *Fascism, Totalitarianism and Political Religion,* 125–59. London and New York: Routledge.

Jack, B. (2015). 'The Rise of the Medical Humanities.' *Times Higher Education,* 22 January. https://www.timeshighereducation.com/features/the-rise-of-the-medical-humanities/2018007 (accessed 20 December 2017).

Jacob, F. (1977). 'Evolution and Tinkering.' *Science* 196(4295): 1161–6.

Jäger, G. (1869). *Darwin'sche Theorie und ihre Stellung zur Moral und Religion.* Stuttgart: Hoffman.

James, W. (1902). *The Varieties of Religious Experience: A Study in Human Nature, Being the Gifford Lectures on Natural Religion Delivered at Edinburgh in 1901–1902.* New York, London and Bombay: Longmans, Green, & Co.

Jameson, F. (1998). *The Cultural Turn: Selected Writings on the Postmodern, 1983–1998.* London and New York: Verso.

Jensen, J. S. (1993). 'Is a Phenomenology of Religion Possible? On the Ideas of a Human and Social Science of Religion.' *Method & Theory in the Study of Religion* 5(2): 109–33.

Jensen, J. S. (2013). 'Normative Cognition in Culture and Religion.' *Journal for the Cognitive Science of Religion* 1(1): 47–70.

Jensen, J. S. (2014). *What Is Religion?* Durham and Bristol, CT: Acumen.

Jensen, J. S. and A. W. Geertz (1991). 'Tradition and Renewal in the Histories of Religions: Some Observations and Reflections.' In A. W. Geertz and J. S. Jensen (eds), *Religion, Tradition, and Renewal*, 11–27. Aarhus: Aarhus University Press.

Jensen, T. and A. W. Geertz (eds) (2015). *NVMEN, the Academic Study of Religion, and the IAHR: Past, Present and Prospects*. Boston and Leiden: Brill.

Jerryson, M., M. Juergensmeyer and M. Kitts (eds) (2013). *The Oxford Handbook of Religion and Violence*. Oxford and New York: Oxford University Press.

Jesi, F. (1973). *Mito*. Milan: Mondadori.

Jesi, F. (2005). *Bachofen. A cura di A. Cavalletti*. Turin: Bollati Boringhieri.

Jesi, F. (2011[1979]). *Cultura di destra. Con tre inediti e un'intervista. A cura di A. Cavalletti*. Rome: nottetempo.

Jones, W. (1788). 'The Third Anniversary Discourse, delivered 2 February 1786.' In *Asiatick Researches: or, Transactions of the Society Instituted in Bengal, for Inquiring into the History and Antiquities, The Arts, Sciences, and Literature, of Asia*. Calcutta: Manuel Cantopher.

Jordan, L. H. and B. Labanca (1909). *The Study of Religion in the Italian Universities*. Oxford: Oxford University Press.

Junginger, H. (ed.) (2008). 'Introduction.' In H. Junginger (ed.), *The Study of Religion Under the Impact of Fascism*, 1–103. Leiden and Boston: Brill.

Juschka, D. M. (2005). 'Religion and Gender.' In J. R. Hinnells (ed.), *The Routledge Companion to the Study of Religion*, 229–42. Abingdon and New York: Routledge.

Juschka, D. M. (2008). 'Deconstructing the Eliadean Paradigm: Symbol.' In W. Braun and R. T. McCutcheon (eds), *Introducing Religion: Essays in Honor of Jonathan Z. Smith*, 163–77. London and Oakville, CT: Equinox.

Kahneman, D. (2011). *Thinking, Fast and Slow*. New York: Farrar, Straus, and Giroux.

Kelemen, D. (1999a). 'Functions, Goals and Intentions: Children's Teleological Reasoning about Objects.' *Trends in Cognitive Sciences* 3(12): 461–8.

Kelemen, D. (1999b). 'Why Are Rocks Pointy? Children's Preference for Teleological Explanations of the Natural World.' *Developmental Psychology* 35(6): 1440–52.

Kelemen, D. (2012). 'Teleological Minds: How Natural Intuitions about Agency and Purpose Influence Learning about Evolution.' In K. Rosengren, S. Brem, E. M. Evans and G. Sinatra (eds), *Evolution Challenges: Integrating Research and Practice in Teaching about Evolution*, 66–92. Oxford: Oxford University Press.

Kelemen, D. and E. Rosset (2009). 'The Human Function Compunction: Teleological Explanation in Adults.' *Cognition* 111(1): 138–43.

Kendal, J., J. J. Tehrani and J. Odling-Smee (2011). 'Human Niche Construction in Interdisciplinary Focus.' *Philosophical Transactions of the Royal Society B* 366(1566): 785–92.

King, R. (2013). 'The Copernican Turn in the Study of Religion.' *Method & Theory in the Study of Religion* 25(2): 137–59.

Kinsley, D. (2002). 'Women's Studies and the History of Religions.' In A. Sharma (ed.), *Methodology in Religious Studies: The Interface with Women's Studies*, 1–16. Albany: State University of New York Press.

Kippenberg, H. G. (2002). *Discovering Religious History in the Modern Age*. Princeton: Princeton University Press.

Kitagawa, J. M. (2005[1987]). *s.v.* 'Eliade, Mircea [First Edition].' In L. Jones (ed.), *Encyclopedia of Religion. Second Edition*, vol. 4, 2753–7. Farmington Hills, MI: Macmillan Reference USA.

Klein-Arendt, R. (2010). *s.v.* 'Frobenius, Leo Viktor.' In F. Abiola Irele and B. Jeyifo (eds), *The Oxford Encyclopedia of African Thought*, vol. I: Abolitionism–Imperialism, 391–3. Oxford and New York: Oxford University Press.

Knott, C. G. (1911). *Life and Scientific Work of Peter Guthrie Tait. Supplementing the Two Volumes of Scientific Papers Published in 1898 and 1900*. Cambridge: Cambridge University Press.

Koertge, N. (2013). 'Belief Buddies versus Critical Communities: The Social Organisation of Pseudoscience.' *Philosophy of Pseudoscience: Reconsidering the Demarcation Problem*, edited by M. Pigliucci and M. Boudry, 165–80. Chicago and London: The University of Chicago Press.

Korte, A. M. (2011). 'Openings: A Genealogical Introduction to Religion and Gender.' *Religion and Gender* 1(1): 1–17.

Krause, J., C. Lalueza-Fox, L. Orlando, W. Enard, R. E. Green, H. A. Burbano, J. J. Hublin, C. Hänni, J. Fortea, M. de la Rasilla, J. Bertranpetit, A. Rosas and S. Pääbo (2007). 'The Derived FOXP2 Variant of Modern Humans Was Shared with Neandertals.' *Current Biology* 17(21): 1908–12.

Kripal, J. J. (2007). *The Serpent's Gift: Gnostic Reflections on the Study of Religion*. Chicago and London: The University of Chicago Press.

Kripal, J. J. (2011a). 'The Future Human: Mircea Eliade and the Fantastic Mutant.' *Archævs* 15(1–2): 187–208.

Kripal, J. J. (2011b). *Mutants and Mystics: Science Fiction, Superhero Comics, and the Paranormal*. Chicago and London: The University of Chicago Press.

Kripal, J. J. (2014). *Comparing Religion: Coming to Terms*. With A. Anzali, A. R. Jain and E. Prophet. Malden, MA and Oxford: Wiley Blackwell.

Kristensen, W. B. (1954[1915]). 'Over wandering van historische gegevens.' In *Symbool en werkelijkheid. Godsdiensthistorische studiën*, 66–84. Arnhem: Van Loghum Slaterus.

Kundt, R. (2015). *Contemporary Evolutionary Theories of Culture and the Study of Religion*. London and New York: Bloomsbury.

Laignel-Lavastine, A. (2002). *Cioran, Eliade, Ionesco: L'oubli du fascisme*. Paris: Presses Universitaires de France.

Lakatos, I. (1989). *The Methodology of Scientific Research Programmes: Philosophical Papers. Volume I*, edited by J. Worrall and G. Currie. Cambridge: Cambridge University Press.

Landy, F. (2008). 'Smith, Derrida and Amos.' In W. Braun and R. T. McCutcheon (eds), *Introducing Religion: Essays in Honor of Jonathan Z. Smith*, 208–30. London and Oakville, CT: Equinox.

Lang, A. (1898). *The Making of Religion*. London: Longmans, Green, and Co.

Lanternari, V. (1960). 'Scienze delle religioni e storicismo.' *Annali della Facoltà di Lettere e Filosofia*, Università di Bari 6: 51–65.

Latour, B. (2004). 'Why Has Critique Run out of Steam? From Matters of Fact to Matters of Concern.' *Critical Inquiry* 30(2): 225–48.

Law, S. (2018). 'The X-claim Argument against Religious Belief.' *Religious Studies* 54(1): 15–35.

Lawson, E. T. (2000). 'Towards a Cognitive Science of Religion.' *Numen* 47(3): 338–49.

Lawson, E. T. (2004). 'The Wedding of Psychology, Ethnography, and History: Methodological Bigamy or Tripartite Free Love?' In H. Whitehouse and L. H. Martin (eds), *Theorizing Religious Past: Archaeology, History, and Cognition*, 1–6. Walnut Creek, CA: AltaMira Press.

Lawson, E. T. and R. N. McCauley (1990). *Rethinking Religion: Connecting Cognition and Culture*. Cambridge: Cambridge University Press.

Leach, E. (2006[1966]). 'Sermons from a Man on a Ladder.' In B. S. Rennie (ed.), *Mircea Eliade: A Critical Reader*, 278–85. London and Oakville, CT: Equinox.

Leeds, A. (1988). 'Darwinian and "Darwinian" Evolutionism in the Study of Society and Culture.' In T. F. Glick (ed.), *The Comparative Reception of Darwinism*, 437–85. Chicago and London: The University of Chicago Press.

Leertouwer, L. (1989). 'Tiele's Strategy of Conquest.' In W. Otterspeer (ed.), *Leiden Eastern Connections, 1850–1940*, 153–67. Leiden: Brill.

Leertouwer, L. (1991). 'Primitive Religion in Dutch Religious Studies.' *Numen* 38(2): 198–213.

Leroi, A. M. (2015). *The Lagoon: How Aristotle Invented Science*. London and New York: Bloomsbury.

Leroi-Gourhan, A. (1964). *Les religions de la préhistoire. Paléolithique*. Paris: Presses Universitaires de France.

Lessa, W. A. (1959). Review of Eliade, M. (1959). *The Sacred and the Profane: The Nature of Religion*. New York: Harcourt, Brace & Co. *American Anthropologist* 61(6): 1146–7.

Leuba, J. H. (1912). *A Psychological Study of Religion: Its Origin, Function, and Future*. New York: Macmillan.

Lévy-Bruhl, L. (1949). *Carnets*. Paris: Presses Universitaires de France.

Lewis, I. M. (1978[1971]). *Ecstatic Religion: An Anthropological Study of Spirit Possession and Shamanism*. Harmondsworth and New York: Penguin.

Lewis-Williams, D. (2010). *Conceiving God: The Cognitive Origin and Evolution of Religion*. London: Thames & Hudson.

Lewontin, R. and R. Levins (2007). *Biology Under the Influence*. New York: Monthly Review Press.

Lincoln, B. (1981). *Emerging from the Chrysalis: Studies in Rituals of Women's Initiation*. Chicago: University of Chicago Press.

Lincoln, B. (ed.) (1986). *Religion, Rebellion, Revolution: An Inter-disciplinary and Cross-cultural Collection of Essays*. New York: St. Martin's Press.

Lincoln, B. (1989). *Discourse and the Construction of Society: Comparative Studies of Myth, Ritual, and Classification*. New York: Oxford University Press.

Lincoln, B. (1991). *Death, War, and Sacrifice: Studies in Ideology and Practice*. Chicago: The University of Chicago Press.

Lincoln, B. (1996). 'Theses on Method.' *Method & Theory in the Study of Religion* 8(3): 225–7.

Lincoln, B. (1999). *Theorizing Myth: Narrative, Ideology, and Scholarship*. Chicago and London: The University of Chicago Press.

Lincoln, B. (2000). 'Culture.' In W. Braun and R. T. McCutcheon (eds), *Guide to the Study of Religion*, 409–22. London: Cassell.

Lincoln, B. (2005). 'Theses on Religion & Violence.' *ISIM Review* 15: 12.

Lincoln, B. (2012). *Gods and Demons, Priests and Scholars: Critical Explanations in the History of Religions*. Chicago and London: The University of Chicago Press.

Lincoln, B. (2015). 'Beginnings of a Friendship.' *Mythos* 9: 13–32.

Livingstone, D. N. (2014). *Dealing with Darwin: Place, Politics, and Rhetoric in Religious Engagements with Evolution*. Baltimore: Johns Hopkins University Press.

Lorenz, K. (1966 [1963]). *On Aggression*, trans. M. K. Wilson. London: Methuen. Vienna: Dr Borotha-Schoeler.

Love, B. J. (ed.) (2006). *Feminists Who Changed America, 1963–1975*. Urbana and Chicago: University of Illinois Press.

Lucas, R. (2005). *s.v.* 'Australian Indigenous Religion: History of Study (Further Considerations).' In L. Jones (ed.), *Encyclopedia of Religion. Second Edition*, vol. 2, 684–92. Detroit: Thomson Gale–MacMillan.

Luhrmann, T. M. (2013). 'Building on William James: The Role of Learning in Religious Experience.' In D. Xygalatas and W. W. McCorkle Jr (2013). *Mental Culture: Classical Social Theory and the Cognitive Science of Religion*, 145–63. Durham and Bristol, CT: Acumen.

Luria, S. E., S. J. Gould and S. Singer (1981). *A View of Life*. Menlo Park, CA: The Benjamin/Cummings Publishing Company.

Lyons, S. L. (1995). 'The Origins of T. H. Huxley's Saltationism: History in Darwin's Shadow.' *Journal of the History of Biology* 28(3): 463–94.

Macalister, A. (1882). *Evolution in Church History*. Dublin: Hodges & Figgis.

MacDonald, I. (2008[2005]). *Revolution in the Head: The Beatles' Records and the Sixties. Second Revised Edition*. London: Vintage Books.

Mackey, J. L. (2009). *Rethinking Roman Religion: Action, Practice, and Belief*. PhD diss., Princeton University.

MacWilliams, M. W. (2005). *s.v.* 'Campbell, Joseph.' In L. Jones (ed.), *Encyclopedia of Religion. Second Edition*, vol. 3, 1377–80. Detroit: Thomson Gale–MacMillan.

Maier, B. (2009). *William Robertson Smith: His Life, His Work and His Times*. Tübingen: Mohr Siebeck.

Malik, K. (2017). 'Outraged Headlines Erupted Last Month When Students at a London University Launched a Campaign to "Decolonise our Minds" of White European Philosophers. But What Is the Truth behind the Headlines?' *The Observer/The New Review*, 19 February: 10–13.

Mandelbrote, S. (2013). 'Early Modern Natural Theologies.' In R. Re Manning with J. H. Brooke and F. Watts (eds), *The Oxford Handbook of Natural Theology*, 75–99. Oxford: Oxford University Press.

Manera, E. (2012). *Furio Jesi. Mito, violenza, memoria*. Rome: Carocci.

Marett, R. R. (1914a[1912]). *Anthropology*. Revised edn. London: Williams and Norgate Ltd., New York: Henry Holt and Co.

Marett, R. R. (1914b[1909]). *The Threshold of Religion*. 3rd edn. London: Methuen.

Maringer, J. (1960[1956]). *The Gods of Prehistoric Man*, trans. M. Ilford. New York: Alfred A. Knopf.

Marsh, S. (2018). 'Jewish School Removed "Homosexual" Mentions from GCSE Textbook.' *The Guardian*, 9 March. https://www.theguardian.com/education/2018/mar/09/yesodey-hatorah-jewish-girls-school-north-london-homosexual-references-textbook (accessed 9 March 2018).

Martin, C. (2014[2012]). *A Critical Introduction to the Study of Religion*. New York and London: Routledge.

Martin, L. H. (1997). 'Biology, Sociology and the Study of Religions: Two Lectures.' *Religio: revue pro religionistiku* 5(1): 21–35.

Martin, L. H. (2000). 'Kingship and the Consolidation of Religio-political Power during the Hellenistic Period.' *Religio: revue pro religionistiku* 8(2): 151–60.

Martin, L. H. (2013). 'The Origins of Religion, Cognition and Culture: The Bowerbird Syndrome.' In A. W. Geertz (ed.), *Origins of Religion, Cognition and Culture*, 178–202. Durham and Bristol, CT: Acumen.

Martin, L. H. (2014). *Deep History, Secular Theory: Historical and Scientific Studies of Religion*. Boston and Berlin: de Gruyter.

Martin, L. H. (2015). 'The Continuing Enigma of "Religion".' *Religion, Brain & Behavior* 5(2): 125–31.

Martin, L. H. (2016). 'Comparative and Historical Studies of Religions: The Return of Science.' In P. Antes, A. W. Geertz and M. Rothstein (eds), *Contemporary Views on*

Comparative Religion: In Celebration of Tim Jensen's 65th Birthday, 33–45. Sheffield and Bristol, CT: Equinox.

Martin, L. H. and I. Pyysiäinen (2013). 'Conclusion: Moving towards a New Science of Religion; or, Have We Already Arrived?' In D. Xygalatas and W. W. McCorkle Jr (eds), *Mental Culture: Classical Social Theory and the Cognitive Science of Religion*, 213–26. Durham and Bristol, CT: Acumen.

Martin, L. H. and D. Wiebe (eds) (2016). *Conversations and Controversies in the Scientific Study of Religion: Collaborative and Co-authored Essays by Luther H. Martin and Donald Wiebe*. Leiden and Boston: Brill.

Martin, L. H. and D. Wiebe (eds) (2017). *Religion Explained? The Cognitive Science of Religion after Twenty-Five Years*. London and New York: Bloomsbury.

Martin, P. S. and R. G. Klein (eds) (1989). *Quaternary Extinctions: A Prehistoric Revolution*. Tucson, AZ: The University of Arizona.

Maryanski, A. (2014). 'The Birth of the Gods: Robertson Smith and Durkheim's Turn to Religion as the Basis of Social Integration.' *Sociological Theory* 32(4): 352–76.

Masse, W. B., E. Wayland Barber, L. Piccardi and P. T. Barber (2007). 'Exploring the Nature of Myth and Its Role in Science.' In L. Piccardi and W. B. Masse (eds), *Myth and Geology*, 9–28. London: The Geological Society.

Massenzio, M. (2005). 'The Italian School of "History of Religions".' *Religion* 35(4): 209–22.

Masuzawa, T. (2000). 'Origin.' In W. Braun and R. T. McCutcheon (eds), *Guide to the Study of Religion*, 209–24. London: Cassell.

Max Müller, F. (1864). *Lectures on the Science of Language Delivered at the Royal Institute of Great Britain in April, May, & June 1861*. 4th edn. London: Longman, Green, Longman, Roberts, & Green.

Max Müller, F. (1872). *Lectures on the Science of Religion. With a Paper on Buddhist Nihilism, and a Translation of the* Dhammapada *or 'Path of Virtue'*. New York: Charles Scribner and Co.

Max Müller, F. (1873). 'Lectures on Mr. Darwin's Philosophy of Language.' Three lectures published in *Fraser's Magazine*, London, May: 525–41; June: 659–79; July: 1–24.

Max Müller, F. (1875). *Chips from a German Workshop*, 4 vols. London: Longmans, Green, & Co.

Max Müller, F. (1878). *Lectures on the Origin and Growth of Religion*. London: Longmans, Green, & Co.

Max Müller, F. (1881). 'Essays on the Science of Religion.' In *Chips from a German Workshop*, vol. 1. New York: Charles Scribner & Sons.

Max Müller, F. (1891). *Lectures on the Science of Language*, 9th edn, 2 vols. London: Longmans, Green, & Co.

Max Müller, F. (1893[1870]). *Introduction to the Science of Religion*. London: Longmans, Green, & Co.

Max Müller, F. (1898a). *Theosophy, or Psychological Religion*. London: Longmans, Green, & Co.

Max Müller, F. (1898b). *Natural Religion*. London: Longmans, Green, & Co.

Max Müller, F. (1898c). *Anthropological Religion*. London: Longmans, Green, & Co.

Max Müller, F. (1899). *Auld Lang Syne*. New York: Scribner.

Max Müller, F. (1909[1856]). *Comparative Mythology: An Essay*. London: Routledge and New York: Dutton & Co.

Mayr, E. (1982). *The Growth of Biological Thought: Diversity, Evolution, and Inheritance*. Cambridge, MA and London: Belknap Press of Harvard University Press.

Mayr, E. (1991). *One Long Argument: Charles Darwin and the Genesis of Modern Evolutionary Thought*. Cambridge, MA: Harvard University Press.

McCauley, R. N. (2011). *Why Religion Is Natural and Science Is Not*. New York: Oxford University Press.

McCauley, R. N. (2017). *Philosophical Foundations of the Cognitive Science of Religion: A Head Start*. With a chapter co-written with E. T. Lawson. London and New York: Bloomsbury.

McCauley, R. N. and E. T. Lawson (2002). *Bringing Ritual to Mind: Psychological Foundations of Cultural Forms*. Cambridge and New York: Cambridge University Press.

McCutcheon, R. T. (1997). *Manufacturing Religion: The Discourse on Sui Generis Religion and the Politics of Nostalgia*. Oxford: Oxford University Press.

McCutcheon, R. T. (2004). 'Critical Trends in the Study of Religion in the United States.' In P. Antes, A. M. Geertz and R. R. Warne (eds), *New Approaches to the Study of Religion. Vol. 1: Regional, Critical, and Historical Approaches*, 317–43. Berlin and New York: de Gruyter.

McCutcheon, R. T. (2008). 'Introducing Smith.' In W. Braun and R. T. McCutcheon (eds), *Introducing Religion: Essays in Honor of Jonathan Z. Smith*, 1–17. London and Oakville, CT: Equinox.

McCutcheon, R. T. (2014). '"Religion" as "*sui generis*".' Interview hosted by T. J. Coleman III. *The Religious Studies Project*, 13 January. http://www.religiousstudiesproject.com/podcast/russell-mccutcheon-on-religion-as-sui-generis/ (accessed 15 May 2017).

McGrath, A. (2015). *Inventing the Universe: Why We Can't Stop Talking about Science, Faith and God*. London: Hodder & Stoughton.

Medawar, P. B. (1986[1984]). *The Limits of Science*. Oxford: Oxford University Press.

Mercier, H. and D. Sperber (2011). 'Why Do Humans Reason? Arguments for an Argumentative Theory.' *Behavioural and Brain Sciences* 34(2): 57–74.

Merkur, D. (1996). 'Psychoanalytic Methods in the History of Religion: A Personal Statement.' *Method & Theory in the Study of Religion* 8(4): 327–43.

Merlo, R. (2011). *Il mito dacico nella letteratura romena dell'Ottocento*. Alessandria: Edizioni dell'Orso.

Mesoudi, A. (2011). *Cultural Evolution: How Darwinian Theory Can Explain Human Culture & Synthesize the Social Science*. Chicago and London: The University of Chicago Press.

Mesoudi, A., K. N. Laland, R. Boyd, B. Buchanan, E. Flynn, R. N. McCauley, J. Renn, V. Reyes-García, S. Shennan, D. Stout and C. Tennie (2013). 'The Cultural Evolution of Technology and Science.' In P. J. Richerson and M. H. Christiansen (eds), *Cultural Evolution: Society, Technology, Language, and Religion*, 193–216. Cambridge, MA and London: The MIT Press.

Mezei, B. M. (2014). 'Memes, Possible Worlds, and Quantum Theory: New Perspectives in the Study of Religion.' In Á. Kovács and J. L. Cox (eds), *New Trends and Recurring Issues in the Study of Religion*, 160–82. Paris: L'Harmattan.

Mikaelson, L. (2005). 'Gendering the History of Religions.' In P. Antes, A. M. Geertz and R. R. Warne (eds), *New Approaches to the Study of Religion. Vol. 1: Regional, Critical, and Historical Approaches*, 295–315. Berlin and New York: de Gruyter.

Milner, R. (2009). *Darwin's Universe: Evolution from A to Z*. Berkeley, Los Angeles and London: University of California Press.

Mincu, M. and R. Scagno (eds) (1986). *Mircea Eliade e l'Italia*. Milan: Jaca Book.

Minois, G. (1998). *Histoire de l'athéisme. Les incroyants dans le monde occidental des origines à nos jours*. Paris: Fayard.

Minois, G. (2012). *Dictionnaire des athées, agnostiques, sceptiques et autres mécréants*. Paris: Albin Michel.

Mischek, U. (2008). 'Antisemitismus und Antijudaismus in den Werken und Arbeiten Pater Wilhelm Schimdts S.V.D. (1868–1954).' In H. Junginger (ed.), *The Study of Religion Under the Impact of Fascism*, 467–90. Leiden and Boston: Brill.

Mishler, B. D. (2000). 'Deep Phylogenetic Relationships among "Plants" and Their Implications for Classification.' *Taxon* 49(4): 661–83.

Molendijk, A. L. (2004). 'Religious Development: C. P. Tiele's Paradigm of Science of Religion.' *Numen* 51(3): 321–51.

Molendijk, A. L. (2005). *The Emergence of the Science of Religion in the Netherlands*. Leiden and Boston: Brill.

Molendijk, A. L. (2016). *Friedrich Max Müller and the Sacred Books of the East*. Oxford: Oxford University Press.

Molendijk, A. L. (2017). 'The Study of Religion in the Netherlands.' *Journal for Theology and the Study of Religion* 71(1): 5–18.

Momigliano, A. (1979[1954]). 'A Hundred Years after Ranke.' In *Primo contributo alla storia degli studi classici e del mondo antico*, 367–73. Rome: Edizioni di Storia e Letteratura.

Momigliano, A. (1993[1971]). *Alien Wisdom: The Limits of Hellenization*. Cambridge: Cambridge University Press.

Momigliano, A. (2005[1987]). *s.v.* 'Historiography.' In L. Jones (ed.), *Encyclopedia of Religion. Second Edition*, vol. 6, 4035–42. Detroit: Thomson Gale–MacMillan.

Momigliano, A. (2016[1974]). 'The Rules of the Game in the Study of Ancient History,' trans. K. W. Yu. *History & Theory* 55(1): 39–45.

Moore, J. R. (1981[1979]). *The Post-Darwinian Controversies: A Study of the Protestant Struggle to Come to Terms with Darwin in Great Britain and in America 1870–1990*. Cambridge: Cambridge University Press.

Moore, J. R. (1989). 'Of Love and Death: Why Darwin "Gave up Christianity".' In J. R. Moore (ed.), *History, Humanity and Evolution: Essays For John C. Greene*, 195–229. Cambridge and New York: Cambridge University Press.

Morgan, T. J. H. and P. L. Harris (2015). 'James Mark Baldwin and Contemporary Theories of Culture and Evolution.' *European Journal of Developmental Psychology* 12(6): 666–77.

Morin, E. (1973). *Le paradigme perdu. La nature humain*. Paris: Éditions du Seuil.

Morris, B. (1987). *Anthropological Studies of Religion: An Introductory Text*. Cambridge: Cambridge University Press.

Morris, D. (1984[1967]). *The Naked Ape*. New York: Dell.

Munz, P. (1985). *Our Knowledge of the Growth of Knowledge: Popper or Wittgenstein?* London: Routledge.

Murphy, A. R. (ed.) (2011). *The Blackwell Companion to Religion and Violence*. Malden, MA and Oxford: Wiley-Blackwell.

Murray, O. (2010). 'In Search of the Key to All Mythologies.' In S. Rebenich, B. von Reibnitz and T. Späth, *Translating Antiquity: Antikebilder im europäischen Kulturtransfer*, 119–29. Basel: Schwabe.

Muthuraj, J. G. (2007). 'The Significance of Mircea Eliade for Christian Theology.' In B. Rennie (ed.), *The International Eliade*, 71–99. Albany: State University of New York.

Nicholls, A. (2014). 'A Germanic Reception in England: Friedrich Max Müller's Critique of Darwin's *Descent of Man*.' In T. F. Glick and E. Shaffer (eds), *The Literary and Cultural Reception of Charles Darwin in Europe*, vol. 3, 78–100. London and New York: Bloomsbury.

Noiret, P. (2017). 'La religiosité au Paléolithique.' *Comptes Rendus Palevol* 16(2): 182–8.

Noll, R. (1994). *The Jung Cult: Origins of a Charismatic Movement*. Princeton: Princeton University Press.

Nongbri, B. (2013). *Before Religion: A History of a Modern Concept*. New Haven and London: Yale University Press.

Norenzayan, A. (2013). *Big Gods: How Religion Transformed Cooperation and Conflict*. Princeton and Oxford: Princeton University Press.

Numbers, R. L. (2006[1992]). *The Creationists: From Scientific Creationism to Intelligent Design. Expanded Edition*. Cambridge, MA: Harvard University Press.

Numbers, R. L. (2009). 'Myth 24: That Creationism Is a Uniquely American Phenomenon.' In R. L. Numbers (ed.), *Galileo Goes to Jail and Other Myths about Science and Religion*, 215–23. Cambridge, MA: Harvard University Press.

Nunn, P. D. (2009). *Vanished Islands and Hidden Continents of the Pacific*. Honolulu: University of Hawai'i Press.

Ohlmarks, Å. (1939). *Studien zum Problem des Schamanismus*. Lund and Copenhagen: C.W.K. Gleerup-Ejnar Munksgaard.

Oişteanu, A. (2010). *Narcotice în cultura română. Istorie, religie şi literatură. Ediţia ilustrată*. Iaşi: Polirom.

Oişteanu, A. (2012[2001]). *Imaginea evreului în cultura română. Studiu de imagologie în context est-central-european. Ediţia a III-a, revăzută, adăugită şi ilustrată*. Iaşi: Polirom.

Olson, C. (1989). 'Theology of Nostalgia: Reflections on the Theological Aspects of Eliade's Work.' *Numen* 36(1): 98–112.

Omodeo, A. (1929). *Tradizioni morali e disciplina storica*. Bari: Laterza.

Orbecchi, M. (2015). *Psicologia dell'anima*. Turin: Bollati Boringhieri.

O'Rourke, M., S. Crowley and C. Gonnerman (2016). 'On the Nature of Cross-disciplinary Integration: A Philosophical Framework.' *Studies in History and Philosophy of Science Part C: Studies in History and Philosophy of Biological and Biomedical Sciences*, 56: 62–70.

Otto, R. (1923[1917]). *The Idea of the Holy*, trans. J. W. Harvey. New York: Oxford University Press.

Oxford Dictionaries (2016a). 'Post-Truth.' https://en.oxforddictionaries.com/definition/post-truth (accessed 20 December 2017).

Oxford Dictionaries (2016b). 'Word of the Year 2016 is . . .' 16 November. https://en.oxforddictionaries.com/word-of-the-year/word-of-the-year-2016 (accessed 20 December 2017).

Paden, W. E. (2016). *New Patterns for Comparative Religion: Passages to an Evolutionary Perspective*. London and New York: Bloomsbury.

Palladino, P. and M. Worboys (1993). 'Science and Imperialism.' *Isis* 84(1): 91–102.

Pearsall, J. (ed.) (1999). *The Concise Oxford Dictionary. Tenth Edition*. Oxford: Oxford University Press.

Penn, D. C. (2011). 'How Folk Psychology Ruined Comparative Psychology and How Scrub Jays Can Save It.' In R. Menzel and J. Fischer (eds), *Animal Thinking: Contemporary Issues in Comparative Cognition*, 253–65. Cambridge, MA and London: The MIT Press.

Penn, D. C., K. J. Holyoak and D. J. Povinelli (2008). 'Darwin's Mistake: Explaining the Discontinuity between Human and Nonhuman Minds.' *Behavioral and Brain Sciences* 31(2): 109–30.

Penner, H. H. (1989). *Impasse and Resolution: A Critique of the Study of Religion*. New York: Lang.

Penner, H. H. and E. A. Yonan (1972). 'Is a Science of Religion Possible?' *The Journal of Religion* 52(2): 107–33.

Pennycook, G., J. A. Cheyne, N. Barr, D. J. Koehler and J. A. Fugelsang (2015). 'On the Reception and Detection of Pseudo-profound Bullshit.' *Judgment and Decision Making* 10(6): 549–63.

Pereltsvaig, A. and M. W. Lewis (2015). *The Indo-European Controversy: Facts and Fallacies in Historical Linguistics*. Cambridge and New York: Cambridge University Press.

Peresani, M., I. Fiore, M. Gala, M. Romandini and A. Tagliacozzo (2011). 'Late Neanderthals and the Intentional Removal of Feathers as Evidenced from Bird Bone Taphonomy at Fumane Cave 44 ky B.P., Italy.' *Proceedings of the National Academy of Sciences* 108(10): 3888–93.

Pernet, H. (2006[1988]). *Ritual Masks: Deceptions and Revelations*, trans. L. Grillo. Eugene, OR: Wipf & Stock.

Pernet, H. (2011). 'Introducere.' In M. Gligor (ed.), *Mircea Eliade – Henry Pernet. Corespondenţă. 1961–1986*, 13–30. Cluj-Napoca: Casa Cărţii de Ştiinţă.

Pernet, H. (2012). 'Lasting Impressions: Mircea Eliade.' In M. Gligor (ed.), *Mircea Eliade between the History of Religions and the Fall into History*, 9–33. Cluj-Napoca: Presa Universitară Clujeană.

Pettazzoni, R. (1911). 'Le superstizioni, Relazione. Primo Congresso di Etnografia Italiana. Roma, 19–24 ottobre 1911.' Rome: n.p.

Pettazzoni, R. (1912a). 'Lo studio delle religioni in Italia.' *Nuova Antologia* 47(243): 107–10.

Pettazzoni, R. (1912b). *La religione primitiva in Sardegna*. Piacenza: Società Editrice Pontremolese.

Pettazzoni, R. (1913). 'La scienza delle religioni e il suo metodo.' *Scientia* 7(13): 239–47.

Pettazzoni, R. (1922). *Dio. Formazione e sviluppo del monoteismo nella storia delle religioni. Vol. I: L'essere celeste nelle credenze dei popoli primitivi*. Rome: Società Editrice Athenaeum (Bologna: Stabilimenti Tipografici Riuniti).

Pettazzoni, R. (1924). *Svolgimento e carattere della storia delle religioni: Lezione inaugurale pronunziata nell'Università di Roma il 17 gennaio 1924*. Bari: Laterza.

Pettazzoni, R. (1927). 'Studi recenti in rapporto con la teoria degli esseri celesti e del monoteismo.' *Studi e Materiali di Storia delle Religioni* 3: 97–113.

Pettazzoni, R. (1933). Review of van der Leeuw, G. (1933). *Phänomenologie der Religion*. Tübingen: Mohr. *Studi e Materiali di Storia delle Religioni* 9: 242–4.

Pettazzoni, R. (1954a). *Essays on the History of Religions*, trans. H. J. Rose. Leiden: Brill.

Pettazzoni, R. (1954b). 'Alle origini della scienza delle religioni.' *Numen* 1(2): 136–17.

Pettazzoni, R. (1956a). 'Das Ende des Urmonotheismus?' *Numen* 3(2): 156–9.

Pettazzoni, R. (1956b[1955]). *The All-Knowing God: Researches into Early Religion and Culture*, trans. H. J. Rose. London: Methuen.

Pettazzoni, R. (1959). 'The Supreme Being: Phenomenological Structure and Historical Development.' In M. Eliade and J. M. Kitagawa (eds), *The History of Religions: Essays in Methodology*, 59–66. Chicago: The University of Chicago Press.

Pharo, L. K. (2011). 'A Methodology for a Deconstruction and Reconstruction of the Concepts "Shaman" and "Shamanism".' *Numen* 58(1): 6–70.

Pickering, M. (2009). *Auguste Comte: An Intellectual Biography. Volume 3*. Cambridge: Cambridge University Press.

Pietsch, T. W. (2012). *Trees of Life: A Visual History of Evolution*. Baltimore: The Johns Hopkins University Press.

Pievani, T. (2011). 'An Evolving Research Programme: The Structure of Evolutionary Theory from a Lakatosian Perspective.' In A. Fasolo (ed.), *The Theory of Evolution and Its Impact*, 211–28. Milan: Springer.

Pievani, T. (2012). 'Many Ways of Being Human, the Stephen J. Gould's Legacy to Palaeoanthropology (2002–2012).' *Journal of Anthropological Sciences* 90: 133–49.

Pievani, T. (ed.) (2013a). *Charles Darwin. Lettere sulla religione.* Turin: Einaudi.

Pievani, T. (2013b). 'Diamond cammina sul filo.' *Doppiozero*, 29 July. http://www.doppiozero.com/materiali/recensioni/diamond-cammina-sul-filo (accessed 20 December 2017).

Pievani, T. and A. Parravicini (2016). 'Multilevel Selection in a Broader Hierarchical Perspective.' In N. Eldredge, T. Pievani, E. Serrelli and I. Temkin (eds), *Evolutionary Theory: A Hierarchical Perspective*, 174–201. Chicago and London: The University of Chicago Press.

Pigliucci, M. (2013). 'The Demarcation Problem: A (Belated) Response to Laudan.' In M. Pigliucci and M. Boudry (eds), *Philosophy of Pseudoscience: Reconsidering the Demarcation Problem*, 9–28. Chicago and London: The University of Chicago Press.

Pigliucci, M. (2015a). 'The Debate about Funding of Basic Scientific Research.' In *Scientia Salon: Philosophy, Science, and All Interesting Things in Between*, 18 June. https://scientiasalon.wordpress.com/2015/06/18/the-debate-about-funding-of-basic-scientific-research/ (accessed 20 December 2017).

Pigliucci, M. (2015b). 'Smolin on Mathematics.' *Scientia Salon: Philosophy, Science, and All Interesting Things in Between*, 21 April. https://scientiasalon.wordpress.com/2015/04/21/smolin-on-mathematics/ (accessed 30 January 2018).

Pigliucci, M. and M. Boudry (eds) (2013). *Philosophy of Pseudoscience: Reconsidering the Demarcation Problem.* Chicago and London: The University of Chicago Press.

Pigliucci, M. and M. Boudry (2014). 'Prove it! The Burden of Proof Game in Science vs. Pseudoscience Disputes.' *Philosophia* 42(2): 487–502.

Pinker, S. (1997). *How the Mind Works.* London: Penguin.

Pinker, S. (2002). *The Blank Slate: The Modern Denial of Human Nature.* London: Penguin.

Pinker, S. (2010). 'The Cognitive Niche: Coevolution of Intelligence, Sociality, and Language.' *Proceedings of the National Academy of Sciences* 107(suppl. 2): 8893–999.

Plantinga, R. J. (1989). 'W. B. Kristensen and the Study of Religion.' *Numen* 36(2): 173–88.

Platvoet, J. (1998a). 'Close Harmonies: The Science of Religion in Dutch *Duplex Ordo* Theology, 1860–1960.' *Numen* 54(2): 115–62.

Platvoet, J. (1998b). 'From Consonance to Autonomy: The Science of Religion in the Netherlands, 1948–1995.' *Method & Theory in the Study of Religion* 10(4): 334–51.

Plotkin, H. (2004). *Evolutionary Thought in Psychology.* Oxford: Blackwell.

Podemann Sørensen, J. (2006). 'The Historical and Comparative Study of Religions: A Rhetorical Approach.' In P. Antes, A. W. Geertz and M. Rothstein (eds), *Contemporary Views on Comparative Religion in Celebration of Tim Jensen's 65th Birthday*, 47–58. Sheffield and Bristol, CT: Equinox.

Popper, K. (1994). *The Myth of the Framework: In Defence of Science and Rationality*, edited by M. Notturno. Abingdon and New York: Routledge.

Prandi, C. (2011). 'La storia delle religioni in Italia tra XX e XXI secolo.' *Humanitas* 66(1): 65–87.

Preus, S. (1987). *Explaining Religion: Criticism and Theory from Bodin to Freud.* New York and London: Yale University Press.

Proctor, R. N. (2008). 'Agnotology: A Missing Term to Describe the Cultural Production of Ignorance (and Its Study).' In R. N. Proctor and L. Schiebinger (eds), *Agnotology: The Making and Unmaking of Ignorance*, 1–35. Stanford: Stanford University Press.

Prothero, D. R. (2013). *Reality Check: How Science Deniers Threaten Our Future.* Bloomington and Indianapolis: Indiana University Press.

Purzycki, B. G. and R. Sosis (2013). 'The Extended Religious Phenotype and the Adaptive Coupling of Ritual and Belief.' *Israel Journal of Ecology & Evolution* 59(2): 99–108.

Purzycki, B. G. and R. A. McNamara (2016). 'An Ecological Theory of Gods' Minds.' In H. De Cruz and R. Nichols (eds), *Advances in Religion, Cognitive Science, and Experimental Philosophy*, 143–67. London and New York: Bloomsbury.

Pyrah, R. (2008). 'Enacting Encyclicals? Cultural Politics and "Clerical Fascism" in Austria, 1933–1938.' In M. Feldman, M. Turda and T. Georgescu (eds), *Clerical Fascism in Interwar Europe*, 157–70. London and New York: Routledge.

Pyysiäinen, I. (2009). *Supernatural Agents: Why We Believe in Souls, Gods, and Buddhas*. Oxford: Oxford University Press.

Pyysiäinen, I. and V. Anttonen (eds) (2002). *Current Approaches in the Cognitive Science of Religion*. London and New York: Continuum.

Rabett, R. J. (2018). 'The Success of Failed *Homo Sapiens* Dispersals Out of Africa and into Asia.' *Nature Ecology & Evolution* 2: 212–19.

Ratnapalan, L. (2008). 'E. B. Tylor and the Problem of Primitive Culture.' *History and Anthropology* 19(2): 131–42.

Raup, D. M. and D. Jablonski (eds) (1986). *Patterns and Processes in the History of Life. Report of the Dahlem Workshop on Patterns and Processes in the History of Life, Berlin 1985, June 16–21*. Berlin, Heidelberg and New York: Springer.

Rennie, B. S. (ed.) (2006). *Mircea Eliade: A Critical Reader*. London and Oakville: Equinox.

Rennie, B. S. (2013). 'Raffaele Pettazzoni from the Perspective of the Anglophone Academy.' *Numen* 60(5–6): 649–75.

Rennie, B. S. (2016). 'Mircea Eliade: Eponym of the Humanities?' *Los Angeles Review of Books*, 13 October. https://lareviewofbooks.org/article/mircea-eliade-eponym-of-the-humanities/#! (accessed 20 December 2017).

Richards, R. J. (1987). *Darwin and the Emergence of Evolutionary Theories of Mind and Behavior*. Chicago and London: The University of Chicago Press.

Richards, R. J. (2013). *Was Hitler a Darwinian? Disputed Questions in the History of Evolutionary Theory*. Chicago and London: The University of Chicago Press.

Richerson, P. J. (2014). Comment on Turchin, P. (2014). 'Cooperation: This Time, between Man and Woman.' *Social Evolution Forum*, 20 October. https://evolution-institute.org/blog/cooperation-this-time-between-man-and-woman/ (accessed 30 January 2018).

Richerson, P. J. and R. Boyd (2005). *Not by Genes Alone: How Culture Transformed Human Evolution*. Chicago and London: The University of Chicago Press.

Richerson, P. J. and R. Boyd (2013). 'Rethinking Paleoanthropology: A World Queerer than We Supposed.' In G. Hatfield and H. Pittman (eds), *Evolution of Mind, Brain and Culture*, 263–302. Philadelphia: University of Pennsylvania Museum of Archaeology and Anthropology.

Richerson, P. J. and L. Newson (2009). 'Is Religion Adaptive? Yes, No, Neutral. But Mostly We Don't Know.' In J. Schloss and M. J. Murray (eds), *The Believing Primate: Scientific, Philosophical, and Theological Reflections on the Origin of Religion*, 100–17. Oxford and New York: Oxford University Press.

Richlin, A. (2009). 'Writing Women into History.' In A. Erskine (ed.), *A Companion to Ancient History*, 146–53. Malden, MA and Oxford: Blackwell.

Richmond, M. L. (2006). 'The 1909 Darwin Celebration Reexamining Evolution in the Light of Mendel, Mutation, and Meiosis.' *Isis* 97(3): 447–84.

Ricketts, M. L. (1966). 'The North American Indian Trickster.' *History of Religions* 5(2): 327–50.

Ricketts, M. L. (1970). *The Sacred Trickster of the American Indians*. Unpublished manuscript. Revised version of the author's 1964 PhD diss. entitled *The Structure and Religious Significance of the Trickster-Culture Hero in the Mythology of the North American Indians*, University of Chicago.

Ricketts, M. L. (1978). 'Mircea Eliade et la mort de Dieu.' In C. Tacou, G. Banu and M.-F. Ionesco (eds), *L'Herne. Mircea Eliade*, 110–19. Paris: Éditions de l'Herne.

Ricketts, M. L. (2000a). 'Eliade and Goethe.' *Archæus* 6(3–4): 283–311.

Ricketts, M. L. (2000b). 'La risposta americana all'opera letteraria di Mircea Eliade.' In J. Ries and N. Spineto (ed.), *Esploratori del pensiero umano. Georges Dumézil e Mircea Eliade*, 369–89. Milan: Jaca Book.

Ricketts, M. L. (2004[1988]). *Rădăcinile româneşti ale lui Mircea Eliade*, 2 vols. Bucharest: Criterion.

Ricketts, M. L. (2007). 'Mircea Eliade, Profesorul meu: Amintiri.' In M. Gligor and M. L. Ricketts (eds), *Întâlniri cu Mircea Eliade*, 209–19. Bucharest: Humanitas.

Ricketts, M. L. (2008). 'Eliade on Diplomatic Service in London.' *Religion* 38(4): 346–54.

Riel-Salvatore, J. (2010). 'A Niche Construction Perspective on the Middle–Upper Paleolithic Transition in Italy.' *Journal of Archaeological Method and Theory* 17(4): 323–55.

Rieppel, O. (2008). 'Re-writing Popper's Philosophy of Science for Systematics.' *History and Philosophy of the Life Sciences* 30(3–4): 293–316.

Ries, J. (2005). *Il mito e il suo significato*. Milan: Jaca Book.

Roberts, D. D. (1987). *Benedetto Croce and the Uses of Historicism*. Berkeley and Los Angeles: University of California Press.

Roberts, J. H. (2009). 'Myth 18: That Darwin Destroyed Natural Theology.' In R. L. Numbers (ed.), *Galileo Goes to Jail and Other Myths about Science and Religion*, 161–9. Cambridge, MA and London: Harvard University Press.

Robinson, A. (2002). *The Life and Work of Jane Ellen Harrison*. Cambridge: Cambridge University Press.

Roebroeks, W., M. J. Sier, T. Kellberg Nielsen, D. D. Loecker, J. M. Parés, C. E. S. Arps and H. J. Mücher (2012). 'Use of Red Ochre by Early Neandertals.' *Proceedings of the National Academy of Sciences* 109(6): 1889–94.

Romero, A. (2009). *Cave Biology: Life in Darkness*. Cambridge: Cambridge University Press.

Rosengren, K. S., S. K. Brem, E. M. Evans and G. M. Sinatra (eds) (2012). *Evolution Challenges: Integrating Research and Practice in Teaching and Learning about Evolution*. Oxford: Oxford University Press.

Rossi, P. (2003). *I segni del tempo. Storia della Terra e storia delle nazioni da Hooke a Vico*. Milan: Feltrinelli.

Roubekas, N. P. (2015). 'Belief in Belief and Divine Kingship in Early Ptolemaic Egypt: The Case of Ptolemy II Philadelphus and Arsinoe II.' *Religio: revue pro religionistiku* 23(1): 3–23.

Roubekas, N. P. (2017). *An Ancient Theory of Religion: Euhemerism from Antiquity to the Present*. New York and Abingdon: Routledge.

Rowland, I. D. (2014). 'Furio Jesi and the *Culture of the Right*.' In R. Bod, J. Maat and T. Weststeijn (eds), *History of the Humanities III: The Modern Humanities*, 293–310. Amsterdam: Amsterdam University Press.

Rudolph, K. and A. Ciattini (2005[1987]). *s.v.* 'Kulturkreislehre.' In L. Jones (ed.), *Encyclopedia of Religion. Second Edition*, vol. 8, 5259–62. Detroit: Thomson Gale–MacMillan.

Ruse, M. (2009). *s.v.* 'Natural Theology.' In M. Ruse and J. Travis (eds), *Evolution: The First Four Billion Years*, 756–62. Cambridge, MA and London: The Belknap Press of Harvard University Press.

Ruse, M. (2013). 'Evolution: From Pseudoscience to Popular Science, from Popular Science to Professional Science.' In M. Pigliucci and M. Boudry (eds), *Philosophy of Pseudoscience: Reconsidering the Demarcation Problem*, 225–44. Chicago and London: University of Chicago Press.

Ruse, M. (2015). *Atheism: What Everyone Needs to Know*. Oxford and New York: Oxford University Press.

Ruse, M. and R. J. Richards (eds) (2009). *The Cambridge Companion to the* Origin of Species. New York: Cambridge University Press.

Russell, E. (2011). *Evolutionary History: Uniting History and Biology to Understand Life on Earth*. Cambridge and New York: Cambridge University Press.

Sagan, C. (1996a). *The Demon-Haunted World: Science as a Candle in the Dark*. New York: Ballantine Books.

Sagan, C. (1996b). 'Carl Sagan's Last Interview with Charlie Rose (Full Interview).' *Youtube*. https://www.youtube.com/watch?v=U8HEwO-2L4w (accessed 20 December 2017).

Saler, B. (2000[1993]). *Conceptualizing Religion: Immanent Anthropologists, Transcendent Natives, and Unbounded Categories*. New York and Oxford: Berghan Books.

Saler, B. (2008). 'Essentialism and Evolution.' In J. Bulbulia, R. Sosis, E. Harris, R. Genet, C. Genet and K. Wyman (eds), *The Evolution of Religion: Studies, Theories, & Critiques*, 379–86. Santa Margarita, CA: Collins Foundation Press.

Saler, B. (2009). *Understanding Religion: Selected Essays*. Berlin and New York: de Gruyter.

Saliba, J. A. (1976) *'Homo Religiosus' in Mircea Eliade: An Anthropological Evaluation*. Leiden: Brill.

Salzman, M. R. (2011[2007]). 'Religious *Koine* and Religious Dissent in the Fourth Century.' In J. Rüpke (ed.), *A Companion to Roman Religion*, 109–25. Malden, MA and Oxford: Wiley-Blackwell.

Sanderson, S. K. (2018). *Religious Evolution and the Axial Age: From Shamans to Priests to Prophets*. London and New York: Bloomsbury.

Sapolsky, R. (2017a). 'Why Your Brain Hates Other People, and How to Make It Think Differently.' *Nautilus* (49), 22 June. http://nautil.us/issue/49/the-absurd/why-your-brain-hates-other-people (accessed 1 July 2017).

Sapolsky, R. (2017b). *Behave: The Biology of Humans at Our Best and Worst*. London: Penguin.

Scagno, R. (2000). 'Alcuni punti fermi sull'impegno politico di Mircea Eliade nella Romania interbellica. Un commento al dossier «Toladot» del 1972.' In J. Ries and N. Spineto (eds), *Esploratori del pensiero umano. Georges Dumézil e Mircea Eliade*, 259–89. Milan: Jaca Book.

Schaller, M. and D. R. Murray (2008). 'Pathogens, Personality, and Culture: Disease Prevalence Predicts Worldwide Variability in Sociosexuality, Extraversion, and Openness to Experience.' *Journal of Personality and Social Psychology* 95(1): 212–21.

Schaller, M. and D. R. Murray (2010). 'Infectious Diseases and the Evolution of Cross-Cultural Differences.' In M. Schaller, A. Norenzayan, S. J. Heine, T. Yamagishi and T. Kameda (eds), *Evolution, Culture, and the Human Mind*, 243–56. New York and London: Psychology Press.

Schilbrack, K. (2005). 'Bruce Lincoln's Philosophy.' *Method & Theory in the Study of Religion* 17(1): 44–58.

Schimmel, A. (1960). 'Summary of the Discussion.' *Numen* 7(2): 235–9.

Schmidt, L. (1952). 'Der "Herr der Tiere" in einigen Sagenlandschaften Europas und Eurasiens.' *Anthropos* 47(3–4): 509–38.

Schmidt, W. (1910). *L'origine de L'idée de Dieu. Étude historico-critique et positive. 1ère Partie: Historico-Critique.* Paris: Picard & Fils.

Schmidt, W. (1913). Review of Pettazzoni, R. (1912). *La Religione primitiva in Sardegna,* 1912. *Anthropos* 8(2–3): 573–6.

Schmidt, W. (1914). Review of Hehn, J., *Die biblische und die babylonische Gottesidee, die israelitische Gottesauffassung im Lichte der altorientalischen Religionsgeschichte* (Leipzig 1913). *Anthropos* (9): 343–8.

Schmidt, W. (1919–1920). 'Die kulturhistorische Methode und die nordamerikanische Ethnologie.' *Anthropos* (14–15): 546–63.

Schmidt, W. (1926–1955). *Der Ursprung der Gottesidee,* 12 vols. Münster: Aschendorff.

Schmidt, W. (1931). *The Origin and Growth of Religion: Facts and Theories,* trans. H. J. Rose. London: Methuen.

Schmidt, W. (1935a). *Rasse und Volk: Ihre allgemeine Bedeutung. Ihre Geltung im deutschen Raum.* Salzburg, Leipzig: Pustet.

Schmidt, W. (1935b). *Der Ursprung der Gottesidee. Eine historisch-kritische und positive Studie. VI: Endsynthese der Religionen der Urvolker – Amerikas, Asiens, Australiens, Afrikas.* Münster: Aschendorff.

Schmidt, W. (1955). *Das Mutterrecht.* Wien-Mödling: Verlag der Missionsdruckerei St Gabriel.

Schmidt, W. (1964). *Wege der Kulturen: Gesammelte Aufsätze.* St Augustin bei Bonn: Verlag des Anthropos-Instituts.

Schmidt, W. and W. Koppers (1924). *Volker und Kulturen. Erster Teil: Gesellschaft und Wirtschaft der Volker.* Regensburg: J. Habbel.

Schwartz, M. A. (2008). 'The Importance of Stupidity in Scientific Research.' *Journal of Cell Science* 121: 1771.

Segal, R. A. (2002a). 'Myth and Politics: A Response to Robert Ellwood.' *Journal of the American Academy of Religion* 70(3): 611–20.

Segal, R. A. (2002b). 'Introduction.' In W. R. Smith, *Religion of the Semites.* New Brunswick and London: Transaction.

Segal, R. A. (2004). *Myth: A Very Short Introduction.* Oxford: Oxford University Press.

Segal, R. A. (2006). 'All Generalizations Are Bad: Postmodernism on Theories.' *Journal of the American Academy of Religion* 74(1): 157–71.

Segal, R. A. (2016). 'Friedrich Max Müller on Religion and Myth.' *Publications of the English Goethe Society* 85(2–3): 135–44.

Sela, Y., T. K. Shackelford and J. R. Liddle (2015). 'When Religion Makes It Worse: Religiously Motivated Violence as a Sexual Selection Weapon.' In D. J. Slone and J. A. Van Slyke (eds), *The Attraction of Religion: A New Evolutionary Psychology of Religion,* 111–32. London and New York: Bloomsbury.

Sepkoski, D. and M. Ruse (eds) (2010). *The Paleobiological Revolution: Essays on the Growth of Modern Paleontology.* Chicago and London: The University of Chicago Press.

Severino, V. S. (2015). 'For a Secular Return to the Sacred: Raffaele Pettazzoni's Last Statement on the Name of the Science of Religions.' *Religion* 45(1): 2–23.

Sfameni Gasparro, G. (2011). *Introduzione alla storia delle religioni.* Rome and Bari: Laterza.

Sfameni Gasparro, G. (2016). 'A Method Without Explanatory Theory: Ugo Bianchi's Historical-Comparative Methodology after Thirty Years.' In P. Antes, A. W. Geertz and M. Rothstein (eds), *Contemporary Views on Comparative Religion: In Celebration of Tim Jensen's 65th Birthday,* 73–86. Sheffield and Bristol, CT: Equinox.

Sharpe, E. J. (1986[1975]). *Comparative Religion: A History. Second Edition*. London: Duckworth & Co.

Sharpe, E. J. (1988). 'Religious Studies, the Humanities, and the History of Ideas.' *Soundings* 71: 245–58.

Shermer, M. (2002). *In Darwin's Shadow. The Life and Science of Alfred Russel Wallace: A Biographical Study on the Psychology of History*. Oxford and New York: Oxford University Press.

Shermer, M. (2013). 'Science and Pseudoscience: The Difference in Practice and the Difference It Makes.' In M. Pigliucci and M. Boudry (eds), *Philosophy of Pseudoscience: Reconsidering the Demarcation Problem*, 203–24. Chicago and London: The University of Chicago Press.

Shermer, M. and A. Grobman (2009). *Denying History: Who Says the Holocaust Never Happened and Why Do They Say It? Updated and Expanded*. Berkeley, Los Angeles and London: University of California Press.

Shettleworth, S. J. (2010). *Cognition, Evolution, and Behavior. Second Edition*. Oxford and New York: Oxford University Press.

Shimron, Y. (2018). 'Religion Historian Jonathan Z. Smith Dies.' *RNS Religion News Service*, 2 January. https://religionnews.com/2018/01/02/religion-historian-jonathan-z-smith-dies/ (accessed 10 February 2018).

Shirokogoroff, S. M. (1935). *Psychomental Complex of the Tungus*. London: Kegan Paul.

Shryock, A. and D. L. Smail (2011a). 'Introduction.' In A. Shryock and D. L. Smail (eds), *Deep History: The Architecture of Past and Present*, 3–20. Berkeley, Los Angeles and London: University of California Press.

Shryock, A. and D. L. Smail (eds) (2011b). *Deep History: The Architecture of Past and Present*. Berkeley, Los Angeles and London: University of California Press.

Shryock, A., T. R. Trautmann and C. Gamble (2011). 'Imagining the Human in the Deep Time.' In A. Shryock and D. L. Smail (eds), *Deep History: The Architecture of Past and Present*, 21–54. Berkeley, Los Angeles and London: University of California Press.

Sidky, H. (2010). 'On the Antiquity of Shamanism and Its Role in Human Religiosity.' *Method & Theory in the Study of Religion* 22(1): 68–92.

Singh, M. (2018). 'The Cultural Evolution of Shamanism.' *Behavioral and Brain Sciences* 41 e66: 1–17. DOI: 10.1017/S0140525X17001893

Sinhababu, S. (2008). 'Full J. Z. Smith Interview.' *The Chicago Maroon*, 2 June. https://www.chicagomaroon.com/2008/06/02/full-j-z-smith-interview/ (accessed 30 January 2018).

Slone, D. J. (ed.) (2006). *Religion and Cognition: A Reader*. London and Oakville, CT: Equinox.

Slone, D. J. (2007[2004]). *Theological Incorrectness: Why Religious People Believe What They Shouldn't*. Oxford and New York: Oxford University Press.

Slone, D. J. (2008). 'The Attraction of Religion: A Sexual Selectionist Account.' In J. Bulbulia, R. Sosis, E. Harris, R. Genet, C. Genet and K. Wyman (eds), *The Evolution of Religion: Studies, Theories, & Critiques*, 181–8. Santa Margarita, CA: Collins Foundation Press.

Slone, D. J. (2013). 'The Opium or the Aphrodisiac of the People? Darwinizing Marx on Religion.' In D. Xygalatas and W. W. McCorkle Jr (eds), *Mental Culture: Classical Social Theory and the Cognitive Science of Religion*, 52–65. Durham and Bristol, CT: Acumen.

Slone, D. J. and J. A. Van Slyke (eds) (2015). *The Attraction of Religion: A New Evolutionary Psychology of Religion*. London and New York: Bloomsbury.

Smail, D. L. (2008). *On Deep History and the Brain*. Berkeley, Los Angeles and London: University of California Press.

Smail, D. L. (2014). 'Retour sur *On Deep History and the Brain.' Tracés. Revue de Sciences humaines* 14. http://traces.revues.org/6014 (accessed 15 May 2017).

Smart, N. (1980). 'Eliade and the *History of Religious Ideas.' The Journal of Religion* 60(1): 67–71.

Smith, C. H. and G. Beccaloni (eds) (2008). *Natural Selection and Beyond: The Intellectual Legacy of Alfred Russel Wallace.* Oxford and New York: Oxford University Press.

Smith, H. W. (2008). *The Continuities of German History: Nation, Religion, and Race across the Long Nineteenth Century.* Cambridge: Cambridge University Press.

Smith, J. Z. (1982). *Imagining Religion: From Babylon to Jonestown.* Chicago and London: The University of Chicago Press.

Smith, J. Z. (1990). *Drudgery Divine: On the Comparison of Early Christianities and the Religions of Late Antiquity.* Chicago: The University of Chicago Press, London: School of Oriental and African Studies, University of London.

Smith, J. Z. (1993). *Map Is Not Territory.* Chicago: The University of Chicago Press.

Smith, J. Z. (2000). 'Classification.' In W. Braun and R. T. McCutcheon (eds), *Guide to the Study of Religions,* 35–44. London and New York: Cassell.

Smith, J. Z. (2004). *Relating Religion: Essays in the Study of Religion.* Chicago and London: The University of Chicago Press.

Smith, W. D. (1991). *Politics and the Sciences of Culture in Germany, 1840–1920.* New York and Oxford: Oxford University Press.

Smith, W. R. (1875). *s.v.* 'Bible.' *Encyclopaedia Britannica,* 9th edn, vol. 3: 634–48. Edinburgh: A. & C. Black.

Smith, W. R. (1880a). *s.v.* 'Hebrew Language and Literature.' *Encyclopaedia Britannica,* 9th edn, vol. 11: 531–8. Edinburgh: A. & C. Black.

Smith, W. R. (1880b). 'Animal Worship and Animal Tribes among the Arabs and in the Old Testament.' *Journal of Philology* 17(9): 75–100.

Smith, W. R. (1912[1869]). 'Christianity and the Supernatural.' In J. S. Black and G. Chrystal (eds), *Lectures and Essays of William Robertson Smith,* 109–36. London: Adam & Charles Black.

Smith, W. R. (1927[1889]). *Lectures on the Religion of the Semites: The Fundamental Institutions. Third edition with an Introduction and Additional Notes by S. A. Cook.* New York: The Macmillan Company, London: A. V. C. Buck Ltd.

Snow, C. P. (1961[1959]). *The Two Cultures and the Scientific Revolution.* New York: Cambridge University Press.

Sober, E. (2008). *Evidence and Evolution: The Logic Behind the Science.* Cambridge: Cambridge University Press.

Sogno, C. (2006). *Q. Aurelius Symmachus: A Political Biography.* Ann Arbor: The University of Michigan Press.

Sokal, A. (2010[2008]). *Beyond the Hoax: Science, Philosophy and Culture.* Oxford and New York: Oxford University Press.

Sokal, A. and J. Bricmont (1998[1997]). *Intellectual Impostures.* London: Profile.

Sommer, M. (2006). 'Mirror, Mirror on the Wall: Neanderthal as Image and "Distortion" in Early 20th-century French Science and Press.' *Social Studies of Science* 36(2): 207–40.

Sørensen, J. (2004). 'Religion, Evolution, and an Immunology of Cultural Systems.' *Evolution and Cognition* 10(1): 61–73.

Sørensen, J. (2007). *A Cognitive Theory of Magic.* Lanham, MD and Plymouth, UK: AltaMira Press.

Spencer, H. (1864). *Principles of Biology,* vol. I. London and Edinburgh: Williams and Norgate.

Sperber, D. (1975). *Rethinking Symbolism*. Cambridge: Cambridge University Press.

Sperber, D. (1996). *Explaining Culture: A Naturalistic Approach*. Oxford: Blackwell.

Sperber, D. (2010). 'The Guru Effect.' *Review of Philosophical Psychology* 1(4): 583–92.

Sperber, D. (2011). 'A Naturalistic Ontology for Mechanistic Explanations in the Social Sciences.' In P. Demeulenaere (ed.), *Analytical Sociology and Social Mechanisms*, 64–77. Cambridge: Cambridge University Press.

Spiegelberg, H. (1982[1960]). *The Phenomenological Movement: A Historical Introduction*. 3rd edn. The Hague: Nijhoff.

Spineto, N. (1994). *Mircea Eliade-Raffaele Pettazzoni. L'histoire des religions a-telle un sens? Correspondance 1926–1959. Texte présenté, établi et annoté par N. Spineto*. Paris: Éditions du Cerf.

Spineto, N. (2006). *Mircea Eliade storico delle religioni*. Brescia: Morcelliana.

Spineto, N. (2009). 'Comparative Studies in the History of Religions Today: Continuity with the Past and New Approaches.' *Historia Religionum: An International Journal* 1: 41–50.

Spineto, N. (2010). 'Studi storico-comparativi.' In A. Melloni (ed.), *Dizionario del sapere storico-religioso del Novecento*, vol. 2, 1256–317. Bologna: il Mulino.

Spineto, N. (2012). *Storia e storici delle religioni in Italia*. Alessandria: Edizioni dell'Orso.

Spineto, N. (2013). 'Gli epistolari come fonte per la storia degli studi storico-religiosi. Questioni di metodo.' *Historia Religionum* 5: 11–18.

Spiro, M. E. (1971). 'Religion: Problems of Definition and Explanation.' In M. Banton (ed.), *Anthropological Approaches to the Study of Religion*, 85–126. London: Tavistock Publications.

Spiro, M. E. (1996). 'Postmodernist Anthropology, Subjectivity, and Science: A Modernist Critique.' *Comparative Studies in Society and History* 38(4): 759–80.

Spöttel, M. (1998). 'German Ethnology and Antisemitism: The Hamitic Hypothesis.' *Dialectical Anthropology* 23(2): 131–50.

Staal, F. (1979). 'The Meaninglessness of Ritual.' *Numen* 26(1): 2–22.

Staal, F. (1988). *Universals: Studies in Indian Logic and Linguistics*. Chicago and London: The University of Chicago Press.

Staal, F. (1989). *Rules Without Meaning: Ritual, Mantras, and the Human Science*. New York and Bern: Peter Lang.

Stausberg, M. (2008). 'Raffaele Pettazzoni and the History of Religions in Fascist Italy (1928–1938).' In H. Junginger (ed.), *The Study of Religion Under the Impact of Fascism*, 365–95. Leiden and Boston: Brill.

Stausberg, M. (ed.) (2009). *Contemporary Theories of Religion: A Critical Companion*. London and New York: Routledge.

Stausberg, M. (2014). 'Advocacy in the Study of Religion/s.' *Religion* 44(2): 220–32.

Stausberg, M. and M. Q. Gardiner (2016). 'Definition.' In M. Stausberg and S. Engler (eds), *The Oxford Handbook of the Study of Religion*, 9–32. Oxford: Oxford University Press.

Stavru, A. (2005). *s.v.* 'Bachofen, J. J.' In L. Jones (ed.), *Encyclopedia of Religion. Second Edition*, vol. 2, 730–3. Detroit: Thomson Gale–MacMillan.

Sterelny, K. (2007[2001]). *Dawkins vs. Gould: Survival of the Fittest. New edition*. Cambridge: Icon Books.

Sterelny, K. (2017). 'Religion Re-explained.' *Religion, Brain & Behaviour*. https://doi.org/10.1080/2153599X.2017.1323789

Stoddart, D. R. (1966). 'Darwin's Impact on Geography.' *Annals of the Association of American Geographers* 56(4): 683–98.

Stone, J. R. (2005). *s.v.* 'Müller, Friedrich Max.' In L. Jones (ed.), *Encyclopedia of Religion. Second Edition*, vol. 8, 6234–7. Detroit: Thomson Gale–MacMillan.

Stott, R. (2012). *Darwin's Ghosts: In Search of the First Evolutionists*. London and New York: Bloomsbury.

Strauss, V. (2015). 'No, Finland Isn't Ditching Traditional School Subjects. Here's What's Really Happening.' *The Washington Post*, 26 March. https://www.washingtonpost.com/news/answer-sheet/wp/2015/03/26/no-finlands-schools-arent-giving-up-traditional-subjects-heres-what-the-reforms-will-really-do/?utm_term=.fd381b0b129f (accessed 20 December 2017).

Strenski, I. (2003). *Theology and the First Theory of Sacrifice*. Leiden and Boston: Brill.

Strenski, I. (2004). 'Ideological Critique in the Study of Religion: Real Thinkers, Real Contexts and a Little Humility.' In P. Antes, A. W. Geertz and R. R. Warne (eds), *New Approaches to the Study of Religion. Vol. 1: Regional, Critical and Historical Approaches*, 271–93. Berlin and New York: de Gruyter.

Strenski, I. (2015). *Understanding Theories of Religion. Second Edition*. Malden, MA and Oxford: Wiley-Blackwell.

Strenski, I. (2016). 'The Magic and Drudgery in J. Z. Smith's Theory of Comparison.' In P. Antes, A. W. Geertz and M. Rothstein (eds), *Contemporary Views on Comparative Religion: In Celebration of Tim Jensen's 65th Birthday*, 7–16. Sheffield and Bristol, CT: Equinox.

Stringer, C. and J. Galway-Witham (2017). 'On the Origin of Our Species.' *Nature* 546(7657): 212–14.

Stroumsa, G. G. (2010). *A New Science: The Discovery of Religion in the Age of Reason*. Cambridge, MA: Harvard University Press.

Stuessy, T. F. (2009). 'Paradigms in Biological Classification (1707–2007): Has Anything Really Changed?' *Taxon* 58(1): 68–76.

Sullivan, H. P. (1970). 'History of Religions: Some Problems and Prospects.' In P. Ramsey and J. F. Wilson (eds), *The Study of Religion in Colleges and Universities*, 246–80. Princeton: Princeton University Press.

Sullivan, L. E. (2005[1987]). *s.v.* 'Supreme Beings.' In L. Jones (ed.), *Encyclopedia of Religion. Second Edition*, vol. 13, 8867–81. Detroit: Thomson Gale–MacMillan.

Sulloway, F. J. (1992[1979]). *Freud, Biologist of the Mind: Beyond the Psychoanalytic Legend. With a New Preface by the Author*. Cambridge, MA and London: Harvard University Press.

Sulloway, F. J. (1998[1996]). *Born to Rebel: Birth Order, Family Dynamics, and Creative Lives*. London: Abacus.

Switek, B. (2012). 'Unless They're Zombies, Fossils Don't Live.' *Laelaps: Wired Science Blogs*, 22 August. http://www.wired.com/2012/08/a-rant-about-living-fossils/ (accessed 20 January 2018).

Tait, P. G. and W. Thomson (1912[1879]). *Treatise on Natural Philosophy*. Cambridge: Cambridge University Press.

Talmont-Kaminski, K. (2013). 'Werewolves in Scientists' Clothing: Understanding Pseudoscientific Cognition.' In M. Pigliucci and M. Boudry (eds), *Philosophy of Pseudoscience: Reconsidering the Demarcation Problem*, 381–96. Chicago and London: The University of Chicago Press.

Tatole, V. (2008). 'Notes on the Reception of Darwin's Theory in Romania.' In E.-M. Engels and T. F. Glick (eds), *The Reception of Charles Darwin in Europe*, vol. 2, 463–79. London and New York: Continuum.

Taves, A. (2009). *Religious Experience Reconsidered: A Building-Block Approach to the Study of Religion and Other Special Things*. Princeton: Princeton University Press.

Taves, A. (2013). 'Building Blocks of Sacralities: A New Basis for Comparison across Cultures and Religions.' In R. F. Paloutzian and C. Park (eds), *Handbook of Psychology of Religion and Spirituality*, 138–64. New York: The Guilford Press.

Taylor, C. (2017). *The Routledge Guidebook to Foucault's The History of Sexuality*. New York and London: Routledge.

Taylor, V. E. (2008). *Religion after Postmodernism: Retheorizing Myth and Literature*. Charlottesville and London: University of Virginia Press.

Thiesse, A.-M. (1999). *La création des identités nationales. Europe XVIIIe–XXe siècle*. Paris: Éditions du Seuil.

Thomassen, E. and T. Jensen (2017). 'Joint Statement on the European Academy of Religion by the European Association for the Study of Religions (EASR) and the International Association for the History of Religions (IAHR).' *European Association for the Study of Religion*, 24 May. http://easr.org/fileadmin/user_upload/content/pdfs/EASR-IAHR_joint_statement_about_the_EuARe.pdf (accessed 20 February 2018).

Thurs, D. P. and R. L. Numbers (2013). 'Science, Pseudoscience, and Science Falsely So-called.' In M. Pigliucci and M. Boudry (eds), *Philosophy of Pseudoscience: Reconsidering the Demarcation Problem*, 121–44. Chicago and London: The University of Chicago Press.

Tiele, C. P. (1860). 'Het onderwijs in de godsdienstgeschiedenis aan de Leidse Hoogeschool.' *De Gids* 24: 815–30.

Tiele, C. P. (1866). 'Theologie en godsdienstwetenschap.' *De Gids* 30(2): 205–44.

Tiele, C.P. (1871). 'Het wezen en de oorsprong van den godsdienst.' *Theologisch Tijdschrift* 5: 373–406.

Tiele, C. P. (1873). 'Over de geschiedenis der oude godsdiensten, haar methode, geest en belang.' *Theologisch Tijdschrift* 7: 573–89.

Tiele, C. P. (1877[1876]). *Outlines of the History of Religion to the Spread of the Universal Religions*, trans. J. Estlin Carpenter. London: Kegan Paul.

Tiele, C. P. (1886). s.v. 'Religions.' *Encyclopaedia Britannica*, 9th edn, vol. 20: 358–71. Edinburgh: A. & C. Black.

Tiele, C. P. (1897–1899). *Elements of the Science of Religion*. 2 vols. Edinburgh and London: Blackwood and Sons.

Tooby, J. and L. Cosmides (1992). 'The Psychological Foundations of Culture.' In J. H. Barkow, L. Cosmides and J. Tooby (eds), *The Adapted Mind: Evolutionary Psychology and the Generation of Culture*, 19–136. Oxford and New York: Oxford University Press.

Tort, P. (2008). *L'effet Darwin. Selection naturelle et naissance de la civilisation*. Paris: Éditions du Seuil.

Travis, J. (2009). s.v. 'Natural History.' In M. Ruse and J. Travis (eds), *Evolution: The First Four Billion Years*, 754–5. Cambridge, MA and London: The Belknap Press of Harvard University Press.

Tremlin, T. (2006). *Minds and Gods: The Cognitive Foundations of Religion*. Oxford: Oxford University Press.

Trivers, R. (2011). *The Folly of Fools: The Logic of Deceit and Self-Deception in Human Life*. New York: Basic Books.

Tuckett, J. (2016a). 'Clarifying Phenomenologies in the Study of Religion: Separating Kristensen and van der Leeuw from Otto and Eliade.' *Religion* 46(1): 75–101.

Tuckett, J. (2016b). 'Clarifying the Phenomenology of Gerardus van der Leeuw.' *Method and Theory in the Study of Religion* 28(3): 227–263.

Tull, H. W. (1991). 'F. Max Müller and A. B. Keith: "Twaddle", the "Stupid" Myth, and the Disease of Indology.' *Numen* 38(1): 27–58.

Tuniz, C., R. Gillespie and C. Jones (2009). *The Bone Readers: Science and Politics in Human Origins Research*. Walnut Creek: Left Coast Press.

Țurcanu, F. (2007[2003]). *Mircea Eliade prizonierul istoriei. Ediția a II-a revăzută*. Bucharest: Humanitas.

Țurcanu, F. (2013). 'Mircea Eliade et Georges Dumézil, «une amitié dans la liberté»?' *Historia Religionum* (5): 109–34.

Turchin, P. (2015). *Ultrasociety: How 10,000 Years of War Made Humans the Greatest Cooperators on Earth*. Chaplin, CT: Beresta Books.

Turner, F. M. (1981). *The Greek Heritage in Victorian Britain*. New Haven: Yale University Press.

Turner, J. (2011). *Religion Enters the Academy: The Origins of the Scholarly Study of Religion in America*. Athens: The University of Georgia Press.

Turner, J. (2014). *Philology: The Forgotten Origins of the Modern Humanities*. Princeton and Oxford: Princeton University Press.

Turner, J. H., A. Maryanski, A. K. Petersen and A. W. Geertz (2018). *The Emergence and Evolution of Religion by Means of Natural Selection*. London and New York: Routledge.

Tylor, E. B. (1866). Review of F. Max Müller's *Lectures on the Science of Language*, H. Wedgwood's *A Dictionary of English Etymology*, and F. W. Farrar's *Chapters on Language*. *The Quarterly Review* 119 (January and April): 394–435.

Tylor, E. B. (1868). Review of F. Max Müller's *Chips from a German Workshop*. 2 vols. *The Fortnightly Review* 9: 225–8.

Tylor, E. B. (1871). *Primitive Culture*. 2 vols. John Murray: London.

University of Chicago Divinity School, The (2012). 'Bruce Lincoln.' https://web.archive.org/web/20120730224549/http://divinity.uchicago.edu/faculty/lincoln-education.shtml (accessed 30 December 2017).

van der Leeuw, G. (1933). *s.n. Sociologus: A Journal of Sociology and Social Psychology* 9: 477–8.

van der Leeuw, G. (1937). *De primitieve Mensch en de Religie*. Wolters: Groningen.

van der Leeuw, G. (1954). 'Confession scientifique faite à l'Université Masaryk de Brno le Lundi 18 Novembre 1946.' *Numen* 1(1): 8–15.

van der Leeuw, G. (1963[1938]). *Religion in Essence and Manifestation: A Study in Phenomenology*, trans. J. E. Turner, with the additions of the second German edition by H. H. Penner. New York: Harper & Row.

Vanhaelemeersch, P. (2007). 'Eliade, "History" and "Historicism".' In B. Rennie (ed.), *The International Eliade*, 151–66. Albany: State University of New York Press.

van Wyhe, J. (ed.) (2002–). *The Complete Work of Charles Darwin Online*. http://darwin-online.org.uk (accessed 20 December 2017).

van Wyhe, J. and P. C. Kjærgaard (2015). 'Going the Whole Orang: Darwin, Wallace and the Natural History of Orang-utans.' *Studies in History and Philosophy of Science Part C: Studies in History and Philosophy of Biological and Biomedical Sciences* (51): 53–63.

Vianu, M. A. and V. Alexandrescu (eds) (1994). *Scrisori către Tudor Vianu II (1936–1949)*. Bucharest: Minerva.

Villar, F. (2009[1991]). *Gli indoeuropei e le origini dell'Europa*. Bologna: il Mulino.

Villareal, L. P. (2008). 'From Bacteria to Belief: Immunity and Security.' In R. D. Sagarin and T. Taylor (eds), *Natural Security: A Darwinian Approach to a Dangerous World*, 42–68. Berkeley, Los Angeles and London: University of California Press.

von Schnurbein, S. (2003). 'Shamanism in the Old Norse Tradition: A Theory between Ideological Camps.' *History of Religions* 43(2): 116–38.

von Stuckrad, K. (2014). *The Scientification of Religion: An Historical Study of Discursive Change, 1800–2000*. Berlin: de Gruyter.

Waardenburg, J. (1999). *Classical Approaches to the Study of Religion: Aims, Methods, and Theories of Research. Introduction and Anthology*. New York and Berlin: de Gruyter.

Waldau, P. (2013). *Animal Studies: An Introduction*. Oxford and New York: Oxford University Press.

Wallace, A. R. (1872). Review of *Primitive Culture: Researches into the Development of Mythology, Philosophy, Religion, Art, and Custom* by Edward B. Tylor, 1871. *Academy* (3)42: 69–71.

Waller, J., M. Edwardsen and M. Hewlett (2005[1987]). *s.v.* 'Evolution: Evolutionism.' In L. Jones (ed.), *Encyclopedia of Religion. Second Edition*, vol. 5, 2913–17. Detroit: Thomson Gale–MacMillan.

Warburg, M. (1989). 'William Robertson Smith and the Study of Religion.' *Religion* 19(1): 41–61.

Warburton, N. (2007). *Thinking from A to Z. Third Edition*. Abingdon and New York: Routledge.

Waterfield, R. (2009[2000]). *The First Philosophers: The Presocratics and the Sophists. A New Translation*. Oxford: Oxford University Press.

Watson, J. B. (1913). 'Psychology as the Behaviorist Views It.' *Psychological Review* 20(2): 158–77.

Wayland Barber, E. and P. T. Barber (2004). *When They Severed Earth from Sky: How the Human Mind Shapes Myth*. Princeton: Princeton University Press.

Wedemeyer, C. K. (2010). 'Introduction I: Two Scholars, a "School", and a Conference.' In C. K. Wedemeyer and W. Doniger (eds), *Hermeneutics, Politics, and the History of Religions: The Contested Legacies of Joachim Wach and Mircea Eliade*, xv–xxvi. Oxford and New York: Oxford University Press.

Wedemeyer, C. K. and W. Doniger (eds) (2010). *Hermeneutics, Politics, and the History of Religions: The Contested Legacies of Joachim Wach and Mircea Eliade*. Oxford and New York: Oxford University Press.

Weinstein, F. (1995). 'Psychohistory and the Crisis of the Social Sciences.' *History and Theory* 34(4): 299–319.

Wenzel, N. G. (2009). 'Postmodernism and Religion.' In P. Clarke (ed.), *The Oxford Handbook of the Sociology of Religion*, 172–93. Oxford: Oxford University Press.

Werblowsky, R. J. Z. (1960). 'Marburg: And After?' *Numen* 7(2): 215–20.

Werblowsky, R. J. Z. (2006[1989]). '*In Nostro Tempore*: On Mircea Eliade.' In B. Rennie (ed.), *Mircea Eliade: A Critical Reader*, 294–301. London and Oakville, CT: Equinox.

Westerink, H. (2010). 'Participation and Giving Ultimate Meaning: Exploring the Entanglement of Psychology of Religion and Phenomenology of Religion in the Netherlands.' *Numen* 57(2): 186–211.

Wheeler-Barclay, M. (2010). *The Science of Religion in Britain, 1860–1915*. Charlottesville: University of Virginia Press.

Whewell, W. (1840). *The Philosophy of the Inductive Sciences, Founded Upon Their History*. 2 vols. London: John W. Parker.

White, L. A. (1938). 'Science Is Sciencing.' *Philosophy of Science* 5(4): 369–89.

White, L. A. (1952). *The Evolution of Culture*. New York: McGraw-Hill.

Whitehouse, H. (2000). *Arguments and Icons: Divergent Modes of Religiosity*. Oxford: Oxford University Press.

Whitehouse, H. (2002). 'Modes of Religiosity: Towards a Cognitive Explanation of the Sociopolitical Dynamics of Religion.' *Method & Theory in the Study of Religion* 143(3–4): 293–315.

Whitehouse, H. (2013a). 'Religion, Cohesion, and Hostility.' In S. Clarke, R. Powell and J. Savulescu (eds), *Religion, Intolerance, and Conflict: A Scientific and Conceptual Investigation*, 36–47. Oxford: Oxford University Press.

Whitehouse, H. (2013b). 'Immortality, Creation and Regulation: Updating Durkheim's Theory of the Sacred.' In D. Xygalatas and W. W. McCorkle Jr, *Mental Culture: Classical Social Theory and the Cognitive Science of Religion*, 66–79. Durham and Bristol, CT: Acumen.

Whitmarsh, T. (2015). *Battling the Gods: Atheism in the Ancient World*. London: Faber and Faber.

Widengren, G. (1967). 'Mircea Eliade Sixty Years Old.' *Numen* 14(3): 165–6.

Wiebe, D. (1999). *The Politics of Religious Studies: The Continuing Conflict with Theology in the Academy*. New York: St Martin's Press.

Wiebe, D. (2000). 'Modernism.' In W. Braun and R. T. McCutcheon (eds), *Guide to the Study of Religion*, 351–64. London and New York: Cassell.

Wiebe, D. (2009). 'Religious Biases in Funding Religious Studies Research?' *Religio: revue pro religionistiku* 17(2): 125–40.

Williamson, P. (2016). 'Take the Time and Effort to Correct Misinformation.' *Nature* 540(7632): 171.

Wilson, D. S. (2002). *Darwin's Cathedral: Evolution, Religion, and the Nature of Society*. Chicago and London: The University of Chicago Press.

Witzel, E. J. M. (2012). *The Origins of the World's Mythologies*. Oxford: Oxford University Press.

Wolfart, J. C. (2000). 'Postmodernism.' In W. Braun and R. T. McCutcheon (eds), *Guide to the Study of Religion*, 380–95. London: Cassell.

Wozniak, R. H. and J. A. Santiago-Blay (2013). 'Trouble at Tyson Alley: James Mark Baldwin's Arrest in a Baltimore Bordello.' *History of Psychology* 16(4): 227–48.

Wunn, I. (2003). 'The Evolution of Religions.' *Numen* 50(4): 387–415.

Xygalatas, D. (2013). 'Special Issue on the Experimental Research of Religion.' *Journal for the Cognitive Science of Religion* 1(2): 137–9.

Xygalatas, D. and W. W. McCorkle Jr (2013). *Mental Culture: Classical Social Theory and the Cognitive Science of Religion*. Durham and Bristol, CT: Acumen.

Yusa, M. (2005[1987]). *s.v.* 'Henotheism.' In L. Jones (ed.), *Encyclopedia of Religion. Second Edition*, vol. 6, 3913–14. Detroit: Thomson Gale–MacMillan.

Zimmer, B. (2010). 'Truthiness.' *The New York Times*, 13 October. http://www.nytimes.com/2010/10/17/magazine/17FOB-onlanguage-t.html (accessed 20 December 2017).

Zimoń, H. (1986). 'Wilhelm Schmidt's Theory of Primitive Monotheism and Its Critique within the Vienna School of Ethnology.' *Anthropos* 81(1–3): 243–60.

Znamenski, A. A. (2003). *Shamanism in Siberia: Russian Records of Indigenous Spirituality*. Dordrecht: Kluwer Academic Publishers.

Znamenski, A. A. (2007). *The Beauty of the Primitive: Shamanism and Western Imagination*. Oxford and New York: Oxford University Press.

Znamenski, A. A. (2009). 'Quest for Primal Knowledge: Mircea Eliade, Traditionalism, and "Archaic Techniques of Ecstasy".' *Shaman* 171(2): 181–204.

Index